1/86

Least Square Estimation
with Applications to
Digital Signal Processing

Least Square Estimation with Applications to Digital Signal Processing

Arthur A. Giordano

Frank M. Hsu

A WILEY-INTERSCIENCE PUBLICATION

JOHN WILEY & SONS

New York • Chichester • Brisbane • Toronto • Singapore

Library of Congress Cataloging in Publication Data:

Giordano, A. A. (Arthur Anthony), 1941–
    Least square estimation with applications to digital
signal processing.

    "A Wiley-Interscience publication."
    Includes index.
    1. Least squares—Data processing. 2. Estimation
theory—Data processing. 3. Signal processing—Digital
techniques.   I. Hsu, F. M. (Frank Ming), 1946–
II. Title.

QA275.G56 1985   621.38'043   84-19496
ISBN 0-471-87857-X

Printed in the United States of America

10 9 8 7 6 5 4 3 2 1

*This book is dedicated
to our families
for their love and patience*

DIANE, LISA, AMANDA
and
HUEY, LINDA, ELLEN, JENNY

Preface

Least squares error techniques were devised independently by Gauss and Legendre in the early 1800s as a method for estimating parameters from noisy measurements. These techniques initiated the development of what is now a voluminous body of mathematical and scientific literature describing investigations on an extensive variety of least square error applications. Each new application seems to spawn its own theoretical formulation. However, in many instances least square error principles with geometrically based foundations provide a unifying thread among seemingly unrelated problems. A dual objective of this text, then, is to establish the mathematical framework of least square error principles, and to subsequently demonstrate the utility and widespread use of these principles in a variety of digital signal processing applications.

Important questions regarding the purpose and objectives of this book require further clarification, specifically:

1. Why is this book important?
2. What can be found in this book that is unavailable elsewhere?
3. What circumstances justify a treatise on this subject at this time?
4. Who is the intended audience?

Although these questions are related, they will be answered individually, beginning with the third question.

Digital computers and real-time digital processors have forced dramatic changes in such diverse scientific disciplines as communications, control, radar, seismology, bioelectronics, etc. Least square error algorithms, which were originally developed to process data, often involve an extensive amount of iterative computation. Thus, special-purpose digital signal processors, programmable digital signal processors and/or digital computers are ideally suited for implementing least square error algorithms. In addition, the research and development of least square error algorithms, which previously had been constrained by a scientist's perseverance in performing hand computations, have been enhanced and extended by the computational power and speed of modern digital signal processors and digital computers. As a result, new applications and new least square error algorithms supporting a variety of applications are now possible that, until the last decade, could not have been conceived.

In response to the second question, this book documents important least square error algorithms in a unified way, with consistent notation. Both deterministic and probabilistic formulations are presented in a geometrical framework. The subject matter treated here is typically scattered in the available literature, or is developed within the context of a narrow specific application. By presenting these algorithms within a single reference source, unique and/or common features associated with the various algorithms can be identified. In Part I, a mathematical formulation of the least square error algorithms is provided. In Part II, certain digital signal processing applications that have achieved widespread use are selected as examples. One goal of this book, then, is to offer the reader an opportunity to comprehend the theory in a manner which permits a transfer of the technology to the specific application at hand.

The importance of the book (by way of an answer to the first question) can be summarized as follows: (1) The generality and widespread applicability of least square error algorithms in digital signal processing is not well known; (2) a consistent unified treatment of least square error algorithms, which permits the reader to implement these algorithms in hardware and/or software, has heretofore not been available. Many numerical examples which can be followed by hand computation are used in the book to help the reader understand detailed computational procedures required for least square error algorithms.

The fourth and final question posed above concerns the definition of the intended audience. This book is specifically written for practicing engineers and scientists involved in digital signal processing and for advanced students interested in digital signal processing. Desirable prerequisites include courses in matrix algebra, probability and stochastic processes, and digital signal processing.

The structure of the book is as follows: Chapters 2–5 (Part I) present the least square error algorithms. The remaining chapters (Part II) cover the digital signal processing applications. In Chapter 2 a Fourier-series expansion using orthogonal functions is presented. The Fourier-series coefficients are derived both by differentiating the mean square error and by applying the orthogonality principle. Subsequently, geometric concepts are presented to introduce least square optimization in Hilbert space. Both the orthogonality principle and the Gram-Schmidt orthogonalization procedure are then used to derive the normal equations. In Chapter 3 the Durbin, Levinson and Burg algorithms are derived. In the derivations a digital communications model is assumed using signals with either known correlation functions or correlation functions which can be estimated from the observed data. The Durbin algorithm provides a recursive solution to the Yule-Walker equations. The Levinson algorithm provides a recursive solution to the normal equations that is referred to as a *Wiener filter*. The Burg algorithm provides a solution using forward and backward error prediction with an implementation which assumes a lattice structure form. In Chapter 4 a general least square lattice algorithm that is recursive in both time and order is described for the nonstationary signal case. In Chapter 5 the Kalman recursive least square estimation algorithm is derived by generalizing the minimum-variance-weighted-least-square method. A computationally stable form of the Kalman algorithm, known as the square root Kalman algorithm, concludes the chapter.

Part II describes applications involving equalization, spectral analysis, digital whitening, adaptive arrays, interchannel interference mitigation, and digital speech processing. The first application, equalization, receives the greatest coverage, since many of the algorithms presented in Part I are utilized. The discussion of equalization algorithms also provides an opportunity to identify a relationship between the Kalman algorithm and generalized least squares lattice structure formulations. Spectral analysis, treated in Chapter 7, considers the application of autoregressive methods, such as the Burg algorithm, for attaining high-resolution spectral estimates. In Chapter 8, which deals with digital whitening, use of the Durbin implementation of a Wiener filter and the Burg implementation of a maximum entropy filter result in substantial performance improvement by suppressing narrowband interference in spread-spectrum communications. The chapters on adaptive arrays and interference mitigation (Chapters 9 and 10) describe techniques for combining multiple signals using minimum mean square error methods to process digital signals efficiently and reduce distortion from interference. Chapter 11, dealing with speech processing, presents applications of linear prediction theory and efficient time-domain waveform coding schemes. These applications represent examples of the use of least square algorithms in digital signal processing and are not intended to be exhaustive.

We would like to express our appreciation to General Telephone and Electronics (GTE) for giving us the opportunity to work on the variety of problems presented in the applications part of the text. We are also indebted to GTE's support during manuscript preparation. We would like to thank numerous colleagues who contributed to the original research at GTE, including P. Anderson, Dr. A. Levesque, J. Lindholm, Dr. H. Nichols, Dr. H. dePedro, M. Sandler, Dr. T. Schonhoff and Dr. John Proakis of Northeastern University. We owe special thanks to Dr. David Freeman, who reviewed the entire manuscript and made many worthwhile suggestions. We are also indebted to L. Carroll for her careful editing and typing of the original manuscript.

ARTHUR A. GIORDANO
FRANK M. HSU

Needham, Massachusetts
January 1985

Contents

Notation

Frequently used symbols required in the text are provided in this section.[*,†] Every attempt has been made to use a consistent notation. Often, subscripts and index parameters are the only means of distinguishing symbols. Unfortunately, very little consistency exists in the literature as a result of the wide variety of researchers and applications. Thus, care should be exercised in comparing algorithms presented here with references available elsewhere.

Among the characters and operators used in the text are

$\mathscr{E}(x)$ ensemble average of random variable x

$V(x)$ variance of random variable x or power of x ($\sigma_x^2 = V(x)$ in Chapters 1–8, 10, 11)

$\hat{x}$ estimate of variable x

$\{x_k\}$ set of variables x_k for k ranging over a specified interval

x^* complex conjugate of variable x

$\boldsymbol{X}^t$ transpose of vector,[†] $\boldsymbol{X}$

$\delta(t)$ impulse function defined by $\int_{-\infty}^{\infty} \delta(t)\, dt = 1$ and $\delta(t) = 0$ for $t \neq 0$

$\delta(l)$ unit sample function $= \begin{cases} 1, & l = 0 \\ 0, & l \neq 0 \end{cases}$

$P_{\min}$ minimum value of the quantity P

[*]Vectors and matrices appear in boldface print
[†]All vectors are defined as column vectors; for example, if $\boldsymbol{X}$ has elements $x_1, \cdots x_N$, then

$$\boldsymbol{X} = \begin{pmatrix} x_1 \\ \vdots \\ x_N \end{pmatrix}.$$

A consistent definition is maintained throughout the text for the following symbols, with the exceptions noted

a_k — prediction error coefficients in Chapters 3–11, scalars in Chapter 2

$a_{M,k}$ — prediction error coefficients with order index

$a_M(m, N)$ — forward prediction error coefficients

$A_M(N)$ — vector of forward prediction error coefficients, that is, $A_M^t(N) = (a_M(1, N), \ldots, a_M(M, N))$

$\overline{A}_M(N)$ — extended vector of forward prediction error coefficients, that is, $\overline{A}_M^t(N) = (1, a_M(1, N), \ldots, a_M(M, N))$

$\alpha_M(N)$ — scalar in lattice structure

α_k — DFE forward coefficients in Chapter 6

b_k — linear weights or prediction coefficients

$b_{M,k}$ — linear weights or prediction coefficients with order index

$b_M(m, N)$ — backward prediction error coefficients

$B_M(N)$ — vector of backward prediction error coefficients, that is, $B_M^t(N) = (b_M(0, N), \ldots, b_M(M - 1, N)$

$\overline{B}_M(N)$ — extended vector of backward prediction error coefficients, that is,
$$\overline{B}_M^t(N) = (1, b_M(0, N), \ldots, b_M(M - 1, N))$$

B — vector of Kalman prediction coefficients, that is,
$$B^t = (b_1, \ldots, b_M)$$

B_k — vector of Kalman prediction coefficients, that is, $B_k^t = (b_{1,k}, \ldots, b_{M,k})$

$\hat{B}_{k,k-1}$ — predicted value of current Kalman coefficient estimate $\hat{B}_k$

β_k — DFE feedback coefficients in Chapter 6

C_k — Fourier-series coefficients

c_k — linear equalizer coefficients in Chapter 6

$\tilde{C}_k$ — vector of feedforward and feedback equalizer coefficients, that is,
$$\tilde{C}_k^t = (\alpha_0, \ldots, \alpha_{M_1}, \beta_1, \ldots, \beta_{M_2}) \text{ (at time instant } k)$$

C_k — vector of linear equalizer coefficients (at time instant k), that is,
$$C_k^t = (c_0, \ldots, c_M)$$

$C_M(N)$ vector of feedforward and feedback equalizer coefficients, that is,

$$C_M^t(N) = (C_M(0, N), \dots, C_M(M, N))$$

$C_{M_T}(N)$ vector of feedforward and feedback lattice equalizer coefficients, that is,

$$C_{M_T}^t(N) = (C_{M_T}(0, N), \dots, C_{M_T}(M, N))$$

δ_k zero mean random vector for Kalman algorithm with M elements

$D_M(k, N)$ lattice structure vector with M elements

D_k square root Kalman $N \times N$ diagonal matrix with elements $d_i(k)$

e_k error signal between received signal and its estimate

ε_k error signal between desired signal or information symbol and its estimate

$\left.\begin{array}{c} e_M^f(k, N) \\ e_M^f(N) \end{array}\right\}$ forward error signal

$\left.\begin{array}{c} e_M^b(k, N) \\ e_M^b(N) \end{array}\right\}$ backward error signal

$\left.\begin{array}{c} e_M(k, N) \\ e_M(N) \end{array}\right\}$ error between information symbol and its estimate

$E_M^f(N)$ minimum forward MSE power vector with $M + 1$ elements, that is, $E_M^f(N) = (f_M(N), 0, \dots, 0)$

$E_M^b(N)$ minimum backward MSE power vector with $M + 1$ elements, that is, $E_M^{bt}(N) = (0, \dots, 0, r_M(N))$

$\underline{e}_M^f(k, N)$ vector of forward error signals for DFE lattice structure, that is,

$$\underline{e}_M^{ft}(k, N) = (e_{1,M}^f(k, N) \, e_{2,M}^f(k, N))$$

$\underline{e}_M^b(k, N)$ vector of backward error signals for DFE lattice structure, that is,

$$\underline{e}_M^{bt}(k, N) = (e_{1,M}^b(k, N) \, e_{2,M}^b(k, N))$$

E Kalman error vector, that is, $E^t = (\varepsilon_1, \dots, \varepsilon_N)$

E_k Kalman error vector, that is, $E_k^t = (\varepsilon_{1k}, \dots, \varepsilon_{Nk})$

E_b/N_0 ratio of signal energy per bit to noise power spectral density or energy contrast ratio

$E(C)$ linear equalizer MSE (ensemble average)

$E(\alpha, \beta)$ DFE MSE (ensemble average)

E time average of continuous squared error signal

$f_m(N)$ minimum forward MSE power

f_k channel coefficients in Chapter 6

$\left.\begin{array}{l} \gamma_M(k, N) \\ \gamma_M(N) \end{array}\right\}$ scalar in lattice structure

g_k crosscorrelation between received and transmitted signals (ensemble average)

                  crosscorrelation between received sample and information symbol in Chapter 6

G_k $M \times N$ Kalman gain matrix

$G_M(N)$ $M + 1$ element Kalman gain vector with order index

Γ_k gradient vector for steepest descent algorithms with $M_1 + M_2 + 1$ elements

I_k kth information symbol

$\tilde{I}_k$ kth information symbol decision

$\left.\begin{array}{l} \hat{I}_M(k, N) \\ \hat{I}_k \end{array}\right\}$ estimate of kth information symbol

$k_M(N)$ scalar in lattice structure

K_k Kalman gain vector, that is, $K_k^t = (K_{1,k}, \ldots, K_{N,k})$

L_f loss function

M order of prediction filter
                  related to number of linear equalizer
                  tap coefficients in Chapter 6, that is, either $M + 1$ or $2M + 1$
                  total number of taps

M_1 number of feedforward DFE coefficients minus one

M_2 number of feedback DFE coefficients

M_T total number of DFE lattice equalizer coefficients, that is, $M_T = M_1 + M_2 + 1$

M_a alphabet size

$\mu_M(N)$ scalar in lattice equalizer structure

N total number of signal samples

n_k additive noise samples

$\eta(t)$ additive channel noise

N_0 noise power spectral density

$\Phi(k, k - 1)$ $M \times M$ transition matrix for Kalman coefficients dynamic model

$p(t)$ transmit filter pulse with symbol duration

P_M ensemble average MSE for either received or desired signals

$P_M(N)$ time average MSE

$P_M^b(N)$ time average backward error power

$P_M^f(N)$ time average forward error power

P_k $M \times M$ error covariance matrix of Kalman prediction coefficients

$P_{k, k-1}$ $M \times M$ Kalman prediction error covariance matrix

$P(f)$ power spectral density of received signal

$q(N)$ zeroth time average autocorrelation element, that is, $q(N) = \rho_N(0, 0)$

$Q_M(N)$ vector of time average autocorrelation coefficients, that is,

$$Q_M^t(N) = (\rho_N(1, 0), \ldots, \rho_N(M, 0))$$

Q_k $M \times M$ covariance matrix of Kalman random vector δ_k

$\rho_N(m, j)$ time average autocorrelation of received signal in prewindowed method (in some instances the weighting coefficient $\omega = 1$)

r_k ensemble average received signal autocorrelation coefficients .

R_M $M \times M$ covariance matrix of received signals (ensemble average)

$R_M(N)$ $M \times M$ time average autocorrelation matrix of received signals

R_ε $N \times N$ Kalman covariance matrix of error signals (ensemble average)

$R_{k\varepsilon}$ $N \times N$ Kalman covariance matrix of error signals (ensemble average)

$r_M(N)$ minimum backward MSE power

s_k desired or transmitted signal samples

$s(t)$ transmitted signal

S Kalman vector of transmitted signal samples, that is, $S^t = (s_1, \ldots, s_N)$

S_k Kalman vector of transmitted signal samples, that is, $S_k^t = (s_{1, k}, \ldots, s_{N, k})$

T symbol interval or duration

T_s sampling interval

T_c chip interval

T_0 period

T_I time interval or integration interval

U_k $N \times N$ square root Kalman upper triangular matrix with elements $u_{i,j}(k)$

$v_M(N)$ Mth time average autocorrelation coefficient, that is,
$v_M(N) = \rho_N(M, M)$

$V_M(N)$ vector of time average autocorrelation coefficients, that is,

$$V_M^t(N) = (\rho_N(0, M), \ldots, \rho_N(M - 1, M))$$

ω weighting coefficient in Chapters 4 and 6

$W_M(N)$ time average crosscorrelation vector between information symbols and received sample, that is,

$$W_M^t(N) = (w_0, \ldots, w_M)$$

$x(t)$ received signal

x_k received or observed signal samples

$\left.\begin{array}{l} X_M(k) \\ X(k) \end{array}\right\}$ received signal vector, that is, $X_M^t(i) = (x_i, \ldots, x_{i-M})$ (In DFE lattice $X_M(i)$ is the vector of received samples and information symbols.)

$X_{M_T}(k)$ vector of received samples and information symbol decisions for DFE lattice with M_T elements

X_M received signal vector, that is, $X_M^t = (x_1, \ldots, x_M)$

$\left.\begin{array}{l} X \\ X_k \end{array}\right\}$ $N \times M$ Kalman matrix of received signal samples

$X_k(i)$ Kalman vector of received signal samples, that is,
$X_k^t(i) = (x_{i,k}, \ldots, x_{i-(M-1),k})$

$\overline{X}_k$ square root Kalman vector of received signal samples, that is,
$\overline{X}_k^t = (x_{1,k}, \ldots, x_{N,k})$

X_k' vector of received signal samples, that is,
$X_k'^t = (x_k, \ldots, x_{k+M})$

$\tilde{X}_k$ vector of received samples and information symbol decisions in Chapter 6

$$\tilde{X}_k^t = (x_k, \ldots, x_{k+M_1}, \tilde{I}_{k-1}, \ldots, \tilde{I}_{k-M_2})$$

NOTES

1. MSE = mean square error
2. DFE = decision feedback equalizer
3. Subscripts on received signal are used to distinguish Kalman and lattice structures. The time index k is used in the Kalman case and the order index M is used in the lattice case.

*Least Square Estimation
with Applications to
Digital Signal Processing*

chapter

1

INTRODUCTION

Least squares theory is extensively utilized in a wide variety of scientific applications. A few specific examples include communications, control, estimation, prediction, numerical analysis, geophysics, and spectral analysis. Due to the large number of possible applications, specialization of least squares theory to one application would be difficult and would limit the utility of the text. However, digital communications has evolved to such pervasive and common usage that special attention to this subject is warranted. Even with this specialization, both the variations of least square algorithms and the variety of applications within the field of digital communications appear to be unbounded. With these preliminary remarks the twofold purpose of the text can be stated:

1. To unify fundamental least squares concepts.
2. To provide an overview of important least square algorithms and selected applications in digital communications.

The least squares concepts presented here are selected on the basis of the principal known applications. Variations of the selected algorithms have been and will continue to be developed. Thus the concepts presented here are not exhaustive but are intended to acquaint the uninitiated reader with the theory.

1.1 BASIS OF LEAST SQUARE THEORY

Estimation problems can be considered to be a subset of the class of approximation problems. The general approximation problem can be roughly stated as the approximation of an unknown quantity from a combination of known quantities. This problem can be formulated either deterministically or probabilistically. A classic example of the deterministic case occurs in Fourier-series expansions, where an arbitrary function is approximated by a combination of elementary functions, for example, sinusoids.[1,2] An example of the probabilistic case occurs in linear prediction, where the next sample in a random sequence is estimated by a linear combination of prior samples. In either the deterministic or the probabilistic case a performance criterion is required that measures the quality of the approximation. One such measure which is simple and leads to practical implementations is the mean square error (MSE). Often this criterion results in optimum or near optimum performance, in comparison with other measures of performance. For example, in the estimation of a signal received in white Gaussian noise, minimization of the mean square error is equivalent to maximum likelihood estimation. This relationship between minimum mean square error and maximum likelihood estimation in additive white Gaussian noise is more completely addressed in Section 1.3. As a result of the widespread use of the MSE criterion in numerous and diverse applications, all of the algorithms emphasized here are based on minimization of the mean square error.

Both deterministic and probabilistic least square algorithms can be presented in terms of geometric principles. As a result, problems of each type, such as classical Fourier-series and least square estimation of random variables, can be formulated as least square optimization problems in Hilbert space.[3,4] This approach is followed in Chapter 2 and is used to introduce the orthogonality principle, an intuitive geometric result of optimization theory that is a direct use of the projection theorem.[3] A simple version of the projection theorem, illustrated in Figure 1-1a, states that the shortest line from a point to a plane in three-dimensional Euclidean space is the perpendicular from the point to the plane. Figure 1-1b illustrates a line outside of a plane which is being approximated by a line within the plane. The error is represented by the length of the line from a point labeled A in Figure 1-1b to any arbitrary point in the plane. The best approximation of the line in the sense of minimum-mean-square error is obtained by dropping a perpendicular from point A to the plane. The orthogonality principle then requires that the error be orthogonal to every line in the plane. This discussion, loosely stated here, is given a firm mathematical basis in Chapter 2, following a presentation of elementary Hilbert space concepts.

This text almost exclusively uses discrete rather than continuous time signal representations. This usage can be traced to the explosive adoption of digital computers and digital processors in virtually unlimited applications. Least square error algorithm developments have followed the recent trend to discrete-time signals. As a result, emphasis is placed on algorithms which are

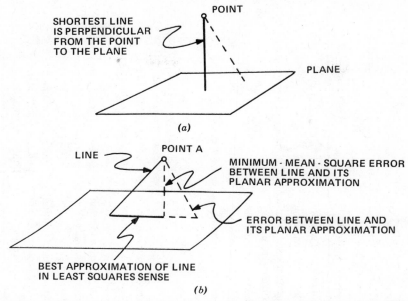

Figure 1-1. (a), Illustration of projection theorem; (b), illustration of orthogonality principle.

suitable for use in digital computers and processors. For example, the traditional Wiener-Hopf conditions expressed as a time-continuous integral are replaced by the normal equations in terms of weights referred to as Wiener filter coefficients. Furthermore, recursive solutions for the filter coefficients that are amenable to numerical computation are desired. An important element of these discrete numerical algorithms is then associated with the computational complexity and associated processing burdens. Often, algorithms with suboptimum performance are used to reduce the computational burden. This tradeoff accounts for the widespread use of the gradient or steepest-descent algorithms in data equalization and adaptive arrays, for example. The relationship between discrete and continuous-time models of a communication system is described in Chapter 6. In the representation, complex low-pass equivalents of real bandpass signals are assumed.

The remaining sections in this chapter present a summary of the least square algorithms and associated applications that are to be developed in the later chapters, and the relationship between least square algorithms and other methods.

1.2   SUMMARY OF LEAST SQUARE ALGORITHMS AND ASSOCIATED APPLICATIONS

A summary of the principal least square algorithms described in the text is given in Table 1-1. All of the algorithms presented in this table implicitly

TABLE 1-1. PRINCIPAL LEAST SQUARE ALGORITHMS

Section	Least Square Algorithm	Assumptions	Solution for Predictors Coefficients	Comments
2.1.4	Durbin	Stationary signal and known autocorrelation coefficients	Durbin recursive solution of Yule-Walker equations	Estimation of received signal
3.2.4	Levinson/Wiener	Stationary signal and known autocorrelation coefficients	Levinson recursive solution of normal equations	Estimation of transmitted signal
3.3	Cholesky Decomposition	Known autocorrelation coefficients	Cholesky recursive solution of normal equations	Estimation of received signal—nonstationary case
3.5.3	Burg	Stationary signal	Burg recursive solution directly from available data	Forward and backward prediction. No correlation coefficient estimation. Related forward and backward coefficients.
4.1.6	Lattice	Prewindowed data	Lattice recursive solution directly from prewindowed data	Forward and backward prediction—nonstationary case
5.3.1	Kalman	Dynamic System Model for predictor coefficients	Kalman recursive solution with error covariance matrix updating	Estimate of transmitted signal—nonstationary case. Sensitive to computer roundoff errors
5.4.3	Square Root Kalman	Dynamic System Model for predictor coefficients	Kalman recursive solution using U-D factorization of error covariance matrix	Similar performance to Kalman with better numerical stability

assume that iterative computation will be performed. These algorithms are considered for both stationary and nonstationary signals, with either known or estimated autocorrelation functions. It is shown that various windowing schemes for the data lead to unique algorithm definitions. Lattice structures that utilize forward and backward prediction and are based on either autocorrelation functions or only available data are extensively described. Kalman algorithms that are based on dynamic system models are presented in which covariance matrix updating is required for adaptive operation. Computationally stable square root Kalman algorithms are introduced for practical digital processing applications implemented with microprocessors or special-purpose digital hardware. In Chapter 6, on equalization, the relationship between the Kalman algorithm and lattice structure formulations is presented.

Table 1-1 provides a summary of principal least square algorithms and indicates the section in which an algorithm summary appears, key assumptions, a description of the solution for the predictor coefficients, and comments. Note that the Cholesky decomposition, lattice, Kalman and square root Kalman algorithms apply to the nonstationary signal case. The Burg and lattice structures are unique in that forward and backward error predictions are accomplished. The Kalman and square root Kalman algorithms are unique in that error covariance matrix updating is performed. In Part 2, the applications section, modifications for adaptive use and variations of these algorithms are described. Typically, lattice and Kalman algorithms offer rapid coefficient convergence with increased computational complexity over simpler least square algorithms such as the gradient or steepest-descent.

Table 1-2 summarizes the application of least square algorithms used in digital communications that are presented here. This table represents only a sample of the applications and is by no means exhaustive. However, a number of different channels are considered, and appropriate algorithms for use on these channels are identified. In the application section it should be clear that the least square algorithm selected is rarely suitable in the general form provided in Part 1. Variations such as the use of diversity, consideration of computational complexity, and specific requirements of the application force the algorithm designer to consider appropriate algorithm modification.

The specific applications described in Part 2 and presented in Table 1-2 include equalization, power spectral analysis, digital whitening, adaptive arrays, interchannel interference mitigation, digital speech processing, and a brief treatment of image processing. The equalization chapter includes a description of wireline and several radio channels, linear and decision feedback equalization implementations, and a variety of adaptive algorithms for coefficient updating that range from steepest descent to Kalman methods. In the power spectral analysis chapter, traditional periodogram techniques based on the Fourier transform, as well as iterative autoregressive methods such as the Burg algorithm, are presented. The bias and variance of the power spectral estimates are important performance measures and are included in the discussion. Digital whitening techniques used to suppress narrowband interference in

TABLE 1-2.   APPLICATIONS OF LEAST SQUARE ALGORITHMS IN DIGITAL COMMUNICATIONS

Application	Approach	Comments
Equalization		
Wireline channel equalization	Linear equalizer	—Transversal filter —Lattice structure
Troposcatter channel equaliztion	Decision feedback equalizer	—Explicit diversity with steepest descent
HF channel equalization	Decision feedback equalizer	—Forward and Backward Transversal Filters with Kalman coefficient updating —Lattice structure
Radio multipath equalization	Linear equalizer	—Parameter estimation via ambiguity function —Nonrecursive and recursive formulations
Power spectral analysis	Periodogram method Autoregressive method	—Weighted and averaged Fast Fourier Transform —Maximum entropy —Iterative solution —Improved spectral resolution
Digital whitening	Wiener and maximum entropy filters	—PN spread spectrum communications —Performance bound for SNR improvement
Adaptive arrays	Interference cancellers and Baseband diversity combining	—Two antennas —Minimum MSE —Square root Kalman
Interchannel interference mitigation	Dual channel combining	—Dual polarization —Minimum MSE
Digital speech processing	Linear prediction	—PARCOR —Pitch estimation
	Time domain waveform coding/adaptive quantization	—PCM, DPCM, DM —APC
Image processing	Two-dimensional waveform coding	—PCM, DPCM, ADPCM

pseudo-noise (PN) spread spectrum applications are described next. Wiener and maximum entropy filters, implemented with the Durbin and Burg algorithms, respectively, are used to enhance the signal-to-noise ratio (SNR) and bit error rate (BER) performance. Adaptive arrays, using interference cancellation and baseband diversity combining, are then presented. For simplicity, only two antennas are assumed and BER performance is computed. Interchannel interference mitigation introduces dual channel combining, where each channel transports different digital data streams. A specific application for this algorithm occurs in dual-polarized microwave radio transmission, where energy from one polarization cross-couples into the other. Digital speech processing, is considered in the last chapter, where a brief introduction to the related field of image processing is also presented. Emphasis is placed on linear prediction techniques based on the use of partial correlation (PARCOR) coefficients in conjunction with pitch estimation. Various time-domain waveform coding techniques, such as pulse code modulation (PCM), differential pulse code modulation (DPCM), delta modulation (DM), and adaptive predictive coding (APC) are considered. The section on speech processing concludes with a general implementation which consists of an adaptive quantizer in addition to an adaptive predictor. The final section, on image processing, provides a two-dimensional generalization of adaptive differential pulse code modulation (ADPCM).

1.3 ESTIMATION BY LEAST SQUARES AND OTHER METHODS

This text is devoted to the important special case of least square estimation. In order to understand the advantages and limitations of least square techniques, it is useful to briefly consider other methods such as maximum a posteriori (MAP) estimation, maximum likelihood (ML) estimation and minimum variance (MV) estimation. A good discussion of these methods is provided in Chapter 6 of Sage and Melsa.[5]

The principal advantage of least square algorithms is that they require very little information on the statistics of the data, and are usually simple to implement. MAP estimation, on the other hand, requires a detailed statistical description of the estimation problem in terms of both the a priori probability density function (pdf) of the random variable to be estimated and the a posteriori pdf of the random variable to be estimated assuming a set of known observations (see Appendix Section A.1). ML estimation assumes that the a priori pdf is unavailable. MV and least squares estimation methods make the fewest assumptions with regard to the statistics of the data. As a result of the relaxed statistical description required for least square methods, a principal disadvantage is that these methods do not always provide the best performance. In spite of this disadvantage, least square algorithms have obtained broad usage. The extensive variety of applications presented in Part 2 is a testimony to this fact.

To illustrate the difference between least square techniques and other methods, a simple parameter estimation example will be given. Suppose a set of real observed samples, x_i, for $i = 1, \ldots, N$, are available. A desired signal estimate, $\hat{s}_i$, is formed by weighting each observed sample x_i by a parameter α, that is,

$$\hat{s}_i = \alpha x_i \tag{1.1}$$

The estimate $\hat{s}_i$ differs from the desired signal s_i by an error, ε_i. In other words, the desired signal can only be approximated as a result of the measurement error, that is,

$$\varepsilon_i = s_i - \hat{s}_i \tag{1.2}$$

Combining (1.1) and (1.2) results in a simple linear parameter estimation problem with a model represented by

$$s_i = \alpha x_i + \varepsilon_i \tag{1.3}$$

Now define a function, $f(\alpha)$, which is the sum of the squares of the error components

$$f(\alpha) = \sum_{i=1}^{N} \varepsilon_i^2$$

$$= \sum_{i=1}^{N} (s_i - \alpha x_i)^2 \tag{1.4}$$

A least squares estimate can now be obtained by differentiating the quadratic function $f(\alpha)$ with respect to α and setting the result to zero:

$$\frac{df(\alpha)}{d\alpha} = \sum_{i=1}^{N} 2(s_i - \alpha x_i)(-x_i)$$

$$= 0$$

When the derivative is set to zero, the parameter α is referred to as the least square estimate and is denoted by $\hat{\alpha}$:

$$\hat{\alpha} = \frac{\sum_{i=1}^{N} s_i x_i}{\sum_{i=1}^{N} x_i^2} \tag{1.5}$$

Alternate methods for determining an estimate of α are now discussed. Assume that the error components are independent, normally distributed random variables each with zero mean and variance, σ_ε^2 (see Appendix Section A.1.14). Here, α is assumed to be a random variable having an a priori pdf, $p(\alpha)$. Given a set of signal samples $\{s_i\}$ for $i = 1, \ldots, N$, the a posteriori pdf, $p(\alpha|s_1, \ldots, s_N)$ can be investigated. A MAP estimate can then be obtained by finding an estimate, $\hat{\alpha}$, which maximizes the conditional pdf $p(\alpha|s_1, \ldots, s_N)$. The MAP estimation rule can then be stated as follows:

1.3.1   MAP Estimation

$$\text{Find } \hat{\alpha} \text{ to maximize } p(\alpha|s_1, \ldots, s_N)$$

Use of Bayes formula allows an equivalent rule to be stated, that is,

$$\text{Find } \hat{\alpha} \text{ to maximize } \left\{ \frac{p(s_1, \ldots, s_N|\alpha)\,p(\alpha)}{p(s_1, \ldots, s_N)} \right\}$$

where $p(s_1, \ldots, s_N)$ is the pdf of the signal samples $\{s_i\}$. Since the pdf $p(s_1, \ldots, s_N)$ does not depend on α, the MAP estimate can be computed from

$$\text{Find } \hat{\alpha} \text{ to maximize } \{ p(s_1, \ldots, s_N|\alpha)\,p(\alpha)\}$$

In the case where the a priori pdf, $p(\alpha)$, is unknown the ML estimate can be computed from:

1.3.2   ML Estimation

$$\text{Find } \hat{\alpha} \text{ to maximize } p(s_1, \ldots, s_N|\alpha)$$

Since the error components are assumed to be independent and normally distributed, the model given by (1.3) allows the conditional pdf $p(s_1, \ldots, s_N|\alpha)$ to be computed for a fixed set of observed samples $\{x_i\}$, for example, (see Appendix Section A.1.14),

$$p(s_1, \ldots, s_N|\alpha) = \frac{1}{\left(\sqrt{2\pi}\,\sigma_\varepsilon\right)^N} \exp\left\{ -\frac{1}{2\sigma_\varepsilon^2} \sum_{i=1}^{N} (s_i - \alpha x_i)^2 \right\} \qquad (1.6)$$

The pdf (1.6) is referred to as the likelihood function. The estimate, $\hat{\alpha}$, of the parameter α can be found by choosing $\hat{\alpha}$ to maximize the likelihood. Note that an equivalent rule is to maximize the logarithm of the likelihood, since the logarithm is a monotonic function. Defining the log-likelihood function

$$l(\alpha) = \ln p(s_1, \ldots, s_N|\alpha)$$

then

$$l(\alpha) = -\frac{N}{2}\ln 2\pi - N \ln \sigma_\varepsilon - \frac{1}{2\sigma_\varepsilon^2} \sum_{i=1}^{N} (s_i - \alpha x_i)^2 \qquad (1.7)$$

The estimate $\hat{\alpha}$ which maximizes the log-likelihood is the same as that which minimizes the sum of squares given in (1.4). Hence, the least square estimate is the same as the maximum likelihood estimate when the error components are independent, normally distributed zero mean random variables, each with variance σ_ε^2.

Now the statistical properties of the estimate given by (1.5) are determined as follows (see Appendix Section A.1.14). For a fixed set of samples $\{x_i\}$, the average value of $\hat{\alpha}$, $\mathscr{E}(\hat{\alpha})$, can be computed as

$$\mathscr{E}(\hat{\alpha}) = \mathscr{E}\left\{ \frac{\sum_{i=1}^{N} s_i x_i}{\sum_{i=1}^{N} x_i^2} \right\}$$

$$= \frac{\sum_{i=1}^{N} \mathscr{E}(s_i) x_i}{\sum_{i=1}^{N} x_i^2}$$

From (1.3) it can be seen that $\mathscr{E}(s_i) = \alpha x_i$ where α is now assumed to be a constant. Thus, the above equation simplifies to

$$\mathscr{E}(\hat{\alpha}) = \alpha \qquad (1.8)$$

Equation (1.8) is, in fact, the definition of an unbiased estimate. Similarly, the variance of $\hat{\alpha}$, $V(\hat{\alpha})$, can be computed:

$$V(\hat{\alpha}) = V\left\{ \frac{\sum_{i=1}^{N} s_i x_i}{\sum_{i=1}^{N} x_i^2} \right\}$$

$$= \frac{\sum_{i=1}^{N} V(s_i) x_i^2}{\left(\sum_{i=1}^{N} x_i^2 \right)^2}$$

From (1.3) it can be seen that $V(s_i) = \sigma_\varepsilon^2$, so that the above equation can be written as

$$V(\hat{\alpha}) = \frac{\sigma_\varepsilon^2}{\sum_{i=1}^{N} x_i^2} \qquad (1.9)$$

It is now shown that (1.9) is also the condition for a minimum variance estimate. An important bound on the variance of a scalar estimate is provided by the Cramer-Rao inequality[5] for an unbiased estimate (see Appendix Section A.1.16.3):

$$V(\hat{\alpha}) \geq \cfrac{-1}{\mathscr{E}\left\{\cfrac{\partial^2 \ln p(s_1, \ldots, s_N | \alpha)}{\partial \alpha^2}\right\}} \tag{1.10}$$

Using (1.7) in (1.10), it can be seen that the minimum variance estimate is achieved when Eq. (1.9) is satisfied. Thus the estimate provided by (1.5) is an unbiased minimum variance estimate in addition to being a least squares estimate. Under very general conditions it can be shown that the ML estimate for large N results in an estimate that is unbiased, normally distributed, and achieves minimum variance[5,6].

This simple example can be generalized by considering the linear filtering problem in which an estimate of the desired signal is formed by a weighted linear addition of the samples x_i according to

$$\hat{s}_i = \sum_{m=1}^{M} b_m x_{i-m}, \qquad i = 1, \ldots, N \tag{1.11}$$

where the coefficients b_i for $i = 1, \ldots, M$ are to be estimated. Least squares optimization then leads to a set of linear equations for the coefficients. Since the estimate in (1.11) is a linear combination of the samples x_i in which the coefficients b_i are obtained by a mean square error criterion, the estimation is termed linear mean square error estimation. (It should be noted that nonlinear least square estimation can be performed by use of a polynomial in x.)

From the above example it can be seen that least square estimation requires knowledge of only second-order moments (see (1.5)). In fact, later chapters will present least square algorithms which are based only on the data itself. The maximum-likelihood method, on the other hand, postulates knowledge of the pdf of the data samples. This additional knowledge was unnecessary in the case of normally distributed errors. However, in general, the additional knowledge leads to nonlinear estimation techniques which provide better performance estimation results.

In this text it is assumed that the reader is familiar with probability and random processes, matrix algebra, elementary digital communications, and sampling theory. To aid the reader, a summary of principal mathematical concepts useful in the study of this text is provided in the Appendix.

REFERENCES

1. B. P. Lathi, *Signals, Systems and Communications*, Wiley, New York, 1965.
2. B. P. Lathi, *An Introduction to Random Signals and Communication Theory*, International Textbook Company, Scranton, Pa., 1968.
3. D. G. Luenberger, *Optimization by Vector Space Methods*, Wiley, New York, 1969.
4. C. Nelson Dorny, *A Vector Space Approach to Models and Optimization*, Krieger, New York, 1980.
5. A. P. Sage and J. C. Melsa, *Estimation Theory with Applications to Communication and Control*, McGraw-Hill, New York, 1971.
6. G. M. Jenkins and D. G. Watts, *Spectral Analysis and its Applications*, Holden-Day, San Francisco, 1969.

part
1

Mathematical Formulations of the Least Square Error Algorithms

Part 1, consisting of Chapters 2–5, provides a mathematical representation of important least square algorithms. Both classical least squares algorithms and numerous least square estimation algorithms are presented. In Part 2 these algorithms are utilized in specific digital signal processing applications.

INTRODUCTION TO LEAST SQUARES THEORY

The mathematics of least squares can be both deterministic and probabilistic. The deterministic case is principally concerned with approximation problems. A classical example of the approximation of a function by a linear combination of orthogonal functions results in a Fourier-series expansion. In the probabilistic case approximation problems reduce to estimation problems. Most of the theory and examples in this text fall in the latter category. For completeness and to motivate the geometric principles of least squares theory, a short treatment of the approximation problem is presented.

2.1 CLASSICAL LEAST SQUARES

This section develops general Fourier-series expansions in terms of orthogonal functions.[1,2] Two methods are used to compute the Fourier coefficients, both based on least squares principles.

In the approximation problem, assume that an arbitrary complex function $f(t)$ is approximated by a complex function $f_a(t)$ over a certain interval $t_1 \leq t \leq t_2$. An error function $f_e(t)$ can then be defined by

$$f_e(t) = f(t) - f_a(t) \tag{2.1}$$

15

A least squares error can be determined by minimizing the quantity E:

$$E = \frac{1}{T_I} \int_{t_1}^{t_2} |f_e(t)|^2 \, dt \tag{2.2}$$

E represents the time average of the magnitude squared error where

$$T_I = t_2 - t_1.$$

To develop a Fourier representation, assume that a set of n functions $f_1(t)$, $f_2(t), \ldots, f_n(t)$ exists that are orthogonal over the interval $t_1 \leq t \leq t_2$:

$$\int_{t_1}^{t_2} f_l(t) f_i{}^*(t) \, dt = \begin{cases} 0, & i \neq l \\ K_i, & i = l \end{cases} \tag{2.3}$$

where K_i is the total energy of $f_i(t)$ in the interval $t_1 \leq t \leq t_2$ and $f_i{}^*(t)$ is the complex conjugate of $f_i(t)$. The approximating function is then represented as a linear combination of the functions $f_i(t)$ as

$$f_a(t) = \sum_{i=1}^{n} C_i f_i(t) \tag{2.4}$$

where the coefficients C_i are to be determined. The coefficients C_i can be obtained by minimizing the squared error E and are referred to as Fourier-series coefficients as n becomes infinite.

Minimization of the squared error, E, can be accomplished either by differentiation or by application of the orthogonality principle. First a derivation based on differentiation is given. Subsequently, the orthogonality principle is used without proof. In a latter portion of this chapter a geometric interpretation of the orthogonality principle will be given.

2.1.1   Derivation of Fourier Series Coefficients by Differentiation

In general, the coefficients C_i are complex and are given by

$$C_i = C_{R_i} + jC_{I_i} \tag{2.5}$$

The least squares error is obtained by setting the derivatives of E with respect to the real and imaginary coefficients to zero:

$$\begin{cases} \dfrac{\partial E}{\partial C_{R_i}} = 0 \\[2mm] \dfrac{\partial E}{\partial C_{I_i}} = 0 \end{cases} \qquad i = 1, \ldots, n \tag{2.6}$$

Using (2.1), (2.2), (2.4) and (2.5), E can be expressed as

$$E = \frac{1}{T_I} \int_{t_1}^{t_2} \left[f(t) - \sum_{i=1}^{n} (C_{R_i} + jC_{I_i}) f_i(t) \right]$$

$$\times \left[f^*(t) - \sum_{k=1}^{n} (C_{R_k} - jC_{I_k}) f_k^*(t) \right] dt \qquad (2.7)$$

Substituting Eq. (2.7) into (2.6) results in

$$\frac{\partial E}{\partial C_{R_l}} = \frac{1}{T_I} \int_{t_1}^{t_2} dt \left\{ -f_l(t) \left[f^*(t) - \sum_{k=1}^{n} C_k^* f_k^*(t) \right] \right.$$

$$\left. -f_l^*(t) \left[f(t) - \sum_{i=1}^{n} C_i f_i(t) \right] \right\} = 0$$

$$\frac{\partial E}{\partial C_{I_l}} = \frac{1}{T_I} \int_{t_1}^{t_2} dt \left\{ -jf_l(t) \left[f^*(t) - \sum_{k=1}^{n} C_k^* f_k^*(t) \right] \right.$$

$$\left. +jf_l^*(t) \left[f(t) - \sum_{i=1}^{n} C_i f_i(t) \right] \right\} = 0$$

Combining the preceding equations yields

$$C_l = \frac{\int_{t_1}^{t_2} f_l^*(t) f(t) \, dt}{\int_{t_1}^{t_2} |f_l(t)|^2 \, dt}, \qquad l = 1, \ldots, n$$

$$= \frac{1}{K_l} \int_{t_1}^{t_2} f_l^*(t) f(t) \, dt, \qquad l = 1, \ldots, n \qquad (2.8)$$

Equation (2.8) is the least square error solution.

An alternate derivation, equivalent to that provided here, can be obtained by differentiating the error, E, with respect to the complex variable, C_i^*, according to

$$\frac{\partial E}{\partial C_i^*} = \frac{\partial E}{\partial C_{R_i}} + j \frac{\partial E}{\partial C_{I_i}}$$

This approach is followed in Section 3.1.2.

2.1.2 Derivation of Fourier-Series Coefficients by Orthogonality Principle

The orthogonality principle requires that each of the functions $f_1(t), f_2(t) \ldots, f_n(t)$ be orthogonal to the error function over the interval $t_1 \le t \le t_2$. Thus

$$\int_{t_1}^{t_2} f_l(t) f_e^*(t)\, dt = 0, \qquad l = 1, \ldots, n \tag{2.9}$$

From Eqs. (2.1), (2.4) and (2.9)

$$\int_{t_1}^{t_2} f_l(t) \left[f^*(t) - \sum_{i=1}^{n} C_i^* f_i^*(t) \right] dt = 0$$

which results in

$$C_l = \frac{\displaystyle\int_{t_1}^{t_2} f_l^*(t) f(t)\, dt}{\displaystyle\int_{t_1}^{t_2} |f_l(t)|^2\, dt}, \qquad l = 1, \ldots, n \tag{2.10}$$

Use of the orthogonality principle leads to the same solution for the Fourier-series coefficients in a more efficient manner.

2.1.3 Evaluation of the Minimum Average Square Error

From (2.1), (2.2) and (2.4), E can be expressed as

$$E = \frac{1}{T_I} \int_{t_1}^{t_2} |f(t) - \sum_{i=1}^{n} C_i f_i(t)|^2\, dt \tag{2.11}$$

Expanding Eq. (2.11) leads to

$$E = \frac{1}{T_I} \int_{t_1}^{t_2} |f(t)|^2\, dt - \sum_{i=1}^{n} C_i \frac{1}{T_I} \int_{t_1}^{t_2} f_i(t) f^*(t)\, dt$$

$$- \sum_{i=1}^{n} C_i^* \frac{1}{T_I} \int_{t_1}^{t_2} f_i^*(t) f(t)\, dt \tag{2.12}$$

$$+ \sum_{i=1}^{n} \sum_{l=1}^{n} C_i C_l^* \frac{1}{T_I} \int_{t_1}^{t_2} f_i(t) f_l^*(t)\, dt$$

Using Eq. (2.10) in (2.12) leads to the minimum average square error

$$E_{\min} = \frac{1}{T_I} \int_{t_1}^{t_2} |f(t)|^2\, dt - \sum_{i=1}^{n} |C_i|^2 \frac{K_i}{T_I} \tag{2.13}$$

2.1.4 Orthogonal Function Expansions

From Eq. (2.13), it can be seen that as n approaches infinity, the minimum average square error approaches zero if the summation

$$\sum_{i=1}^{n} |C_i|^2 K_i$$

approaches the integral

$$\int_{t_1}^{t_2} |f(t)|^2 \, dt$$

Under these conditions the function $f(t)$ can be represented as an infinite series which is said to converge in the mean. The corresponding Fourier series is then given by

$$f(t) = \sum_{i=1}^{\infty} C_i f_i(t) \tag{2.14}$$

where

$$C_i = \frac{\int_{t_1}^{t_2} f_i^*(t) f(t) \, dt}{\int_{t_1}^{t_2} |f_i(t)|^2 \, dt} \qquad i = 1, 2 \ldots$$

Many specific examples of orthogonal functions lead to different Fourier-series representations. Examples of orthogonal functions include trigonometric, exponential, Legendre polynomials, etc. One specific example, given here, assumes that the orthogonal functions $f_i(t)$ are exponentials:

$$f_i(t) = e^{j i \omega_0 t}, \qquad i = 0, \pm 1, \pm 2, \ldots \tag{2.15}$$

These functions are orthogonal over the interval $t_0 \le t \le t_0 + T_0$, where

$$\omega_0 = \frac{2\pi}{T_0}$$

Substituting Eq. (2.15) in (2.14), the Fourier-series expansion becomes

$$f(t) = \sum_{i=-\infty}^{\infty} C_i e^{j i \omega_0 t} \tag{2.16}$$

where

$$C_i = \frac{1}{T_0} \int_{t_0}^{t_0 + T_0} f(t) e^{-j i \omega_0 t} \, dt$$

EXAMPLE

Assume that the orthogonal functions shown in Figure 2-1a are used to determine an approximation of the function shown in Figure 2-1b. Then, using Eqs. (2.4) and (2.8), the Fourier-series coefficients can be computed as follows:

$$\tfrac{1}{4}\int_0^4 |f_i(t)|^2\, dt = 1 \qquad i = 1, 2, 3$$

$$C_1 = \tfrac{1}{4}\left[\int_0^1 1 \cdot \tfrac{17}{4}\, dt + \int_1^2 1 \cdot \tfrac{23}{4}\, dt + \int_2^3 1 \cdot \tfrac{1}{4}\, dt + \int_3^4 1\left(\tfrac{-9}{4}\right) dt\right] = 2$$

$$C_2 = \tfrac{1}{4}\left[\int_0^1 1 \cdot \tfrac{17}{4}\, dt + \int_1^2 1 \cdot \tfrac{23}{4}\, dt + \int_2^3 (-1) \cdot \tfrac{1}{4}\, dt + \int_3^4 (-1) \cdot \left(-\tfrac{9}{4}\right) dt\right] = 3$$

$$C_3 = \tfrac{1}{4}\left[\int_0^1 (-1)\tfrac{17}{4}\, dt + \int_1^2 1 \cdot \tfrac{23}{4}\, dt + \int_2^3 1 \cdot \tfrac{1}{4}\, dt + \int_3^4 (-1) \cdot \left(-\tfrac{9}{4}\right) dt\right] = 1$$

Therefore the approximate function is given by

$$f_a(t) = 2f_1(t) + 3f_2(t) + f_3(t)$$

This approximate function and the corresponding error function are shown in Figures 2-1c and 2-1d, respectively. Using the coefficients C_1, C_2, and C_3, the minimum mean square error can be obtained from (2.11) and is equal to $1/16$. Notice that the error function is orthogonal to the functions $f_i(t)$, $i = 1, 2, 3$ (see Eq. (2.9)).

2.2   LEAST SQUARES OPTIMIZATION IN HILBERT SPACE

Statistical estimation problems can be approached in general as least square optimization problems in Hilbert space. A few simple definitions are required to understand this formulation. A more detailed treatment is available in the references.[3,4]

2.2.1   Hilbert Space Definition

A Hilbert space is a complete inner product space.[5] The meaning of this statement requires an explanation of the terms "complete" and "inner product space." First, it is necessary to define an inner product space (sometimes called a pre-Hilbert space). An inner product space is a vector space S in which pairs of vectors, for instance, x, and y in S, can be used to form an inner product (x, y) that satisfies several axioms. Assume that x, y and z are vectors in S, a and b are scalars, and that the null vector is given by θ. The inner product,

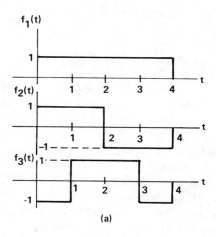

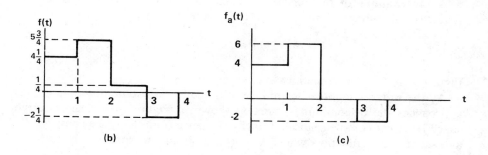

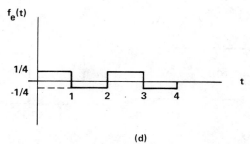

Figure 2-1. (*a*), Orthogonal functions; (*b*), function $f(t)$; (*c*), approximate function $f_a(t)$; (*d*), Error Function $f_e(t)$.

21

(x, y), is a scalar and satisfies the following axioms

$$(x, y) = (y, x)^* \tag{2.17}$$

where $*$ denotes complex conjugate.

$$(ax + by, z) = a(x, z) + b(y, z) \tag{2.18}$$

$$(x, x) \geq 0 \text{ with equality if and only if } x = \theta \tag{2.19}$$

The quantity $\sqrt{(x, x)}$ is denoted $\|x\|$ and is referred to as the norm. It represents the length of the vector x. If $(x, y) = 0$, then x and y are said to be orthogonal. If $(x, y) = 0$ and $\|x\| = \|y\| = 1$, x and y are said to be orthonormal.

In order to define the term "complete," a definition of a Cauchy sequence is required. A sequence y_n, $n = 1, 2, \ldots$, is a Cauchy sequence if the norm $\|y_n - y_m\| \to 0$ as n and $m \to \infty$. An example of a Cauchy sequence is the nth partial sums given in the previous section which led to a Fourier-series expansion. That is, these partial sums become more nearly alike in a mean square error sense as n and m approach infinity. An inner product space is said to be complete if every Cauchy sequence from the space converges in norm to a limit in the space.

As an example of the concept of completeness, consider the case of the space, W, of continuous functions $x_n(t)$ with norm

$$\|x_n\| = \int_0^1 |x_n(t)|\, dt, \qquad n = 1, 2, \ldots$$

Figure 2-2a illustrates functions in the space W.

As the index n approaches infinity, the sequence of continuous functions approaches the discontinuous function shown in Figure 2-2b. Therefore the space W is not complete since a Cauchy sequence from the space converges to a discontinuous function which is not a limit in the space of continuous functions.

Three specific examples of inner product spaces are now presented.

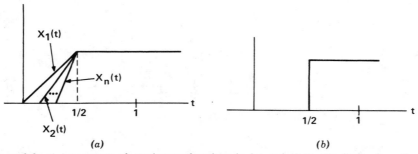

(a)   (b)

Figure 2-2. (a) sequence of continuous functions in interval $(0, 1)$; (b) limit of continuous functions in space W.

EXAMPLE 1:   SPACE OF n-DIMENSIONAL VECTORS.

Let the vector x be a column vector with components $u_1, \ldots, u_n$ and the vector of y be a column vector with components $v_1, \ldots, v_n$. Then the inner product can be expressed as

$$(x, y) = \sum_{i=1}^{n} u_i v_i^* = y^{*t} x \tag{2.20}$$

The norm of x is given by

$$\|x\| = \sqrt{\sum_{i=1}^{n} |u_i|^2} \tag{2.21}$$

EXAMPLE 2:   SPACE OF COMPLEX VALUED CONTINUOUS FUNCTIONS.

Assume the space of complex valued continuous functions is defined on the interval

$$0 \leq t \leq T_I$$

Let $f(t)$ and $g(t)$ be complex valued continuous functions with an inner product.

$$(f, g) = \frac{1}{T_I} \int_0^{T_I} f(t) g^*(t) \, dt \tag{2.22}$$

The norm of f is

$$\|f\| = \sqrt{\frac{1}{T_I} \int_0^{T_I} |f(t)|^2 \, dt} \tag{2.23}$$

EXAMPLE 3:   SPACE OF RANDOM VECTORS.

Let x and y be n-dimensional zero mean random vectors with components $u_1, \ldots, u_n$ and $v_1, \ldots, v_n$ respectively. The inner product can be defined as

$$(x, y) = \mathscr{E}\left\{ \sum_{i=1}^{n} u_i v_i^* \right\} \tag{2.24}$$

where $\mathscr{E}$ denotes an ensemble average over the joint probability density function (pdf) of x and y (see Appendix Section A.1.10). Note that the covariance matrix of x can be expressed as

$$\mathscr{E}\left[xx'^*\right] = \begin{bmatrix} \mathscr{E}\left(u_1u_1^*\right) & \mathscr{E}\left(u_1u_2^*\right) & \cdots & \mathscr{E}\left(u_1u_n^*\right) \\ \vdots & \ddots & & \vdots \\ \mathscr{E}\left(u_nu_1^*\right) & \mathscr{E}\left(u_nu_2^*\right) & \cdots & \mathscr{E}\left(u_nu_n^*\right) \end{bmatrix} \tag{2.25}$$

where t denotes the transpose of the vector x. The norm of x is given by

$$\|x\| = \sqrt{\text{Trace}\,\mathscr{E}\left(xx'^*\right)} \tag{2.26}$$

where the trace

$$\mathscr{E}\left(xx'^*\right) = \sum_{i=1}^{n}\mathscr{E}\left(u_iu_i^*\right)$$

(see Appendix Section A.2.3).

2.2.2 Generalized Orthogonal Function Expansion

Using the Hilbert space definitions given above, a generalized Fourier-series expansion can be obtained. Assume an orthonormal set of vectors $x_1, \ldots, x_n$ exists. Therefore

$$\left(x_ix_l\right) = \begin{cases} 0, & i \neq l \\ 1, & i = l \end{cases} \tag{2.27}$$

An arbitrary vector x in the n-dimensional Hilbert space can then be represented as a linear combination of the orthonormal vectors in terms of the coefficients, C_i, $i = 1, \ldots, n$, as follows:

$$x = \sum_{i=1}^{n} C_ix_i \tag{2.28}$$

Using the inner product axioms, it can be seen that

$$(x, x_l) = \left(\sum_{i=1}^{n} C_ix_i, x_l\right)$$

$$= \sum_{i=1}^{n} C_i(x_i, x_l) \tag{2.29}$$

$$= C_l$$

Therefore the coefficients (x, x_i) are in fact the generalized Fourier-series coefficients. The vector x can then be expressed as

$$x = \sum_{i=1}^{n} (x, x_i) x_i \tag{2.30}$$

If $\{x_i\}$ is an infinite orthonormal set of vectors that span the Hilbert space H, an arbitrary vector x (in some larger space containing H) can be expressed as

$$x = \sum_{n=1}^{\infty} (x, x_i) x_i$$

where the series converges to x (and hence x is in the space H), if and only if

$$\sum_{i=1}^{\infty} |(x, x_i)|^2 < \infty.$$

Using the inner product defined in Example 2 above for complex valued continuous functions, the exponential Fourier series can be defined by means of the orthonormal functions $e^{ji\omega_0 t}$, $i = 0, \pm 1, \pm 2, \ldots$, in the interval $t_0 \leq t \leq t_0 + T_0$ as follows:

$$f_i(t) = e^{ji\omega_0 t}$$

$$(f_i, f_l) = \begin{cases} 1, & i = l \\ 0, & i \neq l \end{cases}$$

$$f(t) = \sum_{n=-\infty}^{\infty} C_n e^{jn\omega_0 t} \tag{2.31}$$

$$C_n = (f, f_n) = \frac{1}{T_0} \int_{t_0}^{t_0 + T_0} f(t) e^{-jn\omega_0 t} \, dt$$

2.3   ORTHOGONALITY PRINCIPLE IN HILBERT SPACE

In this section the orthogonality principle is developed. Prior to presenting the formal theory, a simple geometric interpretation is given.

2.3.1   Geometric Interpretation of Orthogonality Principle

Consider the three dimensional space, V, spanned by the three mutually perpendicular vectors x_1, x_2, x_3 (see Figure 2-3). Let W be the plane or

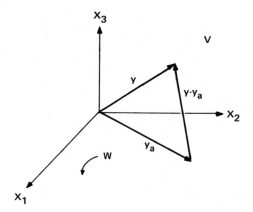

Figure 2-3.   Three-dimensional space V.

subspace of V spanned by the vectors x_1 and x_2. Assume a vector y exists that is in space V but not in W. An important question is which vector, y_a, in W "best" approximates y in V. One definition of "best" is that the length of the error vector $y - y_a$, which is the norm $\| y - y_a \|$, is a minimum. In this example it is clear that by projecting y onto W in a direction perpendicular to W, the length of the error $y - y_a$ is minimized. This result is referred to as the projection theorem. The result can be directly extended in spaces of higher dimension and in infinite dimensional Hilbert space. Since $\| y \|^2 = (y, y)$ involves squaring the length of the vector y, minimizing the norm of the error vector $\| y - y_a \|$ is referred to as a least square error criterion.

This solution is often stated in another form. Notice that the minimum norm $\| y - y_a \|$ is obtained when the error $y - y_a$ is perpendicular to W. Thus, the error must be orthogonal to every vector in W. Since every vector in W can be formed from a linear combination of the basis vectors x_1 and x_2, a simple way of stating the orthogonality principle is that the error must be orthogonal to each of the basis vectors:

$$(y - y_a, x_l) = 0, \qquad l = 1, 2 \tag{2.32}$$

For an n-dimensional space, eq. (2.32) can be directly extended:

$$(y - y_a, x_l) = 0, \qquad l = 1, 2, \ldots, n \tag{2.33}$$

Note that the vector y_a can be expressed as a linear combination of its basis vectors

$$y_a = \sum_{i=1}^{n} a_i x_i \tag{2.34}$$

where a_i, $i = 1, \ldots, n$ are a set of scalars.

From Eqs. (2.33) and (2.34)

$$\left(y - \sum_{i=1}^{n} a_i x_i, x_l \right) = 0, \qquad l = 1, \ldots, n \tag{2.35}$$

or

$$(y, x_l) = \sum_{i=1}^{n} a_i (x_i, x_l), \qquad l = 1, \ldots n \tag{2.36}$$

In matrix form these equations become

$$\begin{bmatrix} (x_1, x_1) & \cdots & (x_n, x_1) \\ \vdots & \ddots & \vdots \\ (x_1, x_n) & \cdots & (x_n, x_n) \end{bmatrix} \begin{pmatrix} a_1 \\ \vdots \\ a_n \end{pmatrix} = \begin{pmatrix} (y, x_1) \\ \vdots \\ (y, x_n) \end{pmatrix} \tag{2.37}$$

Equations (2.37) are referred to as the normal equations.[3,6]

2.3.2 Gram-Schmidt Orthogonalization Procedure

An alternate derivation of the normal equations can be developed by use of the Gram-Schmidt orthogonalization procedure.[3] Assume that the n-dimensional space W is spanned by the vectors $x_1', \ldots, x_n'$, which are not necessarily orthogonal. An orthonormal set of vectors $x_1, \ldots, x_n$ can be obtained by the following steps:

1. Normalize x_1' by setting $x_1 = x_1'/\|x_1'\|$
2. Find the projection of x_2' in the directions of x_1, that is, $(x_2', x_1)x_1$ (see Figure 2-4).
3. Subtract the projection of x_2' on x_1 from x_2' and normalize the result:

$$x_2 = \frac{z_2}{\|z_2\|}$$

where $z_2 = x_2' - (x_2', x_1)x_1$

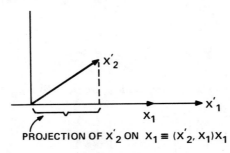

PROJECTION OF X_2' ON $X_1 \equiv (X_2', X_1)X_1$

Figure 2-4. Projection of x_2' on x_1.

4. For the nth vector

$$x_n = \frac{z_n}{\|z_n\|}$$

where

$$z_n = x_n' - \sum_{i=1}^{n-1} (x_n', x_i) x_i.$$

Note that z_n is orthogonal to x_i for $i < n$.

5. Let the vector $y = x_{n+1}'$ where the space V is spanned by $x_1', \ldots, x_{n+1}'$. Then the next vector is given by

$$z_{n+1} = y - \sum_{i=1}^{n} (y, x_i) x_i.$$

Note that z_{n+1} is orthogonal to x_i for $i \le n$.

Step 5 can be reexpressed as

$$\left(y - \sum_{i=1}^{n} (y, x_i) x_i, x_l \right) = 0, \qquad l = 1, \ldots, n \qquad (2.38)$$

Identifying the coefficient $(y, x_i) = a_i$, it can be seen that Eqs. (2.38) are equivalent to Eqs. (2.35). These equations lead to the normal equations given by (2.7). For completeness, the formal theory of optimization in Hilbert space is now presented.

2.3.3 Formal Theory of Optimization in Hilbert Space

Consider the linear algebraic equations

$$a_{11} u_1 + \cdots + a_{1n} u_n = v_1$$
$$\vdots \qquad \ddots \qquad \vdots \qquad \vdots$$
$$a_{m1} u_1 + \cdots + a_{mn} u_n = v_m \qquad (2.39)$$

If the set of vectors $x_1, \ldots, x_n$ and y are defined by

$$x_1 = \begin{pmatrix} a_{11} \\ \vdots \\ a_{m1} \end{pmatrix} \cdots x_n = \begin{pmatrix} a_{1n} \\ \vdots \\ a_{mn} \end{pmatrix}, \qquad y = \begin{pmatrix} v_1 \\ \vdots \\ v_m \end{pmatrix}$$

then y can be expressed as a linear combination of the x_i's, $i = 1, \ldots, n$ according to

$$y = x_1 u_1 + \cdots + x_n u_n \qquad (2.40)$$

Equations (2.39) can also be expressed in matrix notation by

$$A x = y$$

where

$$A = \begin{pmatrix} a_{11} & \cdots & a_{1n} \\ \vdots & \ddots & \vdots \\ a_{m1} & & a_{mn} \end{pmatrix} \qquad \text{and} \qquad x = \begin{pmatrix} u_1 \\ \vdots \\ u_n \end{pmatrix}$$

(see Appendix Section A.2).

Matrix A can be interpreted to be a "bounded linear" operator that transforms the vector x in a Hilbert space W to a vector y in a Hilbert space V. The term "bounded" implies that $\|Ax\| \le M\|x\|$ where M is a constant. The term "linear" implies that if x and z are vectors in W and a and b are scalars, then $A(ax + bz) = aAx + bAz$ where Ax and Az are vectors in V.

Three possible solutions for $Ax = y$ exist:
1. The equation has a unique solution $x = A^{-1}y$ where A^{-1} is the inverse of A.
2. The equation has no solution since there are in fact more equations than unknowns. This case is the overdetermined case.
3. The equation has many solutions since there are in fact fewer equations than unknowns. This case is the underdetermined case.

A unique solution exists when the homogeneous equation $Ax = \theta$ has only the solution $x = \theta$, the null vector. In this case matrix A is nonsingular and its determinant is nonzero. If $Ax = \theta$ has solutions other than $x = \theta$, either case 2 or case 3 exists.

The above geometric example (see Section 2.3) was an example of case 2. Assume that vectors $x_1, x_2, \ldots, x_n$ form a basis for the Hilbert space W. Therefore, they represent an independent set of vectors in that space. Assuming that vectors $x_1, \ldots, x_n, x_{n+1}$ form a basis for the Hilbert space V, then W is a subspace of V. An approximation of vector y in V can be found and is represented by vector y_a in W. This approximation is given by

$$y_a = \sum_{i=1}^{n} \tilde{u}_i x_i$$

where $\tilde{u}_i$ are a set of scalars. Assume that the operator A transforms the n-dimensional vector $x_a^t = (\tilde{u}_1, \ldots, \tilde{u}_n)$ into vectors in W such that

$$y_a = A x_a = \sum_{i=1}^{n} \tilde{u}_i x_i$$

Then the solution which minimizes the norm $\| y - y_a \|$ or $\| y - A x_a \|$ in V is obtained from the normal equations.

$$(y - y_a, x_l) = 0, \qquad l = 1, \ldots n$$

$$\left(y - \sum_{i=1}^{n} \tilde{u}_i x_i, x_l \right) = 0$$

$$\begin{pmatrix} (x_1, x_1) & \cdots & (x_n, x_1) \\ \vdots & \ddots & \vdots \\ (x_1, x_n) & & (x_n, x_n) \end{pmatrix} \begin{pmatrix} \tilde{u}_1 \\ \vdots \\ \tilde{u}_n \end{pmatrix} = \begin{pmatrix} (y, x_1) \\ \vdots \\ (y, x_n) \end{pmatrix}$$

Note that A can be written as $A = (x_1, \ldots, x_n)$ and

$$A^{t*} = \begin{pmatrix} x_1^{t*} \\ \vdots \\ x_n^{t*} \end{pmatrix}$$

Using the inner product, that is, $(x, y) = y^{t*}x$, for n-dimensional vectors x and y, the normal equations can be expressed as

$$A^{t*}A x_a = A^{t*}y$$

and can be obtained by premultiplying $A x_a = y$ by A^{t*}. Then if $A^{t*}A$ is invertible, the solution for x_a is

$$x_a = (A^{t*}A)^{-1}A^{t*}y$$

Case 3 is often referred to as the dual problem of case 2. When many solutions for x exist for a consistent but underdetermined set of equations, a least squares solution can be obtained by choosing the solution with minimum norm $\|x\|$. In this case the solution takes the form[3]

$$x_a = A^{t*}(AA^{t*})^{-1}y$$

Case 2. Examples.

EXAMPLE 1:   LINEAR REGRESSION

Problem.   An experiment is conducted in which the input to a system denoted as w produces a corresponding output z according to the transformation $z = f(w)$ (see Figure 2-5). For discrete values of the input denoted by w_i, the corresponding output produced is $z_i = f(w_i)$, $i = 1, \ldots, m$. A plot of the input-output data is shown in Figure 2-6.

Solution Using Direct Differentiation.   In linear regression the problem is to determine a linear function $f_a(w)$ which yields a least squares error solution. The function $f_a(w)$ is given by

$$f_a(w) = c_0 + c_1 w$$

where c_0 and c_1 are coefficients to be determined that produce the least squares error solution. Let the individual deviations, δ_i, be defined as

$$\delta_i = f(w_i) - f_a(w_i) \qquad i = 1, \ldots, m$$

The least squares error solution is obtained by choosing the coefficients c_0 and

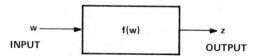

Figure 2-5.   System for linear regression example.

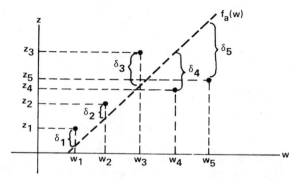

Figure 2-6.   Plot of input-output data.

c_1 to minimize S, the sum of the squares of the deviations:

$$S = \sum_{i=1}^{m} \delta_i^2$$

$$= \sum_{i=1}^{m} [f(w_i) - f_a(w_i)]^2$$

$$= \sum_{i=1}^{m} [z_i - (c_0 + c_1 w_i)]^2$$

Setting the derivatives of S with respect to c_0 and c_1 results in

$$\frac{\partial S}{\partial c_0} = \sum_{i=1}^{m} 2[z_i - (c_0 + c_1 w_i)](-1) = 0$$

$$\frac{\partial S}{\partial c_1} = \sum_{i=1}^{m} 2[z_i - (c_0 + c_1 w_i)](-w_i) = 0$$

Rewriting these equations leads to

$$\sum_{i=1}^{m} z_i = mc_0 + \left(\sum_{i=1}^{m} w_i \right) c_1$$

$$\sum_{i=1}^{m} w_i z_i = \left(\sum_{i=1}^{m} w_i \right) c_0 + \left(\sum_{i=1}^{m} w_i^2 \right) c_1$$

The solution for the coefficients is then

$$c_0 = \frac{(\sum_{i=1}^{m} z_i)(\sum_{i=1}^{m} w_i^2) - (\sum_{i=1}^{m} w_i)(\sum_{i=1}^{m} w_i z_i)}{\Delta}$$

$$c_1 = \frac{m \sum_{i=1}^{m} w_i z_i - (\sum_{i=1}^{m} w_i)(\sum_{i=1}^{m} z_i)}{\Delta}$$

where

$$\Delta = m \sum_{i=1}^{m} w_i^2 - \left(\sum_{i=1}^{m} w_i \right)^2$$

As a specific numerical example assume eight values for w_i, z_i as shown in Table 2-1.

Then

$$\sum_{i=1}^{8} w_i = 56,$$

$$\sum_{i=1}^{8} z_i = 40,$$

$$\sum_{i=1}^{8} w_i^2 = 524,$$

$$\sum_{i=1}^{8} w_i z_i = 364$$

and

$$\Delta = 8(524) - (56)^2 = 1056.$$

Therefore,

$$c_0 = \frac{40(524) - 56(364)}{1056} = \frac{6}{11}$$

and

$$c_1 = \frac{8(364) - 56(40)}{1056} = \frac{7}{11}$$

Therefore,

$$f_a(w) = \tfrac{7}{11}w + \tfrac{6}{11}$$

TABLE 2-1.   ASSUMED INPUT AND OUTPUT VALUES

w_i	z_i
1	1
3	2
4	4
6	4
8	5
9	7
11	8
14	9

Solution Using Optimization in Hilbert Space.   This problem can also be solved via optimization in Hilbert space. First the linear algebraic equations are formed from the data:

$$u_1 + u_2 w_1 = z_1$$

$$\vdots \qquad \vdots$$

$$u_1 + u_2 w_m = z_m$$

In matrix notation these equations are

$$A x = y$$

where

$$A = \begin{pmatrix} 1 & w_1 \\ \vdots & \vdots \\ 1 & w_m \end{pmatrix}, \quad x = \begin{pmatrix} u_1 \\ u_2 \end{pmatrix}, \quad \text{and } y = \begin{pmatrix} z_1 \\ \vdots \\ z_m \end{pmatrix}$$

Note that y can be expressed as

$$y = x_1 u_1 + x_2 u_2$$

where

$$x_1 = \begin{pmatrix} 1 \\ \vdots \\ 1 \end{pmatrix}, \quad \text{and} \quad x_2 = \begin{pmatrix} w_1 \\ \vdots \\ w_m \end{pmatrix}$$

The approximation y_a is given by $y_a = A x_a = \tilde{u}_1 x_1 + \tilde{u}_2 x_2$ where $c_0 = \tilde{u}_1$ and $c_1 = \tilde{u}_2$. Using the orthogonality principle and an inner product definition for vectors yields

$$(y - y_a, x_l) = 0, \qquad l = 1, 2$$
$$(y - (x_1 \tilde{u}_1 + x_2 \tilde{u}_2), x_l) = 0, \qquad l = 1, 2$$

or

$$\begin{pmatrix} (x_1, x_1) & (x_2, x_1) \\ (x_1, x_2) & (x_2, x_2) \end{pmatrix} \begin{pmatrix} \tilde{u}_1 \\ \tilde{u}_2 \end{pmatrix} = \begin{pmatrix} (y, x_1) \\ (y, x_2) \end{pmatrix}$$

The inner products are given by

$$(x_1, x_1) = \sum_{i=1}^{m} 1^2 = m$$

$$(x_2, x_2) = \sum_{i=1}^{m} w_i^2$$

$$(x_1, x_2) = (x_2, x_1) = \sum_{i=1}^{m} w_i$$

$$(y, x_1) = \sum_{i=1}^{m} z_i$$

$$(y, x_2) = \sum_{i=1}^{m} w_i z_i$$

The normal equations then become

$$
\begin{pmatrix}
m & \sum_{i=1}^{m} w_i \\
\sum_{i=1}^{m} w_i & \sum_{i=1}^{m} w_i^2
\end{pmatrix}
\begin{pmatrix}
\tilde{u}_1 \\
\tilde{u}_2
\end{pmatrix}
=
\begin{pmatrix}
\sum_{i=1}^{m} z_i \\
\sum_{i=1}^{m} w_i z_i
\end{pmatrix}
$$

These equations are identical to those determined previously by direct differentiation.

It should be noted that linear regression often gives a poor fit to experimental data. For example the data in Figure 2-7 would yield a better fit if a

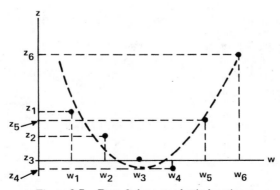

Figure 2-7. Data fit by a quadratic function.

quadratic approximating function is chosen. In general, an nth degree polynomial would be used, resulting in a polynomial regression problem.

EXAMPLE 2:   FUNCTION APPROXIMATION.

Problem.   Find a linear approximation of the function e^w that minimizes the least square error over the interval $(0, 1)$.

Solution Using Direct Differentiation.   The approximate function, $f_a(w)$, is given by

$$f_a(w) = c_0 + c_1 w$$

where c_0 and c_1 are the coefficients to be computed by use of least squares. The average squared error, E, is given by

$$E = \int_0^1 [e^w - (c_0 + c_1 w)]^2 \, dw$$

Differentiating E with respect to c_0 and c_1 yeilds

$$\frac{\partial E}{\partial c_0} = \int_0^1 2(-1)[e^w - (c_0 + c_1 w)] \, dw = 0$$

$$\frac{\partial E}{\partial c_1} = \int_0^1 2[e^w - (c_0 + c_1 w)](-w) \, dw = 0$$

Following integration these equations become

$$e - 1 = c_0 + \frac{c_1}{2}$$

$$1 = \frac{c_0}{2} + \frac{c_1}{3}$$

The solution is $c_1 = 6(3 - e) \approx 1.69$ and $c_0 = 4e - 10 \approx 0.873$.

Solution Using Optimization in Hilbert Space.   The function $f(w)$ is approximated by $f_a(w) = \tilde{u}_1 x_1 + \tilde{u}_2 x_2$ where $x_1 = 1$ and $x_2 = w$. The orthogonality principle can be applied with an inner product defined for continuous functions f and z according to

$$(f, z) = \int_0^1 f(w) z(w) \, dw$$

so that

$$(f - f_a, x_i) = 0 \qquad i = 1, 2$$

These equations can be reexpressed as

$$(f - \tilde{u}_1 x_1 - \tilde{u}_2 x_2, x_i) = 0 \qquad i = 1, 2$$

The normal equations are given by

$$\begin{pmatrix} (x_1, x_1) & (x_2, x_1) \\ (x_1, x_2) & (x_2, x_2) \end{pmatrix} \begin{pmatrix} \tilde{u}_1 \\ \tilde{u}_2 \end{pmatrix} = \begin{pmatrix} (f, x_1) \\ (f, x_2) \end{pmatrix}$$

The inner products can be computed as follows:

$$(x_1, x_1) = \int_0^1 1 \, dw = 1$$

$$(x_2, x_1) = (x_1, x_2) = \int_0^1 w \, dw = \tfrac{1}{2}$$

$$(x_2, x_2) = \int_0^1 w^2 \, dw = \tfrac{1}{3}$$

$$(f, x_1) = \int_0^1 e^w \, dw = e - 1$$

$$(f, x_2) = \int_0^1 w e^w \, dw = 1$$

Substituting the inner products yields the normal equations previously given in the solution using direct differentiation.

EXAMPLE 3: ESTIMATION OF A RANDOM VARIABLE.

Problem. A sequence of stationary random variables w_i is known to have the following moments (typically these moments would require estimation):

$$\mathscr{E}(w_i) = 1$$

$$\mathscr{E}(w_i^2) = 4$$

$$\mathscr{E}(w_i w_{i-1}) = 3$$

$$\mathscr{E}(w_i w_{i-2}) = 2$$

An estimate of the random variable w_i denoted by $\hat{w}_i$ is obtained from a weighted combination of the two previous random variables according to

$$\hat{w}_i = c_0 + c_1 w_{i-1} + c_2 w_{i-2}$$

Find the coefficients that minimize the least square error

$$\mathscr{E}\left\{(w_i - \hat{w}_i)^2\right\}$$

Solution. The normal equations can be determined by using the inner product, that is, $(x_j, x_k) = \mathscr{E}(x_j x_k)$, $j \geq 1, k \geq 1$, defined for real random variables x_j and x_k. Note that $x_1 = 1$, $x_2 = w_{i-1}$, and $x_3 = w_{i-2}$. The normal equations are then

$$\begin{pmatrix} \mathscr{E}(x_1^2) & \mathscr{E}(x_2 x_1) & \mathscr{E}(x_3 x_1) \\ \mathscr{E}(x_1 x_2) & \mathscr{E}(x_2^2) & \mathscr{E}(x_3 x_2) \\ \mathscr{E}(x_1 x_3) & \mathscr{E}(x_2 x_3) & \mathscr{E}(x_3^2) \end{pmatrix} \begin{pmatrix} c_0 \\ c_1 \\ c_2 \end{pmatrix} = \begin{pmatrix} \mathscr{E}(w_i x_1) \\ \mathscr{E}(w_i x_2) \\ \mathscr{E}(w_i x_3) \end{pmatrix}$$

Due to stationarity

$$\mathscr{E}\left(w_i^2\right) = \mathscr{E}\left(w_{i-1}^2\right) = \mathscr{E}\left(w_{i-2}^2\right)$$

$$\mathscr{E}\left(w_i w_{i-1}\right) = \mathscr{E}\left(w_{i-1} w_{i-2}\right)$$

The inner products are then given by

$$\mathscr{E}\left(x_1^2\right) = 1; \; \mathscr{E}\left(x_2^2\right) = \mathscr{E}\left(w_{i-1}^2\right) = 4; \; \mathscr{E}\left(x_3^2\right) = \mathscr{E}\left(w_{i-2}^2\right) = 4$$

$$\mathscr{E}(x_1 x_2) = \mathscr{E}(x_2 x_1) = \mathscr{E}(w_{i-1}) = 1$$

$$\mathscr{E}(x_1 x_3) = \mathscr{E}(x_3 x_1) = \mathscr{E}(w_{i-2}) = 1$$

$$\mathscr{E}(x_2 x_3) = \mathscr{E}(x_3 x_2) = \mathscr{E}(w_{i-1} w_{i-2}) = 3$$

$$\mathscr{E}(w_i x_1) = \mathscr{E}(w_i) = 1$$

$$\mathscr{E}(w_i x_2) = \mathscr{E}(w_i w_{i-1}) = 3$$

$$\mathscr{E}(w_i x_3) = \mathscr{E}(w_i w_{i-2}) = 2$$

Substituting the inner products into the normal equations yeilds

$$\begin{pmatrix} 1 & 1 & 1 \\ 1 & 4 & 3 \\ 1 & 3 & 4 \end{pmatrix} \begin{pmatrix} c_0 \\ c_1 \\ c_2 \end{pmatrix} = \begin{pmatrix} 1 \\ 3 \\ 2 \end{pmatrix}$$

Solving for the coefficients results in

$$\begin{pmatrix} c_0 \\ c_1 \\ c_2 \end{pmatrix} = \begin{pmatrix} 2/5 \\ 4/5 \\ -1/5 \end{pmatrix}$$

The estimate $\hat{w}_i$ is then

$$\hat{w}_i = \tfrac{2}{5} + \tfrac{4}{5}w_{i-1} - \tfrac{1}{5}w_{i-2}$$

REFERENCES

1. B. P. Lathi, *Signals, Systems and Communications*, Wiley, New York, 1965.
2. B. P. Lathi, *An Introduction to Random Signals and Communication Theory*, International Textbook Company, Scranton, Pa., 1968.
3. D. G. Luenberger, *Optimization by Vector Space Methods*, Wiley, New York, 1969.
4. C. Nelson Dorny, *A Vector Space Approach to Models and Optimization*, Krieger, New York, 1980.
5. S. K. Berberian, *Introduction to Hilbert Space*, Oxford Univ. Press, 1961.
6. D. A. S. Fraser, *Statistics, an Introduction*, Wiley, New York, 1958.

3

LEAST SQUARE ESTIMATION ALGORITHMS—PART 1

In this chapter several recursive algorithms for performing least square estimation are described. Several algorithms presented are referred to as the Durbin, Levinson, Cholesky and Burg algorithms. The Durbin algorithm permits the received signal to be estimated from a sequence of previous received signal samples. The Levinson algorithm is used to compute an estimate of the transmitted signal from a sequence of received signal samples. Both the Durbin and Levinson algorithms require estimates of the autocorrelation of the received signal. The Durbin algorithm can be extended using forward and backward and error prediction. The Burg algorithm uses forward and backward error prediction based only on available received signal samples and does not require autocorrelation function estimates. The Cholesky algorithm is a generalization of the Durbin algorithm for the nonstationary case. This chapter represents the first of three parts on least square estimation algorithms.

3.1 DURBIN ALGORITHM

3.1.1 Model Description

In this section the Durbin Algorithm (DA) for estimating and predicting the observed signal from a sequence of observations is presented.[1,2,3] In order to

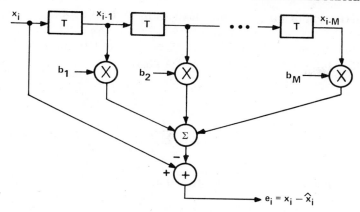

Figure 3-1.   Linear prediction filter shown as a tapped-delay line.

develop the algorithm, it is assumed that an all-pole model is used to generate the signal. Estimation of the signal is then accomplished by means of linear prediction. In the following description the observed signal is the received signal obtained by adding white noise to the transmitted signal. An estimate of the received signal is obtained from delayed versions of the received signal linearly weighted by the prediction coefficients. Minimization of the mean square error between the received signal and its estimate leads to a set of equations whose solution is the prediction coefficients.

Let x_i and s_i denote the complex low-pass equivalent of the received and transmitted signals taken at the ith sample, respectively.* The received signal is expressed as

$$x_i = s_i + n_i, \qquad i = 1, 2, \ldots, N \tag{3.1}$$

where n_i is assumed to be stationary white noise and N is the number of signal samples. Assuming that the signal is stationary, x_i may be predicted from $x_{i-1}, x_{i-2}, \ldots, x_{i-M}$. That is,

$$\hat{x}_i = \sum_{m=1}^{M} b_m x_{i-m} \tag{3.2}$$

where b_m are the coefficients of the linear predictor, M is the order of filter, and $\hat{x}_i$ is the predicted value of x_i. The linear prediction can be accomplished by use of a transversal filter (tapped-delay line), depicted in Figure 3-1, where T denotes the symbol or sample interval. (See Section 6.1).

---

*A real bandpass signal, $x_B(t)$, whose power is concentrated about a carrier frequency f_0, can be expressed in terms of its complex low-pass equivalent signal, $x(t)$, which is nonzero at baseband. Thus $x_B(t) = \mathrm{Re}\{x(t)e^{j2\pi f_0 t}\}$, where Re denotes the real part.

3.1.2   Derivation of Yule-Walker Equations by Differentiation

The coefficients in Eq. (3.2) are determined by minimizing the mean square error (MSE) P_M, defined as an ensemble average:

$$P_M = \mathcal{E}\left[|e_i|^2\right]$$

$$= \mathcal{E}\left[\left|x_i - \sum_{m=1}^{M} b_m x_{i-m}\right|^2\right] \tag{3.3}$$

where e_i is the error signal at the ith sample

$$e_i = x_i - \hat{x}_i \tag{3.4}$$

and $\mathcal{E}$ is the ensemble average operator. Since the estimate of the received signal attempts to remove the predictable portion of the received signal, the error signal is approximately a white noise process. Equation (3.4) can be expressed as

$$e_i = \sum_{m=0}^{M} a_m x_{i-m} \tag{3.5}$$

where a_m are the prediction error filter coefficients given by

$$a_0 \equiv 1, a_m = -b_m, \qquad m \neq 0 \tag{3.6}$$

Substituting (3.6) into (3.3) leads to

$$P_M = \sum_{m=0}^{M} \sum_{j=0}^{M} a_m a_j^* \mathcal{E}\left[x_{i-m} x_{i-j}^*\right]$$

$$= \sum_{m=0}^{M} \sum_{j=0}^{M} a_m a_j^* r_{j-m} \tag{3.7}$$

where

$$r_{j-m} = \mathcal{E}\left[x_{i-m} x_{i-j}^*\right] \tag{3.8}$$

is the ensemble average autocorrelation coefficients. Minimization of P_M with respect to the predictor coefficients a_l^* leads to the set of linear equations

$$\frac{\partial}{\partial a_l^*} P_M = 2 \sum_{m=0}^{M} a_m r_{l-m} = 0, \qquad l = 1, \ldots, M \tag{3.9}$$

Equation (3.9) can be reexpressed in a form known as the Yule-Walker equations:[1,4,5]

$$\sum_{m=1}^{M} b_m r_{l-m} = r_l \qquad l = 1, \ldots, M \qquad (3.10)$$

3.1.3 Derivation of Yule-Walker Equations by Orthogonality Principle

Minimization of P_M with respect to the predictor coefficients $\{a_l^*\}$ can also be accomplished by using the orthogonality principle. The orthogonality principle requires the input signal to be orthogonal to the error signal:

$$\mathscr{E}\left[x_{i-l} e_i^*\right] = 0, \qquad l = 1, \ldots, M \qquad (3.11)$$

Equation (3.11) leads to

$$\sum_{m=0}^{M} a_m^* r_{m-l} = 0, \qquad l = 1, \ldots, M \qquad (3.12)$$

Since r_{m-l} is equal to r_{l-m}^* for a stationary signal process, Eq. (3.12) may be written as

$$\sum_{m=0}^{M} a_m r_{l-m} = 0, \qquad l = 1, \ldots, M \qquad (3.13)$$

Equation (3.13) is identical to Eq. (3.10), which is obtained by the minimum mean square error (MMSE) criterion. The minimum mean square error $P_{M_{\min}}$ can be computed in terms of the optimum filter coefficients $\{a_m\}$:

$$P_{M_{\min}} = \mathscr{E}\left[|e_i|^2\right]$$

$$= \sum_{j=0}^{M} a_j^* \left(\sum_{m=0}^{M} a_m r_{j-m} \right)$$

$$= \sum_{m=0}^{M} a_m r_{-m} \qquad (3.14)$$

Equation (3.14) is obtained by noting that the term $\sum_{m=0}^{M} a_m r_{j-m}$ is identical to zero except for $j = 0$ as shown in (3.13). Combining (3.13) and (3.14) results in

$$\sum_{m=0}^{M} a_m r_{l-m} = P_{M_{\min}} \delta(l) \qquad l = 0, 1, \ldots, M \qquad (3.15)$$

where $\delta(l)$ is the unit sample function defined as

$$\delta(l) = \begin{cases} 1 & l = 0 \\ 0 & l \neq 0 \end{cases} \tag{3.16}$$

The equations in (3.10) or (3.15) are the Yule-Walker equations.

Several methods exist for estimating the autocorrelation coefficients from the signal directly. However, in this development the autocorrelation coefficients will be assumed known, so that errors in estimating the autocorrelation coefficients are neglected. The prediction coefficients can be obtained directly by solving the linear simultaneous equations in (3.10). The number of multiplications in this method is proportional to M^3. The prediction coefficients can also be solved efficiently by means of the Durbin algorithm. The computational complexity (the number of multiplications and additions) is then reduced and is proportional to M^2.

3.1.4 Durbin Recursive Solution

The Durbin algorithm is an order-recursive method for solving (3.10). That is, it solves for the coefficients of an M-order predictor recursively from the coefficients of an $(M - 1)$-order predictor. Introducing another subscript in the prediction coefficients to indicate the order results in

$$\sum_{j=1}^{M} b_{M,j} r_{l-j} = r_l, \qquad l = 1, \dots, M \tag{3.17}$$

or

$$
\begin{aligned}
b_{M,M} r_0 &\quad + b_{M,M-1} r_1 + \cdots + &\quad b_{M,1} r_{M-1} = r_M \\
b_{M,M} r_1^* &\quad + b_{M,M-1} r_0 + \cdots + &\quad b_{M,1} r_{M-2} = r_{M-1} \\
&\vdots &\vdots \quad \vdots \\
b_{M,M} r_{M-1}^* + b_{M,M-1} r_{M-2}^* + \cdots &+ b_{M,1} r_0 &\quad = r_1
\end{aligned}
\tag{3.18}
$$

where r_l^* is equal to r_{-l} for a stationary process.

The M equations in (3.18) take the form

$$R_M \begin{pmatrix} b_{M,M} \\ \vdots \\ b_{M,1} \end{pmatrix} = \begin{pmatrix} r_M \\ \vdots \\ r_1 \end{pmatrix} \tag{3.19}$$

where

$$R_M = \begin{pmatrix} r_0 & r_1 & \cdots & r_{M-1} \\ r_1^* & r_0 & \cdots & r_{M-2} \\ \vdots & & \ddots & \vdots \\ r_{M-1}^* & r_{M-2}^* & \cdots & r_0 \end{pmatrix} \tag{3.20}$$

is a Toeplitz matrix,[6] in which all elements in each diagonal are equal. R_M is called the Mth-order covariance matrix. Equation (3.18) consists of $M - 1$ equations of the form

$$R_{M-1} \begin{pmatrix} b_{M,M-1} \\ \vdots \\ b_{M,1} \end{pmatrix} = \begin{pmatrix} r_{M-1} - b_{M,M} r_1^* \\ \vdots \\ r_1 - b_{M,M} r_{M-1}^* \end{pmatrix} \qquad (3.21)$$

where R_{M-1} is the $(M - 1)$-th order covariance matrix. Multiplying (3.21) by R_{M-1}^{-1} yields

$$\begin{pmatrix} b_{M,M-1} \\ \vdots \\ b_{M,1} \end{pmatrix} = R_{M-1}^{-1} \begin{pmatrix} r_{M-1} \\ \vdots \\ r_1 \end{pmatrix} - b_{M,M} R_{M-1}^{-1} \begin{pmatrix} r_1^* \\ \vdots \\ r_{M-1}^* \end{pmatrix} \qquad (3.22)$$

Using the Toeplitz properties of a covariance matrix, Eq. (3.22) reduces to

$$\begin{pmatrix} b_{M,M-1} \\ \vdots \\ b_{M,1} \end{pmatrix} = \begin{pmatrix} b_{M-1,M-1} \\ \vdots \\ b_{M-1,1} \end{pmatrix} - b_{M,M} \begin{pmatrix} b_{M-1,1}^* \\ \vdots \\ b_{M-1,M-1}^* \end{pmatrix} \qquad (3.23)$$

The procedure for demonstrating the equivalence between Eqs. (3.22) and (3.23) is more explicitly presented in Section 3.2.4 (see Eqs. (3.67) and (3.68)). Equation (3.23) can now be written as

$$b_{M,k} = b_{M-1,k} - b_{M,M} b_{M-1,M-k}^* \qquad k = 1, \ldots, M - 1. \qquad (3.24)$$

The coefficient $b_{M,M}$ is termed the reflection coefficient (RC) or partial correlation coefficient (PARCOR) (see Chapter 11). The reflection coefficient must be computed in order to make the recursive procedure complete. The first equation in (3.18) can be rewritten in vector form as follows:

$$b_{M,M} r_0 + (r_1, r_2, \ldots, r_{M-1}) \begin{pmatrix} b_{M,M-1} \\ \vdots \\ b_{M,1} \end{pmatrix} = r_M \qquad (3.25)$$

When (3.23) is substituted into (3.25), the solution for the reflection coefficient becomes

$$b_{M,M} = \frac{r_M - (r_1, r_2, \ldots, r_{M-1}) \begin{pmatrix} b_{M-1,M-1} \\ \vdots \\ b_{M-1,1} \end{pmatrix}}{r_0 - (r_1, r_2, \ldots, r_{M-1}) \begin{pmatrix} b_{M-1,1}^* \\ \vdots \\ b_{M-1,M-1}^* \end{pmatrix}} \qquad (3.26)$$

Knowledge of $b_{M-1,k}$ for $k = 1,\ldots, M-1$ from the previous iteration allows the computation of $b_{M,M}$. Subsequently, $b_{M,k}$ for $k = 1,\ldots, M-1$ is computed from (3.24) and (3.26). Equation (3.24) is called the Levinson recursion.[7]

Omitting the order iteration index, the denominator in (3.26) can be written as

$$r_0 - (r_1, r_2, \ldots, r_{M-1}) \begin{pmatrix} b_1^* \\ \vdots \\ b_{M-1}^* \end{pmatrix} = \sum_{m=0}^{M-1} a_m^* r_m \qquad (3.27)$$

Since the received signal is assumed to be stationary and the minimum mean square error (MMSE) is always positive, P_{M-1} may be expressed as

$$P_{M-1_{\min}} = \sum_{m=0}^{M-1} a_m r_{-m}$$

$$= \sum_{m=0}^{M-1} a_m^* r_m \qquad (3.28)$$

From (3.27) and (3.28), the denominator of (3.26) can be identified as $P_{M-1_{\min}}$. Equation (3.26) can then be rewritten as

$$a_{M,M} = \frac{-r_M - (r_1, r_2, \ldots, r_{M-1}) \begin{pmatrix} a_{M-1,M-1} \\ \vdots \\ a_{M-1,1} \end{pmatrix}}{P_{M-1_{\min}}} \qquad (3.29)$$

Adding the order iteration index to the prediction coefficients allows $P_{M_{\min}}$ to be expressed as

$$P_{M_{\min}} = \sum_{m=0}^{M} a_{M,m}^* r_m \qquad (3.30)$$

Expanding (3.30) and combining the result with (3.24), (3.28) and (3.29) yields

$$P_{M_{\min}} = a_{M,M}^* \left(r_M + \sum_{m=1}^{M-1} a_{M-1,M-m} r_m \right) + \sum_{m=1}^{M-1} a_{M-1,m}^* r_m + r_0$$

$$= P_{M-1_{\min}} (1 - |a_{M,M}|^2) \qquad (3.31)$$

where $a_{M,0} = 1$ and $a_{M,m} = -b_{M,m}$, $m = 1, 2, \ldots, M$. Equation (3.31) reveals that the MMSE, $P_{M_{\min}}$, can be computed recursively from $P_{M-1_{\min}}$ and $a_{M,M}$.

The initial value of $P_{M_{\min}}$ is $P_{0_{\min}} = r_0$. The Levinson recursion relationship in (3.24) can be rewritten as

$$a_{M,k} = a_{M-1,k} + a_{M,M}a^*_{M-1,M-k}, \qquad k = 1,\ldots, M-1 \qquad (3.32)$$

by noting that $a_{M,k} = -b_{M,k}$ for $k = 1, 2, \ldots, M-1$. Equations (3.24) (or (3.32)), (3.29), and (3.31) are the essential equations required for computing the prediction coefficients recursively (order iterative method). The computation procedure based on these equations is called the Durbin algorithm (DA).[1]

EXAMPLE

Suppose the autocorrelation coefficients are known as $r_n = \rho^n$, for $n = 0, 1, 2, 3$. The prediction coefficients can be computed by the direct method (solution of simultaneous equations) or the iterative method (Durbin algorithm). The results based on these two methods are identical.

Proof. Let's consider the direct method first by forming the simultaneous equations using (3.10)

$$\begin{pmatrix} 1 & \rho & \rho^2 \\ \rho^* & 1 & \rho \\ \rho^{2*} & \rho^* & 1 \end{pmatrix} \begin{pmatrix} b_3 \\ b_2 \\ b_1 \end{pmatrix} = \begin{pmatrix} \rho^3 \\ \rho^2 \\ \rho \end{pmatrix} \qquad (3.33)$$

A direct solution of the matrix equations results in

$$\begin{pmatrix} b_3 \\ b_2 \\ b_1 \end{pmatrix} = \dfrac{1 - |\rho|^2}{\begin{vmatrix} 1 & \rho & \rho^2 \\ \rho^* & 1 & \rho \\ \rho^{2*} & \rho^* & 1 \end{vmatrix}} \begin{pmatrix} 1 & -\rho & 0 \\ -\rho^* & 1 + |\rho|^2 & -\rho \\ 0 & -\rho^* & 1 \end{pmatrix} \begin{pmatrix} \rho^3 \\ \rho^2 \\ \rho \end{pmatrix} \qquad (3.34)$$

where the determinant becomes

$$\begin{vmatrix} 1 & \rho & \rho^2 \\ \rho^* & 1 & \rho \\ \rho^{2*} & \rho^* & 1 \end{vmatrix} = \left(1 - |\rho|^2\right)^2 \qquad (3.35)$$

Substituting Eq. (3.35) into Eq. (3.34) gives

$$\begin{pmatrix} b_3 \\ b_2 \\ b_1 \end{pmatrix} = \begin{pmatrix} 0 \\ 0 \\ \rho \end{pmatrix} \qquad (3.36)$$

On the other hand, if the iterative Durbin algorithm is used, the first order iteration steps from (3.29), (3.31) and (3.32) are given by

$$a_{1,1} = \frac{-a_{0,0}r_1}{P_{0_{min}}} = -\frac{r_1}{r_0} = -\rho$$

$$a_{1,0} = 1 \qquad\qquad\qquad\qquad\qquad\qquad (3.37)$$

and

$$P_{1_{min}} = P_{0_{min}}(1 - |a_{1,1}|^2) = r_0(1 - |a_{1,1}|^2) \qquad (3.38)$$

$$= 1 - |\rho|^2 \qquad\qquad\qquad\qquad (3.39)$$

where $a_{0,0} = 1$. The second order iteration steps are

$$a_{2,2} = \frac{-a_{1,0}r_2 - a_{1,1}r_1}{P_{1_{min}}} = \frac{-r_2 + \rho r_1}{1 - |\rho|^2} \qquad (3.40)$$

$$= \frac{-\rho^2 + \rho^2}{1 - |\rho|^2} = 0 \qquad\qquad\qquad (3.41)$$

$$a_{2,1} = a_{1,1} + a_{2,2}a_{1,1}^* = -\rho \qquad\qquad (3.42)$$

$$a_{2,0} = 1$$

and

$$P_{2_{min}} = P_{1_{min}}(1 - |a_{2,2}|^2) = 1 - |\rho|^2 \qquad (3.43)$$

The third order iteration steps are

$$a_{3,3} = \frac{-a_{2,0}r_3 - a_{2,1}r_2 - a_{2,2}r_1}{P_{2_{min}}} = \frac{-r_3 + \rho r_2}{1 - |\rho|^2}$$

$$= \frac{-\rho^3 + \rho^3}{1 - |\rho|^2} = 0 = -b_3 \qquad\qquad (3.44)$$

$$a_{3,2} = a_{2,2} + a_{3,3}a_{2,1}^* = 0 = -b_2 \qquad\qquad (3.45)$$

$$a_{3,1} = a_{2,1} + a_{3,3}a_{2,2}^* = -\rho = -b_1 \qquad\qquad (3.46)$$

From the above computation it can be seen that the solution $b_1 = \rho$, $b_2 = 0$, $b_3 = 0$ from the Durbin algorithm is identical to that obtained from the direct method.

3.2 LEVINSON ALGORITHM

3.2.1 Model Description

In this section we present the Levinson recursive algorithm.[7,8] The Levinson algorithm is similar to the Durbin algorithm, except that the former estimates and predicts the transmitted signal while the latter estimates and predicts the received signal. The received signal is assumed to be stationary and to have been generated by an all-pole model. Therefore the estimation can be accomplished by means of linear prediction.

Let x_i and s_i denote the lowpass equivalent of the received and transmitted signals taken at the ith sample, respectively. The signal can be expressed as

$$x_i = s_i + n_i \qquad i = 1, 2, \ldots, N \tag{3.47}$$

where n_i is assumed to be stationary white noise and N is the number of samples. Let us predict the transmitted signal s_i from delayed versions of the received signal x_{i-m}, $m = 1, 2, \ldots, M$:

$$\hat{s}_i = \sum_{m=1}^{M} b_m x_{i-m} \tag{3.48}$$

where $\hat{s}_i$ is the estimated signal, $\{b_m\}$ are the prediction coefficients, and M is the order of the prediction filter.

3.2.2 Derivation of Normal Equations by Differentiation

The prediction coefficients can be determined by minimizing the mean square error P_M, defined as

$$P_M = \mathscr{E}\left[|\varepsilon_i|^2\right]$$

$$= \mathscr{E}\left[\left|s_i - \sum_{m=1}^{M} b_m x_{i-m}\right|^2\right] \tag{3.49}$$

where ε_i is the error signal at the ith sample

$$\varepsilon_i = s_i - \hat{s}_i \tag{3.50}$$

Minimizing P_M with respect to the prediction coefficients $\{b_l^*\}$ results in

$$\frac{\partial}{\partial b_l^*} P_M = -2\mathscr{E}\left[x_{i-l}^*\left(s_i - \sum_{m=1}^{M} b_m x_{i-m}\right)\right]$$

$$= -2\left[g_l - \sum_{m=1}^{M} b_m r_{l-m}\right] \tag{3.51}$$

$$= 0, \qquad l = 1, \ldots, M$$

where

$$g_l = \mathscr{E}\left[s_i x_{i-l}^*\right] \tag{3.52}$$

is the crosscorrelation between the received and transmitted signals and r_l is the autocorrelation coefficient defined in (3.8). Equations (3.51) can be written as

$$g_l = \sum_{m=1}^{M} b_m r_{l-m}, \qquad l = 1, 2, \ldots, M \tag{3.53}$$

Equations (3.53) are called the normal equations. The coefficients b_m, resulting from the solution of these equations, are referred to as Wiener filter coefficients.[8]

3.2.3   Derivation of the Normal Equations by the Orthogonality Principle

The normal equations can also be derived by assuming that the received and error signals are statistically orthogonal to each other:

$$\mathscr{E}\left[x_{i-l}\varepsilon_i^*\right] = 0 \qquad l = 1, 2, \ldots, M \tag{3.54}$$

Substituting Eq. (3.50) into Eqs. (3.54) and noting that the signal is stationary, gives

$$\mathscr{E}\left[x_{i-l}s_i^*\right] = \sum_{m=1}^{M} b_m^* \mathscr{E}\left[x_{i-l}x_{i-m}^*\right] \tag{3.55}$$

or

$$g_l^* = \sum_{m=1}^{M} b_m^* r_{m-l}, \qquad l = 1, 2, \ldots, M \tag{3.56}$$

Since $r_{m-l} = r_{l-m}^*$, the normal equations (3.53) are easily obtained from Eq. (3.56).

The minimum MSE, $P_{M_{\min}}$, can be obtained easily by use of the orthogonality principle in the expansion of Eq. (3.49):

$$P_M = \mathscr{E}\left[|\varepsilon_i|^2\right]$$

$$= \mathscr{E}\left[\left(s_i - \sum_{m=1}^{M} b_m x_{i-m}\right)\varepsilon_i^*\right] \tag{3.56a}$$

Using the orthogonality principle (Eq. 3.54), Eq. (3.56a) becomes

$$P_{M_{\min}} = \mathscr{E}\left(s_i \varepsilon_i^*\right)$$

$$= \mathscr{E}\left(s_i\left(s_i^* - \sum_{m=1}^{M} b_m^* x_{i-m}^*\right)\right) \tag{3.56b}$$

Letting $\sigma_s^2 = \mathscr{E}(s_i s_i^*)$ and using Eq. (3.52) in (3.56b) yields

$$P_{M_{\min}} = \sigma_s^2 - \sum_{m=1}^{M} b_m g_m^* \tag{3.56c}$$

since $P_{M_{\min}}$ is real.

3.2.4 Levinson Recursive Solution

The prediction coefficients $\{b_m\}$ can be obtained directly by solving the simultaneous normal equations (3.53) or recursively using the Levinson algorithm. The Levinson algorithm is an order-recursive algorithm. Adding a new subscript M to indicate the Mth recursion step, Eqs. (3.53) are given by

$$\sum_{m=1}^{M} b_{M,m} r_{l-m} = g_l \qquad l = 1, \ldots, M \tag{3.57}$$

or

$$
\begin{aligned}
b_{M,M} r_0 &+ b_{M,M-1} r_1 + \cdots + & b_{M,1} r_{M-1} &= g_M \\
b_{M,M} r_{-1} &+ b_{M,M-1} r_0 + \cdots + & b_{M,1} r_{M-2} &= g_{M-1} \\
&\;\;\vdots & &\;\;\vdots \\
b_{M,M} r_{1-M} &+ b_{M,M-1} r_{2-M} + \cdots + b_{M,1} r_0 & &= g_1
\end{aligned}
\tag{3.58}
$$

Equations (3.58) differ from Eqs. (3.18) only in the right hand side. Similarly, the M equations take the form

$$\boldsymbol{R}_M \begin{pmatrix} b_{M,M} \\ \vdots \\ b_{M,1} \end{pmatrix} = \begin{pmatrix} g_M \\ \vdots \\ g_1 \end{pmatrix} \tag{3.59}$$

where $\boldsymbol{R}_M$ is given in Eq. (3.20). The last $(M-1)$ equations in (3.58) take the form

$$\boldsymbol{R}_{M-1} \begin{pmatrix} b_{M,M-1} \\ \vdots \\ b_{M,1} \end{pmatrix} = \begin{pmatrix} g_{M-1} - b_{M,M} r_1^* \\ \vdots \\ g_1 - b_{M,M} r_{M-1}^* \end{pmatrix} \tag{3.60}$$

Multiplying Eq. (3.60) by $\boldsymbol{R}_{M-1}^{-1}$ yields

$$
\begin{aligned}
\begin{pmatrix} b_{M,M-1} \\ \vdots \\ b_{M,1} \end{pmatrix} &= \boldsymbol{R}_{M-1}^{-1} \begin{pmatrix} g_{M-1} \\ \vdots \\ g_1 \end{pmatrix} - b_{M,M} \boldsymbol{R}_{M-1}^{-1} \begin{pmatrix} r_1^* \\ \vdots \\ r_{M-1}^* \end{pmatrix} \\
&= \begin{pmatrix} b_{M-1,M-1} \\ \vdots \\ b_{M-1,1} \end{pmatrix} - b_{M,M} \boldsymbol{R}_{M-1}^{-1} \begin{pmatrix} r_1^* \\ \vdots \\ r_{M-1}^* \end{pmatrix}
\end{aligned}
\tag{3.61}
$$

where the first-column vector in the right-hand side is obtained from the solution to the $(M - 1)$ equations using the form given in Eq. (3.59). Define the intermediate vector

$$\begin{pmatrix} \phi^*_{M-1,1} \\ \vdots \\ \phi^*_{M-1,M-1} \end{pmatrix} \equiv \boldsymbol{R}^{-1}_{M-1} \begin{pmatrix} r^*_1 \\ \vdots \\ r^*_{M-1} \end{pmatrix} \tag{3.62}$$

Equation (3.61) is then given by

$$\begin{pmatrix} b_{M,M-1} \\ \vdots \\ b_{M,1} \end{pmatrix} = \begin{pmatrix} b_{M-1,M-1} \\ \vdots \\ b_{M-1,1} \end{pmatrix} - b_{M,M} \begin{pmatrix} \phi^*_{M-1,1} \\ \vdots \\ \phi^*_{M-1,M-1} \end{pmatrix} \tag{3.63}$$

Equation (3.63) may be expressed in general at the kth step as

$$b_{M,k} = b_{M-1,k} - b_{M,M}\phi^*_{M-1,M-k} \qquad k = 1, \ldots, M - 1 \tag{3.64}$$

The reflection coefficient $b_{M,M}$ can be obtained from the first equation in Eqs. (3.58) as

$$b_{M,M}r_0 + (r_1, r_2, \ldots, r_{M-1}) \begin{pmatrix} b_{M,M-1} \\ \vdots \\ b_{M,1} \end{pmatrix} = g_M \tag{3.65}$$

Substituting Eq. (3.63) into Eq. (3.65) yields

$$b_{M,M} = \frac{g_M - (r_1, r_2, \ldots, r_{M-1}) \begin{pmatrix} b_{M-1,M-1} \\ \vdots \\ b_{M-1,1} \end{pmatrix}}{r_0 - (r_1, r_2, \ldots, r_{M-1}) \begin{pmatrix} \phi^*_{M-1,1} \\ \vdots \\ \phi^*_{M-1,M-1} \end{pmatrix}} \tag{3.66}$$

Knowledge of $b_{M-1,k}$ for $k = 1, \ldots, M - 1$ from the previous iteration and from the intermediate vector in (3.62) allows the computation of $b_{M,M}$ and subsequently $b_{M,k}$. To complete the Levinson recursion, computation of the intermediate vector given in Eq. (3.62) is required. Equation (3.62) may be written as

$$r_0\phi_{M-1,1} + r^*_1\phi_{M-1,2} + \cdots + r^*_{M-2}\phi_{M-1,M-1} = r_1$$

$$r_1\phi_{M-1,1} + r_0\phi_{M-1,2} + \cdots + r^*_{M-3}\phi_{M-1,M-1} = r_2$$

$$\vdots \qquad\qquad\qquad\qquad\qquad \vdots$$

$$r_{M-2}\phi_{M-1,1} + r_{M-3}\phi_{M-1,2} + \cdots + r_0\phi_{M-1,M-1} = r_{M-1} \tag{3.67}$$

Reversing the order of equations and terms in each equation, Eqs. (3.67) become

$$r_0\phi_{M-1,M-1} + \cdots + r_{M-2}\phi_{M-1,1} = r_{M-1}$$

$$\vdots \qquad\qquad\qquad \vdots$$

$$r_{M-2}^*\phi_{M-1,M-1} + \cdots + r_0\phi_{M-1,1} = r_1 \tag{3.68}$$

or

$$\begin{pmatrix} \phi_{M-1,M-1} \\ \vdots \\ \phi_{M-1,1} \end{pmatrix} = R_{M-1}^{-1} \begin{pmatrix} r_{M-1} \\ \vdots \\ r_1 \end{pmatrix} \tag{3.69}$$

Equations. (3.68) or (3.69) have the same form as the Durbin recursion for $M - 1$ equations. (see (3.19), (3.24) and (3.26)).
The vector

$$\begin{pmatrix} \phi_{M-1,M-1} \\ \vdots \\ \phi_{M-1,1} \end{pmatrix}$$

can be solved using the Durbin recursion as follows:

$$\phi_{M-1,k} = \phi_{M-2,k} - \phi_{M-1,M-1}\phi_{M-2,M-1-k}^*, \qquad k = 1, 2, \ldots, M - 2$$

$$\tag{3.70}$$

$$\phi_{M-1,M-1} = \frac{r_{M-1} - (r_1, r_2, \ldots, r_{M-2})\begin{pmatrix} \phi_{M-2,M-2} \\ \vdots \\ \phi_{M-2,1} \end{pmatrix}}{r_0 - (r_1, r_2, \ldots, r_{M-2})\begin{pmatrix} \phi_{M-2,1}^* \\ \vdots \\ \phi_{M-2,M-2}^* \end{pmatrix}} \tag{3.71}$$

Equations (3.70), (3.71), (3.64), and (3.66) constitute the Levinson recursion algorithm.[7,8] A table illustrating the recursive equations for 1, 2, 3 in (3.57) or 4 equations is shown on the following page.

EXAMPLE OF LEVINSON RECURSIVE ALGORITHM

Let $r_n = \rho^n$ and $g_n = q^{n-1}$ for $n = 1, 2, 3$ in (3.57). Show that the predictor coefficients obtained from the Levinson recursive algorithm are equivalent to those obtained by solving the simultaneous equations.

TABLE 3-1. LEVINSON RECURSION FOR 1, 2, 3, AND 4 EQUATIONS

Number Equations	Normal Equations	Solution of Normal Equations
1	$b_{1,1}r_0 = g_1$	$b_{1,1} = \dfrac{g_1}{r_0}$
2	$b_{2,2}r_0 + b_{2,1}r_1 = g_2$ $b_{2,2}r_1^* + b_{2,1}r_0 = g_1$	$\phi_{1,1} = \dfrac{r_1}{r_0}$ $b_{2,2} = \dfrac{g_2 - r_1 b_{1,1}}{r_0 - r_1 \phi_{1,1}^*}$ $b_{2,1} = b_{1,1} - b_{2,2}\phi_{1,1}^*$
3	$b_{3,3}r_0 + b_{3,2}r_1 + b_{3,1}r_2 = g_3$ $b_{3,3}r_1^* + b_{3,2}r_0 + b_{3,1}r_1 = g_2$ $b_{3,3}r_2^* + b_{3,2}r_1^* + b_{3,1}r_0 = g_1$	$\phi_{2,2} = \dfrac{r_2 - r_1 \phi_{1,1}}{r_0 - r_1 \phi_{1,1}^*}$ $\phi_{2,1} = \phi_{1,1} - \phi_{2,2}\phi_{1,1}^*$ $b_{3,3} = \dfrac{g_3 - r_1 b_{2,2} - r_2 b_{2,1}}{r_0 - r_1 \phi_{2,1}^* - r_2 \phi_{2,2}^*}$ $\begin{pmatrix} b_{3,1} \\ b_{3,2} \end{pmatrix} = \begin{pmatrix} b_{2,1} \\ b_{2,2} \end{pmatrix} - b_{3,3}\begin{pmatrix} \phi_{2,2}^* \\ \phi_{2,1}^* \end{pmatrix}$
4	$b_{4,4}r_0 + b_{4,3}r_1 + b_{4,2}r_2 + b_{4,1}r_3 = g_4$ $b_{4,4}r_1^* + b_{4,3}r_0 + b_{4,2}r_1 + b_{4,1}r_2 = g_3$ $b_{4,4}r_2^* + b_{4,3}r_1^* + b_{4,2}r_0 + b_{4,1}r_1 = g_2$ $b_{4,4}r_3^* + b_{4,3}r_2^* + b_{4,2}r_1^* + b_{4,1}r_0 = g_1$	$\phi_{3,3} = \dfrac{r_3 - (r_1, r_2)\begin{pmatrix} \phi_{2,2} \\ \phi_{2,1} \end{pmatrix}}{r_0 - (r_1, r_2)\begin{pmatrix} \phi_{2,1}^* \\ \phi_{2,2}^* \end{pmatrix}}$ $\begin{pmatrix} \phi_{3,2} \\ \phi_{3,1} \end{pmatrix} = \begin{pmatrix} \phi_{2,2} \\ \phi_{2,1} \end{pmatrix} - \phi_{3,3}\begin{pmatrix} \phi_{2,1}^* \\ \phi_{2,2}^* \end{pmatrix}$ $b_{4,4} = \dfrac{g_4 - (r_1, r_2, r_3)\begin{pmatrix} b_{3,3} \\ b_{3,2} \\ b_{3,1} \end{pmatrix}}{r_0 - (r_1, r_2, r_3)\begin{pmatrix} \phi_{3,1}^* \\ \phi_{3,2}^* \\ \phi_{3,3}^* \end{pmatrix}}$ $\begin{pmatrix} b_{4,1} \\ b_{4,2} \\ b_{4,3} \end{pmatrix} = \begin{pmatrix} b_{3,1} \\ b_{3,2} \\ b_{3,3} \end{pmatrix} - b_{4,4}\begin{pmatrix} \phi_{3,3}^* \\ \phi_{3,2}^* \\ \phi_{3,1}^* \end{pmatrix}$

Problem.   Solve the simultaneous equations

$$\begin{pmatrix} 1 & \rho & \rho^2 \\ \rho^* & 1 & \rho \\ \rho^{2*} & \rho^* & 1 \end{pmatrix} \begin{pmatrix} b_3 \\ b_2 \\ b_1 \end{pmatrix} = \begin{pmatrix} q^2 \\ q \\ 1 \end{pmatrix} \tag{3.72}$$

Using matrix inversion, the solution becomes

$$\begin{pmatrix} b_3 \\ b_2 \\ b_1 \end{pmatrix} = \begin{pmatrix} 1 & \rho & \rho^2 \\ \rho^* & 1 & \rho \\ \rho^{2*} & \rho^* & 1 \end{pmatrix}^{-1} \begin{pmatrix} q^2 \\ q \\ 1 \end{pmatrix}$$

$$= \frac{1}{1 - |\rho|^2} \begin{pmatrix} 1 & -\rho & 0 \\ -\rho^* & 1 + |\rho|^2 & -\rho \\ 0 & -\rho^* & 1 \end{pmatrix} \begin{pmatrix} q^2 \\ q \\ 1 \end{pmatrix}$$

$$= \begin{pmatrix} \dfrac{q^2 - \rho q}{1 - |\rho|^2} \\ \dfrac{q(1 + |\rho|^2) - \rho - q^2\rho^*}{1 - |\rho|^2} \\ \dfrac{1 - q\rho^*}{1 - |\rho|^2} \end{pmatrix} \tag{3.73}$$

On the other hand, if the recursive Levinson algorithm is used, the first order iteration steps are given by

$$b_{1,1} = \frac{g_1}{r_0} = 1 \tag{3.74}$$

$$\phi_{1,1} = \frac{r_1}{r_0} = \rho \tag{3.75}$$

Then the second order iteration steps become

$$b_{2,2} = \frac{g_2 - r_1 b_{1,1}}{r_0 - r_1 \phi_{1,1}^*} = \frac{q - \rho}{1 - |\rho|^2} \tag{3.76}$$

$$b_{2,1} = b_{1,1} - b_{2,2}\phi_{1,1}^* = 1 - \frac{q - \rho}{1 - |\rho|^2}\rho^* = \frac{1 - q\rho^*}{1 - |\rho|^2} \tag{3.77}$$

$$\phi_{2,2} = \frac{r_2 - r_1 \phi_{1,1}}{r_0 - r_1 \phi_{1,1}^*} = \frac{\rho^2 - \rho^2}{1 - |\rho|^2} = 0 \tag{3.78}$$

$$\phi_{2,1} = \phi_{1,1} - \phi_{2,2}\phi_{1,1}^* = \rho \tag{3.79}$$

Subsequently, the third order iteration steps follow

$$b_{3,3} = \frac{g_3 - r_1 b_{2,2} - r_2 b_{2,1}}{r_0 - r_1 \phi_{2,1}^* - r_2 \phi_{2,2}^*} = \frac{q^2 - \rho\left(\dfrac{q - \rho}{1 - |\rho|^2}\right) - \rho^2\left(\dfrac{1 - q\rho^*}{1 - |\rho|^2}\right)}{1 - |\rho|^2}$$

$$= \frac{q^2 - \rho q}{1 - |\rho|^2} \tag{3.80}$$

$$b_{3,1} = b_{2,1} - b_{3,3}\phi_{2,2}^* = \frac{1 - q\rho^*}{1 - |\rho|^2} \tag{3.81}$$

$$b_{3,2} = b_{2,2} - b_{3,3}\phi_{2,1}^* = \frac{q - \rho}{1 - |\rho|^2} - \frac{q^2 - \rho q}{1 - |\rho|^2}\rho^*$$

$$= \frac{q(1 + |\rho|^2) - \rho - q^2\rho^*}{1 - |\rho|^2} \tag{3.82}$$

By comparing (3.73) with (3.80), (3.81) and (3.82), it can be seen that the predictor coefficients obtained from the direct method and Levinson recursive method are identical.

EXAMPLE 2:   LEVINSON RECURSIVE ALGORITHM

Suppose two unknown samples, s_1 and s_0, are transmitted through a channel modeled as a tapped delay line as shown in Figure 3-2. Then the output signal samples $\{\tilde{y}_k\}$ are given by

$$\begin{cases} \tilde{y}_0 = f_0 s_0 + n_0 \\ \tilde{y}_1 = f_0 s_1 + f_1 s_0 + n_1 \\ \tilde{y}_2 = f_1 s_1 + n_2 \end{cases} \tag{3.83}$$

where $\{n_k\}$ are unknown additive noise samples. The least squares solutions

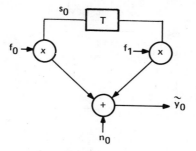

Figure 3-2. A tapped delay line.

for s_0 and s_1, denoted by $\hat{s}_0$ and $\hat{s}_1$, respectively, is now obtained by ignoring the noise and rewriting (3.83) as

$$Ax = y$$

where

$$x = \begin{pmatrix} s_0 \\ s_1 \end{pmatrix}, \quad A = \begin{pmatrix} f_0 & 0 \\ f_1 & f_0 \\ 0 & f_1 \end{pmatrix} \quad \text{and} \quad y = \begin{pmatrix} \tilde{y}_0 \\ \tilde{y}_1 \\ \tilde{y}_2 \end{pmatrix}$$

Defining

$$x_1 = \begin{pmatrix} f_0 \\ f_1 \\ 0 \end{pmatrix} \quad \text{and} \quad x_2 = \begin{pmatrix} 0 \\ f_0 \\ f_1 \end{pmatrix}$$

an approximation, y_a, for y is $y_a = \hat{s}_0 x_1 + \hat{s}_1 x_2$. Using the inner product (2.20), the least squares solutions for s_0 and s_1 given the output signals are determined from the following equations (see Chapter 2).

$$\begin{pmatrix} \hat{s}_1 \\ \hat{s}_0 \end{pmatrix} = \begin{bmatrix} r_0 & r_1 \\ r_1^* & r_0 \end{bmatrix}^{-1} \begin{pmatrix} g_2 \\ g_1 \end{pmatrix} \tag{3.84}$$

where

$$\begin{cases} r_0 = |f_0|^2 + |f_1|^2 \\ r_1 = f_0^* f_1 \\ g_1 = f_0^* \tilde{y}_0 + f_1^* \tilde{y}_1 \\ g_2 = f_0^* \tilde{y}_1 + f_1^* \tilde{y}_2 \end{cases}$$

Equations (3.84) can be solved using the Levinson recursive equations, that is,

$$\begin{cases} b_{1,1} = \dfrac{g_1}{r_0} \\[2mm] \phi_{1,1} = \dfrac{r_1}{r_0} \\[2mm] b_{2,2} = \dfrac{g_2 - r_1 b_{1,1}}{r_0 - r_1 \phi_{1,1}^*} = \hat{s}_1 \\[2mm] b_{2,1} = b_{1,1} - b_{2,2}\phi_{1,1}^* = \hat{s}_0 \end{cases} \tag{3.85}$$

For example, if s's, f's and $\tilde{y}$'s are assigned as follows:

$$\begin{cases} s_0 = 1, s_1 = -1, f_0 = 1, f_1 = 1 \\ \tilde{y}_0 = 0.8, \tilde{y}_1 = 0.1, \tilde{y}_2 = -1.1 \end{cases}, \tag{3.86}$$

then r's, and g's are given by

$$\begin{cases} r_0 = 2, r_1 = 1, \\ g_1 = 0.9, g_2 = -1 \end{cases} \tag{3.87}$$

The unknowns, $\hat{s}_0$ and $\hat{s}_1$, are computed as follows

$$\begin{cases} b_{1,1} = \dfrac{0.9}{2} \\[2mm] \phi_{1,1} = \tfrac{1}{2} \\[2mm] b_{2,2} = \dfrac{-1 - \dfrac{0.9}{2}}{2 - 0.5} = -0.966 = \hat{s}_1 \\[2mm] b_{2,1} = \dfrac{0.9}{2} + \dfrac{2.9}{6} = 0.933 = \hat{s}_0 \end{cases}$$

The numerical values selected for the received signal correspond to a high signal to noise ratio condition. As a result the least square solutions ignoring the noise are close to the actual transmitted values.

3.3   CHOLESKY DECOMPOSITION ALGORITHM

In the derivation of the Yule-Walker equations given in (3.10) the signal x_i is assumed to be stationary. As a result of the stationarity assumption, the covariance matrix, R_M, given in (3.20), is a Toeplitz matrix. The Toeplitz form of R_M then led to the simple Durbin recursive algorithm for the predictor coefficients. If the signal, x_i, is not stationary, the Durbin algorithm is not applicable and a method referred to as Cholesky decomposition can be used.[9] The development of this algorithm is similar to that followed above for the stationary case and is described below.

Assume that the error signal, e_i, at sample i is given by (3.5) and (3.6). The mean square error is again represented by (3.3) and for the nonstationary case leads to a generalization of (3.7):

$$P_M = \sum_{m=0}^{M} \sum_{j=0}^{M} a_m a_j^* \mathscr{E}\left[x_{i-m} x_{i-j}^*\right] \tag{3.88}$$

If the ensemble average autocorrelation coefficients are defined in the nonstationary signal case by r_{mj} where

$$r_{mj} = \mathscr{E}\left[x_{i-m} x_{i-j}^*\right], \tag{3.89}$$

then the mean square error can be expressed as

$$P_M = \sum_{m=0}^{M} \sum_{j=0}^{M} a_m a_j^* r_{mj}$$

Minimization of P_M with respect to the predictor coefficients a_i^* leads to the normal equations

$$\sum_{m=1}^{M} a_m r_{ml} = -r_{0l}, \qquad l = 1, 2, \ldots, M \tag{3.90}$$

Equations (3.90) can be expressed in matrix form as

$$\boldsymbol{R}_M \boldsymbol{A}_M = -\boldsymbol{F}_M \tag{3.91}$$

where

$$\boldsymbol{R}_M = \begin{pmatrix} r_{11} & r_{12} & \cdots & r_{1M} \\ r_{21} & r_{22} & \cdots & r_{2M} \\ \vdots & & & \\ r_{M1} & r_{M2} & \cdots & r_{MM} \end{pmatrix} \tag{3.92}$$

$$\boldsymbol{A}_M^t = (a_1, a_2, \ldots, a_M) \tag{3.93}$$

$$\boldsymbol{F}_M^t = (r_{01}, r_{02}, \ldots, r_{0M}) \tag{3.94}$$

From (3.89) it can be seen that

$$r_{mj} = r_{jm}^*$$

so that the covariance matrix in (3.92) is complex conjugate symmetric. As a result, the covariance matrix, R_M, can be decomposed as follows (see Appendix Sections A.2.5 and A.2.6):

$$R_M = S_M S_M^{*t} \tag{3.95}$$

where

$$S_M = \begin{pmatrix} s_{11} & & & \phi \\ s_{21} & s_{22} & & \\ \vdots & & \ddots & \\ s_{M1} & s_{M2} & \cdots & s_{MM} \end{pmatrix} \tag{3.96}$$

is a lower triangular matrix and ϕ represents zero elements. This formulation is referred to as a square root factorization. Defining the received signal vector by $X_M^t = (x_1, x_2, \ldots, x_M)$, the elements of the triangular matrices can be obtained by examining the quadratic form $X_M^t R_M X_M^*$. This form can be expanded as follows:

$$X_M^t R_M X_M^* = \sum_{i=1}^{M} \sum_{j=1}^{M} r_{ij} x_i x_j^*$$

$$= r_{11}|x_1|^2 + \sum_{j=2}^{M} r_{1j} x_1 x_j^* + \sum_{i=2}^{M} r_{i1} x_i x_1^* + \sum_{i=2}^{M} \sum_{j=2}^{M} r_{ij} x_i x_j^*$$

$$= \left| r_{11}^{1/2} x_1 + \sum_{j=2}^{M} \frac{r_{j1}}{r_{11}^{1/2}} x_j \right|^2 - \left| \sum_{j=2}^{M} \frac{r_{j1}}{r_{11}^{1/2}} x_j \right|^2 + \sum_{i=2}^{M} \sum_{j=2}^{M} r_{ij} x_i x_j^* \tag{3.97}$$

Equation (3.97) can be rewritten as

$$X_M^t R_M X_M^* = |y_1|^2 + \sum_{i=2}^{M} \sum_{j=2}^{M} \left(r_{ij} x_i x_j^* - s_{i1} s_{j1}^* x_i x_j^* \right) \tag{3.98}$$

where

$$\begin{cases} s_{11} = r_{11}^{1/2} \\ s_{j1} = \dfrac{r_{j1}}{s_{11}}, \qquad j = 2, \ldots, M \\ y_1 = \displaystyle\sum_{j=1}^{M} s_{j1} x_j \end{cases} \tag{3.99}$$

Equation (3.98) can be rewritten by defining the quantity $r_{ij}^{(2)}$ as follows

$$r_{ij}^{(2)} \equiv r_{ij} - s_{i1}s_{j1}^*, \qquad i = 2, \ldots, M; \quad j = 2, \ldots, M$$

so that

$$X_M^t R_M X_M^* = |y_1|^2 + \sum_{i=2}^{M} \sum_{j=2}^{M} r_{ij}^{(2)} x_i x_j^* \tag{3.100}$$

By repeating the above procedure for another $(M - 1)$ steps, Eq. (3.100) can be reduced to

$$X_M^t R_M X_M^* = \sum_{i=1}^{M} |y_i|^2 \tag{3.101}$$

where

$$y_i = \sum_{j=1}^{M} s_{ji} x_j, \qquad i = 1, \ldots, M \tag{3.102}$$

$$s_{jj} = \sqrt{r_{jj}^{(j)}}, \qquad j = 1, \ldots, M \tag{3.103}$$

$$r_{ij}^{(1)} \equiv r_{ij} \tag{3.104}$$

$$s_{kj} = \frac{r_{kj}^{(j)}}{s_{jj}}, \qquad k = j + 1, \ldots, M \tag{3.105}$$

$$r_{ik}^{(j+1)} \equiv r_{ik}^{(j)} - s_{ij}s_{kj}^*, \qquad \begin{matrix} k = j + 1, \ldots, M \\ i = k, \ldots, M \end{matrix} \tag{3.106}$$

Equation (3.101) can be expressed in matrix form as

$$X_M^t R_M X_M^* = Y_M^t Y_M^* \tag{3.107}$$

where

$$Y_M^t = (y_1, y_2, \ldots, y_M) \tag{3.108}$$

From (3.102) it can be seen that Y_M can be expressed in matrix notation, that is,

$$Y_M = S_M^t X_M$$

Therefore (3.107) becomes

$$X_M^t R_M X_M^* = X_M^t S_M S_M^{*t} X_M^*$$ (3.109)

where

$$R_M = S_M S_M^{*t}$$

and the elements of the triangular matrix S_M are given in (3.103) to (3.106). Substituting (3.95) into (3.91) yields a pair of equations

$$S_M^{*t} A_M = C_M$$ (3.110)

$$S_M C_M = -F_M$$ (3.111)

where an intermediate vector, C_M, has the form

$$C_M^t = (c_1, c_2, \ldots, c_M)$$ (3.112)

By examining the lower triangular form of S_M, Eq. (3.111) can be solved recursively. For example, from the first row of (3.111)

$$c_1 = \frac{-r_{01}}{s_{11}}$$

From the second row

$$s_{21} c_1 + s_{22} c_2 = -r_{02}$$

which results in

$$c_2 = \frac{-1}{s_{22}} (r_{02} + c_1 s_{21}).$$

In general the jth row results in the solution for c_j, given by

$$c_j = -\frac{1}{s_{jj}} \left(r_{0j} + \sum_{i=1}^{j-1} c_i s_{ji} \right), \qquad j = 1, 2, \ldots, M$$ (3.113)

Similarly, Eq. (3.110) can be solved recursively, yielding

$$a_j = \frac{1}{s_{jj}} \left(c_j - \sum_{i=j+1}^{M} s_{ij}^* a_i \right), \qquad j = M, M-1, \ldots, 1$$ (3.114)

Equations (3.103), (3.104), (3.105), (3.106), (3.113), and (3.114) constitute an algorithm termed the Cholesky decomposition algorithm.[9]

EXAMPLE

Suppose a covariance matrix R_2 is defined as

$$R_2 = \begin{pmatrix} 1 & j \\ -j & 2 \end{pmatrix}$$

The elements of S can be found by using (3.103) to (3.106) as

$$s_{11} = r_{11}^{1/2} = 1$$

$$s_{21} = \frac{r_{21}^{(1)}}{s_{11}} = \frac{r_{21}}{s_{11}} = \frac{-j}{1} = -j$$

$$r_{22}^{(2)} = r_{22}^{(1)} - s_{21}s_{21}^{*}$$

$$= r_{22} - s_{21}s_{21}^{*}$$

$$= 2 - (-j)(j) = 1$$

$$s_{22} = \sqrt{r_{22}^{(2)}} = 1$$

Note that

$$S_2 S_2^{*t} = \begin{pmatrix} 1 & 0 \\ -j & 1 \end{pmatrix}\begin{pmatrix} 1 & j \\ 0 & 1 \end{pmatrix} = \begin{pmatrix} 1 & j \\ -j & 2 \end{pmatrix} = R_2$$

Assume $F_2^t = (1 \quad 2)$ then

$$c_1 = -\tfrac{1}{1} = -1$$

$$c_2 = -\tfrac{1}{1}(2 - 1(-j)) = -j - 2$$

$$a_2 = \frac{-j - 2}{1} = -j - 2$$

$$a_1 = \tfrac{1}{1}(-1 - j(-j - 2)) = -2 + 2j$$

3.4   AUTOCORRELATION COEFFICIENTS COMPUTATIONAL METHODS

In the preceding sections ensemble averages are used to define correlation coefficients. Predictor coefficients are then obtained by assuming that the correlation coefficients are known. If only data samples are provided, however, correlation coefficients are generally estimated by using time averages in place of the ensemble averages. Once time averages are introduced, various windows can be applied to the data to restrict the amount of data used in the estimate. Certain window functions commonly employed result in correlation coefficient estimates that do not yield Toeplitz covariance matrices. Several windowing techniques used for correlation coefficient estimation are outlined below; they are the correlation method, the covariance method, the prewindowed method, and the postwindow method.[10]

Prior to presenting these methods, a form of the Yule-Walker equations is derived using time averages instead of ensemble averages. From Eqs. (3.2) and (3.4)

$$\hat{x}_i = \sum_{m=1}^{M} b_m x_{i-m} \tag{3.115}$$

$$e_i = x_i - \hat{x}_i, \tag{3.116}$$

the time average MSE, $\tilde{P}_M$, can be written as

$$\tilde{P}_M = \sum_i |e_i|^2$$

$$= \sum_i \left| x_i - \sum_{m=1}^{M} b_m x_{i-m} \right|^2. \tag{3.117}$$

Minimization of $\tilde{P}_M$ by differentiation or use of the orthogonality principle leads to

$$\sum_{m=1}^{M} b_m \sum_i x_{i-m} x_{i-l}^* = \sum_i x_i x_{i-l}^*, \qquad l = 1, 2, \ldots, M. \tag{3.118}$$

Suppose x_i is available for all i and $x_i = 0$ for $i \le 0$. Then

$$\rho_{ml} = \sum_{i=1}^{\infty} x_{i-m} x_{i-l}^* \qquad \text{for} \quad l > 0, \quad m > 0 \tag{3.119}$$

is a time average autocorrelation coefficient. Here, the ensemble average autocorrelation coefficients, r_{ml}, are replaced by time average autocorrelation coefficients, ρ_{ml}. It is easy to verify that

$$\rho_{ml} = \rho_{m+1, l+1} \qquad \text{for } l > 0, \quad m > 0 \tag{3.120}$$

Note that the definition of the data in (3.119) results in a single subscripted variable $\rho_{l-m} = \rho_{ml}$. The correlation matrix, R, associated with Eqs. (3.120), is a Toeplitz matrix. The efficient Durbin (or Levinson) method can be used to solve for the filter coefficients $\{b_m, \ m = 1, \ldots, M\}$. Furthermore, the method will lead to a stable solution.

Now suppose that x_i is available for $i = 1, 2, \ldots, N$ and that $x_i = 0$ outside this range. Four different windows can be applied to the data:

A.   Correlation Method.

Let the window be defined in the range $1 \le i \le N + M$. Then

$$\rho_{ml} = \sum_{i=1}^{N+M} x_{i-m} x_{i-l}^*$$

$$= x_1 x_{1+m-l}^* + x_2 x_{2+m-l}^* + \cdots + x_{N+l-m} x_N^* \qquad \text{for } m \ge l. \quad (3.121)$$

It can be verified that

$$\rho_{ml} = \rho_{m+1, l+1} = \rho_{l-m} \text{ for } \quad l = 1, 2, \ldots, M \quad m = 1, 2, \ldots, M.$$

Hence, the correlation matrix R is Toeplitz and consequently, the Durbin (or Levinson) algorithm can be used to achieve a stable solution.

B.   Covariance Method.

Let the window be defined in the range $M + 1 \le i \le N$. Then

$$\rho_{ml} = \sum_{i=M+1}^{N} x_{i-m} x_{i-l}^*$$

$$= x_{M+1-m} x_{M+1-l}^* + x_{M+2-m} x_{M+2-l}^* + \cdots + x_{N-m} x_{N-l}^*. \quad (3.122)$$

Equation (3.122) shows that

$$\rho_{ml} \ne \rho_{l-m} \qquad\qquad\qquad (3.123)$$

Since (3.123) implies that R is not a Toeplitz matrix, the Durbin (or Levinson) method cannot be used to solve the filter coefficients. If ρ_{ml} is set equal to ρ_{l-m}, then the solution may not be stable.

C.   Prewindowed Method.

Let the window be defined in the range $1 \le i \le N$. Then

$$\rho_{ml} = \sum_{i=1}^{N} x_{i-m} x_{i-l}^* \qquad\qquad\qquad (3.124)$$

and R is not a Toeplitz matrix.

D. Postwindow Method.

Let the window be defined in the range $M + 1 \leq i \leq M + N$. Then

$$\rho_{ml} = \sum_{i=M+1}^{M+N} x_{i-m} x_{i-l}^* \tag{3.125}$$

and R is not a Toeplitz matrix.

It has now been shown that the correlation method allows time averages to be used in the solution of the normal equations (e.g., Levinson) where the Toeplitz property of the matrix R is required. In Chapter 4 a weighted form of the prewindowed method is utilized to develop the least square lattice structure. An alternate formulation can be obtained using the postwindow method but is not presented in this text. In Chapter 11 the covariance method is applied to digital speech processing.

3.5 FORWARD AND BACKWARD PREDICTION

3.5.1 Model Description

The one-way predictor shown in Sections 3.1 and 3.2 can be extended to forward and backward predictors. Two different forward and backward prediction algorithms are presented in this section. The first algorithm is a direct extension of the Durbin algorithm, in which the autocorrelation coefficients are given. The second one, originally developed by Burg, estimates the predictor coefficients directly from the available data without making any assumption about the unknown data. The latter algorithm is called the Burg algorithm.[11,12,13] In both algorithms the signal is assumed to be stationary. This assumption leads to the result that the covariance matrix is Toeplitz. Consequently, the Levinson recursion is used in both algorithms for estimating the predictor coefficients recursively.

As shown in previous sections, the error signal e_i formed at each filter iteration from an $(M + 1)$-tap filter with prediction filter coefficients $(1, a, \ldots, a_M)$ can be expressed as (see Figure 3-3)

$$e_i = \sum_{m=0}^{M} a_m x_{i-m} \qquad i = M + 1, \ldots, N \tag{3.126}$$

where $a_0 = 1$ and $x_1, x_2, \ldots, x_N$ are the given data samples. Defining the total estimated error power, $P_{M,N}$, as a time average of all possible error terms given in (3.126), results in

$$P_{M,N} = \sum_{i=M+1}^{N} |e_i|^2 \tag{3.127}$$

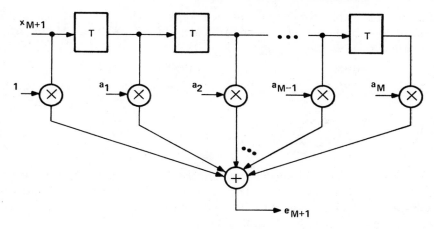

Figure 3-3.   Prediction filter.

Generally, the unknown data is assumed to have a zero time average. In (3.126) and (3.127), this assumption is not required. Since the data is stationary, the prediction coefficients $\{a_m\}$ can be recursively estimated using the Levinson recursion (3.32)

$$a_{M,k} = a_{M-1,k} + a_{M,M}a^*_{M-1,M-k}, \qquad k = 1,\ldots, M-1 \quad (3.128)$$

where $a_{M,M}$ is known as the reflection coefficient. Substituting (3.128) into (3.127), $P_{M,N}$ can be expressed as

$$P_{M,N} = \sum_{i=M+1}^{N} \left| \sum_{m=0}^{M-1} a_{M-1,m}x_{i-m} + a_{M,M} \sum_{m=1}^{M} a^*_{M-1,M-m}x_{i-m} \right|^2$$

$$(3.129)$$

Define the forward error, $e^f_{i,M-1}$, and backward error, $e^b_{i-1,M-1}$, at the $(M-1)$-th step as

$$e^f_{i,M-1} = \sum_{m=0}^{M-1} a_{M-1,m}x_{i-m} \qquad i = M+1,\ldots,N \quad (3.130)$$

$$e^b_{i-1,M-1} = \sum_{m=1}^{M} a^*_{M-1,M-m}x_{i-m} \qquad i = M+1,\ldots,N \quad (3.131)$$

Then, (3.129) can be rewritten as

$$P_{M,N} = \sum_{i=M+1}^{N} \left| e^f_{i,M-1} + a_{M,M}e^b_{i-1,M-1} \right|^2 \quad (3.132)$$

The total estimated power $P_{M,N}$ at step M is seen to be a function of the forward and backward error signals at step $M - 1$. At step M, the error equations are given by

$$e^f_{i,M} = \sum_{m=0}^{M} a_{M,m} x_{i-m} \qquad i = M + 1, \ldots, N \qquad (3.133)$$

$$e^b_{i,M} = \sum_{m=0}^{M} a^*_{M,M-m} x_{i-m} \qquad i = M + 1, \ldots, N \qquad (3.134)$$

where the backward filter coefficients are the reversed complex conjugates of the forward filter coefficients. The relationship between the forward and backward coefficients can also be found by minimizing the forward and backward mean square errors defined as

$$P^f_M = \mathscr{E}\left[|e^f_{i,M}|^2\right] \qquad (3.135)$$

$$P^b_M = \mathscr{E}\left[|e^b_{M+i,M}|^2\right] \qquad (3.136)$$

where

$$e^f_{i,M} = \sum_{m=0}^{M} a_{M,m} x_{i-m}, \qquad i = M + 1, \ldots, N \qquad (3.137)$$

$$e^b_{M+i,M} = \sum_{m=0}^{M} c_{M,m} x_{i+m}, \qquad i = 1, \ldots, N - M \qquad (3.138)$$

Note that $a_{M,m}$ and $c_{M,m}$ denote the forward and backward error filter coefficients, respectively. Minimization of P^f_M with respect to $a^*_{M,m}$ leads to

$$\sum_{m=0}^{M} a_{M,m} r_{l-m} = 0, \qquad l = 1, \ldots, M \qquad (3.139)$$

Similarly, minimization of P^b_M with respect to $c_{M,m}$ yields

$$\sum_{m=0}^{M} c^*_{M,m} r_{l-m} = 0, \qquad l = 1, \ldots, M \qquad (3.140)$$

Equation (3.140) is identical to Eq. (3.139), except that the complex conjugate prediction coefficients are used. Hence, $c^*_{M,m} = a_{M,m}$, that is, the backward predictor is just the complex conjugate of the forward predictor so that (3.138) becomes

$$e^b_{M+i,M} = \sum_{m=0}^{M} a^*_{M,m} x_{i+m}, \qquad i = 1, \ldots, N - M \qquad (3.141)$$

Since x_i is stationary, the optimum weight $a^*_{M,m}$ is independent of m. By replacing m by $M - m$, the backward prediction error can be rewritten as

$$e^b_{M+i,M} = \sum_{m=0}^{M} a^*_{M,M-m} x_{M+i-m} \qquad i = 1, \ldots, N - M \qquad (3.142)$$

or

$$e^b_{i,M} = \sum_{m=0}^{M} a^*_{M,M-m} x_{i-m} \qquad i = M + 1, \ldots, N \qquad (3.143)$$

Equation (3.143) is identical to Eq. (3.134). This result could have been anticipated since the received signal is assumed to be stationary.

3.5.2 Lattice Structure with Known Autocorrelation Coefficients

The prediction coefficients shown in (3.137) and (3.143) can be recursively computed according to the Durbin algorithm if the autocorrelation coefficients are given. A lattice structure can be formed from the forward and backward predictions.[14] The forward predictor in (3.137) may be written as

$$e^f_{i,M} = \sum_{m=0}^{M-1} a_{M-1,m} x_{i-m} + a_{M,M} \sum_{m=1}^{M} a^*_{M-1,M-m} x_{i-m} \qquad (3.144)$$

by substituting (3.128) into (3.137). Equation (3.144) can be written as

$$e^f_{i,M} = e^f_{i,M-1} + a_{M,M} e^b_{i-1,M-1} \qquad (3.145)$$

where $e^f_{i,M-1}$ and $e^b_{i-1,M-1}$ are given by (3.130) and (3.131). Similarly, the backward predictor in (3.143) can be written as

$$e^b_{i,M} = \sum_{m=1}^{M} a^*_{M-1,M-m} x_{i-m} + a^*_{M,M} \sum_{m=0}^{M-1} a_{M-1,m} x_{i-m}$$

$$= e^b_{i-1,M-1} + a^*_{M,M} e^f_{i,M-1} \qquad (3.146)$$

by substituting

$$a^*_{M,M-m} = a^*_{M-1,M-m} + a^*_{M,M} a_{M-1,m} \qquad m = 1, \ldots, M - 1 \qquad (3.147)$$

into (3.143). Equations (3.145) and (3.146) can be given a graphical interpretation. In Figure 3-4, L Lattice sections connected in cascade are depicted. Each section (Fig. 3-4a) consists of one delay element, two multipliers and two adders. The input to the system so formed equals $x_i = e^f_{i,0} = e^b_{i,0}$, and the two

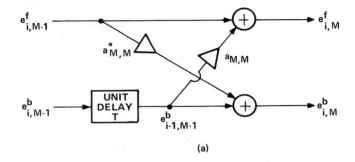

(a)

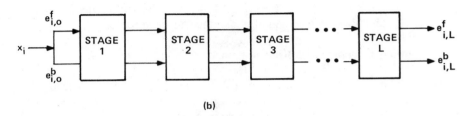

(b)

Figure 3-4. Lattice structure.

outputs equal forward and backward predictor errors at step L. An alternative form for the reflection coefficient $a_{M,M}$ is derived here. The forward MSE is denoted as

$$P_M^f = \mathscr{E}\left[\left|e_{i,M}^f\right|^2\right] = \mathscr{E}\left[\left|e_{i,M-1}^f + a_{M,M}e_{i-1,M-1}^b\right|^2\right] \qquad (3.148)$$

Minimization of (3.148) with respect to $a_{M,M}^*$ leads to

$$a_{M,M} = -\frac{\mathscr{E}\left[e_{i,M-1}^f e_{i-1,M-1}^{b*}\right]}{\mathscr{E}\left[\left|e_{i-1,M-1}^b\right|^2\right]} \qquad (3.149)$$

The above equation can be written in the symmetrical form

$$a_{M,M} = \frac{\mathscr{E}\left[e_{i,M-1}^f e_{i-1,M-1}^{b*}\right]}{\frac{1}{2}\mathscr{E}\left[\left|e_{i,M-1}^f\right|^2 + \left|e_{i-1,M-1}^b\right|^2\right]} \qquad (3.150)$$

This is a consequence of the fact that the forward and backward MSE's are equal. (see (3.130) and (3.131).)

3.5.3   Burg Algorithm

Burg developed an algorithm for estimating the prediction coefficients directly from the data.[11,12] Burg makes no assumptions about the unavailable data except that the signal is treated as a stationary process. No estimation of the autocorrelation coefficients is required in this algorithm. Using (3.145) and (3.146), the time average total estimated forward and backward error powers at step M can be expressed as

$$P_{M,N}^f = \sum_{i=M+1}^{N} |e_{i,M}^f|^2$$

$$= \sum_{i=M+1}^{N} |e_{i,M-1}^f + a_{M,M} e_{i-1,M-1}^b|^2 \qquad (3.151)$$

$$P_{M,N}^b = \sum_{i=M+1}^{N} |e_{i,M}^b|^2$$

$$= \sum_{i=M+1}^{N} |e_{i-1,M-1}^b + a_{M,M}^* e_{i,M-1}^f|^2 \qquad (3.152)$$

where the running index i is limited to $M + 1 \leq i \leq N$. Note that only the available data are used. Since x_i is stationary, an estimate of the total power can be obtained by averaging the forward and backward power:

$$P_{M,N} = \tfrac{1}{2}[P_{M,N}^f + P_{M,N}^b]$$

$$= \tfrac{1}{2} \sum_{i=M+1}^{N} \left\{ |e_{i,M-1}^f + a_{M,M} e_{i-1,M-1}^b|^2 + |e_{i-1,M-1}^b + a_{M,M}^* e_{i,M-1}^f|^2 \right\}$$

$$(3.153)$$

To find the best least squares estimate, the total average error power $P_{M,N}$ is minimized with respect to the coefficient $a_{M,M}$:

$$\frac{\partial P_{M,N}}{\partial a_{M,M}} = 0 \qquad (3.154)$$

The solution of (3.154) is found to be

$$a_{M,M} = \frac{-2\sum_{i=M+1}^{N} e_{i-1,M-1}^{b*} e_{i,M-1}^f}{\sum_{i=M+1}^{N} \left(|e_{i-1,M-1}^b|^2 + |e_{i,M-1}^f|^2 \right)} \qquad (3.155)$$

Define D_{M-1} to be the denominator of (3.155) at step $M - 1$:

$$D_{M-1} = \sum_{i=M+1}^{N} \left(|e_{i-1,M-1}^b|^2 + |e_{i,M-1}^f|^2 \right)$$

$$= 2\tilde{P}_{M-1,N_{\min}} \qquad (3.156)$$

where $\tilde{P}_{M-1, N_{\min}}$ is the average error power at step $M-1$. In general, the use of time averages results in unequal forward and backward error powers. A recursive estimation of the denominator of $a_{M,M}$ can be obtained as follows. Substituting the recursive equations for $e^b_{i-1, M-1}$ and $e^f_{i, M-1}$ into (3.156), D_{M-1} can be written

$$D_{M-1} = \sum_{i=M+1}^{N} \left(\left| e^b_{i-2, M-2} + a^*_{M-1, M-1} e^f_{i-1, M-2} \right|^2 \right)$$

$$+ \sum_{i=M+1}^{N} \left(\left| e^f_{i, M-2} + a_{M-1, M-1} e^b_{i-1, M-2} \right|^2 \right)$$

$$= \sum_{i=M}^{N-1} \left(\left| e^b_{i-1, M-2} + a^*_{M-1, M-1} e^f_{i, M-2} \right|^2 \right)$$

$$+ \sum_{i=M+1}^{N} \left(\left| e^f_{i, M-2} + a_{M-1, M-1} e^b_{i-1, M-2} \right|^2 \right) \qquad (3.157)$$

Adding two end terms to (3.157) allows D_{M-1} to be rewritten as

$$D_{M-1} = - \left| e^b_{N-1, M-2} + a^*_{M-1, M-1} e^f_{N, M-2} \right|^2$$

$$- \left| e^f_{M, M-2} + a_{M-1, M-1} e^b_{M-1, M-2} \right|^2$$

$$+ \sum_{i=M}^{N} \left\{ \left| e^b_{i-1, M-2} + a^*_{M-1, M-1} e^f_{i, M-2} \right|^2 \right.$$

$$+ \left| e^f_{i, M-2} + a_{M-1, M-1} e^b_{i-1, M-2} \right|^2 \right\}$$

$$= - \left| e^b_{N, M-1} \right|^2 - \left| e^f_{M, M-1} \right|^2$$

$$+ \sum_{i=M}^{N} \left\{ \left(1 + \left| a_{M-1, M-1} \right|^2 \right) \left(\left| e^b_{i-1, M-2} \right|^2 + \left| e^f_{i, M-2} \right|^2 \right) \right.$$

$$+ 2 a_{M-1, M-1} e^b_{i-1, M-2} e^{f*}_{i, M-2}$$

$$+ 2 a^*_{M-1, M-1} e^{b*}_{i-1, M-2} e^f_{i, M-2} \right\} \qquad (3.158)$$

Substituting the following equation for $a_{M-1, M-1}$:

$$a_{M-1, M-1} = - \frac{2\sum_{i=M}^{N} e_{i-1, M-2}^{b*} e_{i, M-2}^{f}}{\sum_{i=M}^{N} \left(\left| e_{i-1, M-2}^{b} \right|^2 + \left| e_{i, M-2}^{f} \right|^2 \right)}$$

$$= - \frac{2\sum_{i=M}^{N} e_{i-1, M-2}^{b*} e_{i, M-2}^{f}}{D_{M-2}} \qquad (3.159)$$

into (3.158), allows D_{M-1} to be expressed as

$$D_{M-1} = -\left| e_{N, M-1}^{b} \right|^2 - \left| e_{M, M-1}^{f} \right|^2 + \left(1 - |a_{M-1, M-1}|^2 \right) D_{M-2} \quad (3.160)$$

To begin the computation of the Burg algorithm, M is set to zero. The initial values of $a_{0,0}$, $e_{i,0}^{f}$ and $e_{i,0}^{b}$ are given by

$$a_{0,0} = 1, \qquad e_{i,0}^{f} = e_{i,0}^{b} = x_i \qquad (3.161)$$

At step $M = 1$, $a_{1,1}$ is computed according to

$$a_{1,1} = \frac{-2\sum_{i=2}^{N} x_{i-1}^{*} x_i}{D_0} \qquad (3.162)$$

where

$$D_0 = 2\tilde{P}_{0, N_{\min}} = \sum_{i=2}^{N} \left(|x_i|^2 + |x_{i-1}|^2 \right) \qquad (3.163)$$

A lattice structure in the form given by Figure 3-4 can be constructed for the Burg algorithm.

A Numerical Example

In this section the Durbin and Burg algorithms will be used to illustrate the computational procedures and their numerical results. First a preliminary comparison for the first coefficient will be made. Assume that x_i is available for $i = 1, 2, \ldots, N$ and that $x_i = 0$ outside this range. The correlation method given in Section 3.4 can then be used to compute the first and second autocorrelation coefficients:

$$\rho_0 = \sum_{i=1}^{N} |x_i|^2$$

$$\rho_1 = \sum_{i=1}^{N-1} x_{i+1} x_i^*$$

Using time averages in place of ensemble averages, the Durbin algorithm leads to the solution (see (3.26))

$$b_{1,1} = \frac{\rho_1}{\rho_0} = \frac{\sum_{i=1}^{N-1} x_{i+1} x_i^*}{\sum_{i=1}^{N} |x_i|^2}$$

If the Burg algorithm is used, the solution becomes (see (3.162) and (3.163))

$$b_{1,1} = \frac{2\sum_{i=2}^{N} x_i x_{i-1}^*}{\sum_{i=2}^{N} \left(|x_i|^2 + |x_{i-1}|^2 \right)}$$

From the above coefficients it is clear that the solutions using the Durbin and Burg algorithms are different, resulting in a difference in performance. The following simple numerical example will illustrate the difference explicitly.

Assume that the samples shown in Figure 3-5 are used to compute the prediction coefficients, the MSEs, and the prediction sample values. The available samples are assumed to be $x_1 = 1$, $x_2 = 2$, $x_3 = 3$, $x_4 = 4$ and $x_5 = 5$. The Durbin and Burg algorithms are now used in the estimation of the samples and a comparison in estimation accuracy will be made.

A. Durbin Algorithm.

Let $M = 3$ and $N = 5$. The autocorrelation coefficients can be obtained by using the correlation method given in Section 3.4:

$$\begin{cases} \rho_0 = \sum_{i=1}^{8} x_i^2 = 55 \\[2mm] \rho_1 = \sum_{i=1}^{8} x_{i-1} x_i = 40 \\[2mm] \rho_2 = \sum_{i=1}^{8} x_{i-2} x_i = 26 \\[2mm] \rho_3 = \sum_{i=1}^{8} x_{i-3} x_i = 14 \end{cases}$$

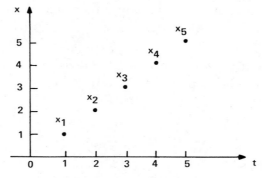

Figure 3-5.   Data samples for estimation.

The prediction coefficients and MSEs can be computed and are given by

$$a_{1,1} = -\frac{\rho_1}{\rho_0} = -0.72727$$

$$a_{1,0} = 1$$

$$\tilde{P}_{1_{min}} = \rho_0\left(1 - a_{1,1}^2\right) = 25.909$$

$$a_{2,2} = \frac{-\rho_2 - a_{1,1}\rho_1}{\tilde{P}_{1_{min}}} = 0.11929$$

$$a_{2,1} = a_{1,1} + a_{2,2}a_{1,1} = -0.81403$$

$$\tilde{P}_{2_{min}} = \tilde{P}_{1_{min}}\left(1 - a_{2,2}^2\right) = 25.5403$$

$$a_{3,3} = \frac{-a_{2,0}\rho_3 - a_{2,1}\rho_2 - a_{2,2}\rho_1}{\tilde{P}_{2_{min}}} = 0.093702$$

$$a_{3,2} = a_{2,2} + a_{3,3}a_{2,1} = 0.04301$$

$$a_{3,1} = a_{2,1} + a_{3,3}a_{2,2} = -0.80285$$

$$\tilde{P}_{3_{min}} = \tilde{P}_{2_{min}}\left(1 - a_{3,3}^2\right) = 25.316$$

where $\tilde{P}_{i_{min}}$ for $i = 1, 2, 3$ denotes the minimum ensemble average MSE estimated by means of a time average. Note that the index $N = 5$ in the power recursions is suppressed for simplicity. From the above computations, the predicted sample x_5 can be obtained by a third order linear prediction as

$$\hat{x}_5 = \sum_{i=1}^{3} b_i x_{5-i} = -\sum_{i=1}^{3} a_{3,i} x_{5-i} = 2.895$$

If a second order linear prediction is used, $\hat{x}_5$ and $\hat{x}_1$ can be obtained by forward and backward predictions, respectively:

$$\hat{x}_5 = \sum_{i=1}^{2} b_i x_{5-i} = -\sum_{i=1}^{2} a_{2,i} x_{5-i} = 2.8983$$

$$\hat{x}_1 = \sum_{i=1}^{2} b_i x_{1+i} = 1.27$$

B.   Burg Algorithm.

The prediction coefficients, the MSE's and prediction values can also be obtained by using the Burg algorithm. In the first step, the prediction coefficient, $b_{1,1}$, is computed as

$$b_{1,1} = \frac{2\sum_{i=2}^{5} x_i x_{i-1}}{D_0} = \frac{80}{84} = 0.9524$$

where

$$D_0 = \sum_{i=2}^{5} \left(x_i^2 + x_{i-1}^2 \right) = 84$$

The second order reflection coefficient, $a_{2,2}$, is then given by

$$a_{2,2} = -\frac{2\sum_{i=3}^{5} e_{i-1,1}^b e_{i,1}^f}{D_1} = 0.9575$$

where

$$e_{2,1}^f = x_2 + a_{1,1} x_1 = 1.0476$$

$$e_{3,1}^f = x_3 + a_{1,1} x_2 = 1.0952$$

$$e_{4,1}^f = x_4 + a_{1,1} x_3 = 1.1428$$

$$e_{5,1}^f = x_5 + a_{1,1} x_4 = 1.1908$$

$$e_{2,1}^b = x_1 + a_{1,1} x_2 = -0.9048$$

$$e_{3,1}^b = x_2 + a_{1,1} x_3 = -0.8572$$

$$e_{4,1}^b = x_3 + a_{1,1} x_4 = -0.8096$$

$$e_{5,1}^b = x_4 + a_{1,1} x_5 = -0.762$$

and

$$D_1 = -e_{5,1}^{b^2} - e_{2,1}^{f^2} + \left(1 - a_{1,1}^2\right)D_0 = 6.13$$

The prediction coefficient, $a_{2,1}$, is obtained as

$$a_{2,1} = a_{1,1} + a_{2,2}a_{1,1} = -0.9524(1.9575) = -1.8637$$

Next, the MSEs are evaluated where the index $N = 5$ in the power recursions is suppressed:

$$\begin{cases} \tilde{P}_{0_{min}} = \tfrac{1}{2}D_0 = 42 \\ \tilde{P}_{1_{min}} = \tfrac{1}{2}D_1 = 3.065 \\ \tilde{P}_{2_{min}} = \tfrac{1}{2}D_2 = 0.17415 \end{cases}$$

where

$$\begin{cases} e_{5,2}^b = e_{4,1}^b + a_{2,2}e_{5,1}^f = 0.3306 \\ e_{3,2}^f = e_{3,1}^f + a_{2,2}e_{2,1}^b = 0.2289 \\ D_2 = -\left(e_{5,2}^b\right)^2 - \left(e_{3,2}^f\right)^2 + \left(1 - a_{2,2}^2\right)D_1 = 0.3483 \end{cases}$$

If a second order linear prediction is used, $\hat{x}_5$ and $\hat{x}_1$ are given by

$$\begin{cases} \hat{x}_5 = -\sum_{i=1}^{2} a_{2,i}x_{5-i} = 4.5823 \\ \hat{x}_1 = -\sum_{i=1}^{2} a_{2,i}x_{1+i} = 0.8549 \end{cases}$$

The MSEs and predicted sample values of the Durbin and Burg algorithms are listed in Table 3-2. It is shown clearly that the Durbin algorithm performs poorly relative to the Burg algorithm in this example. The Durbin algorithm has larger mean square errors (MSE's) and larger prediction errors than the Burg algorithm. Data samples used for linear prediction in the Burg and Durbin algorithms are indicated in Figure 3-6. The Burg algorithm uses only the available samples without making any assumption about the unavailable data. The Durbin algorithm assumes unavailable data samples are zero (i.e., $x_{-2} = x_{-1} = x_6 = x_7 = x_8 = 0$. Because the Durbin algorithm makes more assumptions about the unavailable data, its performance is worse than that of the Burg algorithm. For large data records where end effects contribute little to the final results, the performance between the two algorithms is similar. Thus for short records the Burg algorithm is preferred.

TABLE 3-2.   THE MSE's AND PREDICTED SAMPLE VALUES OF THE DURBIN AND BURG ALGORITHMS

	Durbin Algorithm	Burg Algorithm	Actual Sample Value
$\tilde{P}_{1_{min}}$	25.909	3.065	—
$\tilde{P}_{2_{min}}$	25.5403	0.17415	—
$\hat{x}_1$	1.27	0.8549	1
$\hat{x}_5$	2.8983	4.5823	5

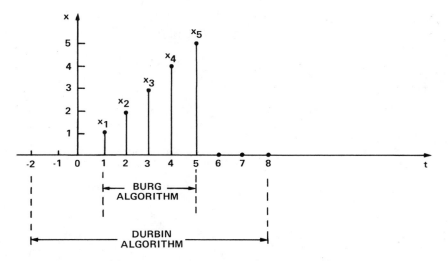

Figure 3-6.   Data samples used for linear prediction in the Burg and Durbin algorithms.

REFERENCES

1. G. E. Box and G. M. Jenkins, *Time Series Analysis, Forecasting and Control*, Holden-Day, San Francisco, 1970.
2. J. Durbin, The Fitting of Time Series Models, *Rev. Int. Inst. Statistics 28*, 1960, p. 233.
3. J. Durbin, Efficient Estimation of Parameters in Moving-Average Models, *Biometrika 46*, 1959, pp. 306–316.
4. G. U. Yule, On a Method of Investigating Periodicities in Desturbed Series with Special Reference to Wölfer's Sunspot Numbers, *Phil. Trans. Roy. Soc. A 226*, 1927, pp. 267–288.
5. G. Walker, On Periodicity in Series of Related Terms, In *Proc. Royal Soc. A131*, 1931, p. 518.

6. G. M. Jenkins and D. G. Watts, *Spectral Analysis and Its Applications*, Holden-Day, San Francisco, 1969.

7. N. Levinson, The Wiener RMS (Root Mean Square) Error Criterion in Filter Design and Prediction, *J. Math. Physics 25*, 1947, pp. 261–278.

8. N. Wiener, *Extrapolation, Interpolation and Smoothing of Stationary Time Series*, Wiley, New York, 1949.

9. G. J. Bierman, *Factorization Methods for Discrete Sequential Estimation*, Academic, New York, 1977.

10. M. Morf, B. Dickenson, T. Kailath and A. Viera, Efficient Solution of Covariance Equations for Linear Predictions, *IEEE Trans. Acoustics, Speech and Signal Processing ASSP-25*, Oct. 1977, pp. 429–433.

11. J. P. Burg, Maximum Entropy Spectral Analyses, Ph.D. Dissertation, Geophysics Dept., Stanford University, 1975.

12. J. P. Burg, Maximum Entropy Spectral Analysis, In *Modern Spectral Analysis*, (D. G. Childers, Ed), IEEE Press, New York, 1978, pp. 34–41.

13. T. J. Ulrych and T. N. Bishop, Maximum Entropy Spectral Analysis and Autoregressive Decomposition, *Rev. Geophysics 13*, 1975, pp. 182–200.

14. J. D. Markel and A. H. Gray, Jr., *Linear Prediction of Speech*, Springer-Verlag, New York, 1976.

chapter
4

LEAST SQUARE ESTIMATION
ALGORITHMS–PART 2

4.1 RECURSIVE LEAST SQUARES LATTICE ALGORITHMS

In the previous chapter algorithms that accomplish linear prediction were developed. These algorithms are implemented either with a linear transversal filter or with a lattice filter. In the case of the transversal filter the prediction coefficients are the tap coefficients of the filter and are computed iteratively. At each iteration the order of the filter is fixed and all of the tap coefficients change. In the case of the lattice filter, the total length of the filter increases by a complete section at each iteration. However, the parameters of each section are computed without affecting the parameters determined for prior sections.

The algorithms developed in this section deal with the nonstationary signal case. As a result, the algorithms are recursive in both order and time. In the stationary case described in Chapter 3, the lattice algorithms described are recursive in order but not in time. An important attribute of lattice filters is that the numbers of filter coefficients are dynamically assigned and are thus insensitive to round-off errors.

4.1.1 Model Description

The algorithms shown in the previous chapter were developed by minimizing the average error power, using the Levinson algorithm as a recursion constraint for obtaining the coefficients. This implies that the covariance matrix is Toeplitz. The algorithm to be presented in this chapter solves the unconstrained least squares problems using the prewindowed method.[1,2,3,4,5] In the prewindowed method, the data samples are $x_1, \ldots, x_N$, and are zero outside this interval. Let the forward error signal $e_M^f(i, N)$ be defined as

$$e_M^f(i, N) = \sum_{m=0}^{M} a_M(m, N) x_{i-m}, \qquad i = 1, \ldots, N \qquad (4.1)$$

where $a_M(m, N)$ is the forward prediction coefficient, M denotes the order of filter (Mth order) and N denotes the time index of filter. It is assumed that

$$x_0 = x_{-1} = \cdots = x_{1-M} = 0 \qquad (4.2)$$

and

$$a_M(0, N) = 1 \qquad (4.3)$$

in the prewindowed least squares method. The mean square error for the forward filter is then defined to be the time average:

$$P_M^f(N) = \sum_{i=1}^{N} \omega^{N-i} \left| e_M^f(i, N) \right|^2 \qquad (4.4)$$

where ω is a weighting factor with $0 < \omega < 1$. Minimization of $P_M^f(N)$ with respect to $a_M(j, N)$ leads to

$$\sum_{j=0}^{M} a_M(j, N) \rho_N(m, j) = 0, \qquad m = 1, \ldots, M \qquad (4.5)$$

where

$$\rho_N(m, j) = \sum_{i=1}^{N} \omega^{N-i} x_{i-m}^* x_{i-j} \qquad (4.6)$$

(see Section 3.3). The minimum mean square error power, for the forward filter, is denoted by $f_m(N)$, and is found by substituting (4.1) and (4.5) into (4.4):

$$f_M(N) = \sum_{j=0}^{M} a_M(j, N) \rho_N(0, j) \qquad (4.7)$$

In the development of the Yule-Walker equations in Section 3.1.3, using ensemble averages rather than time averages, a single equation (Eq. 3.15) concisely expressed the results. In a similar fashion, equations (4.5) and (4.7) can be combined resulting in

$$\sum_{j=0}^{M} a_M(j, N)\rho_N(m, j) = f_M(N)\delta(m), \qquad m = 0,\ldots, M \qquad (4.8)$$

where $\delta(0) = 1$ and $\delta(m) = 0$ for $m \neq 0$. Equations (4.8) can be written in matrix form as

$$\boldsymbol{R}_{M+1}(N)\overline{\boldsymbol{A}}_M(N) = \boldsymbol{E}_M^f(N) \qquad (4.9)$$

where

$$\boldsymbol{A}_M(N) = \begin{pmatrix} a_M(1, N) \\ \vdots \\ a_M(M, N) \end{pmatrix} \qquad (4.10)$$

$$\overline{\boldsymbol{A}}_M(N) = \begin{pmatrix} 1 \\ a_M(1, N) \\ \vdots \\ a_M(M, N) \end{pmatrix} = \begin{pmatrix} 1 \\ \boldsymbol{A}_M(N) \end{pmatrix} \qquad (4.11)$$

$$\boldsymbol{E}_M^f(N) = \begin{pmatrix} f_M(N) \\ 0 \\ \vdots \\ 0 \end{pmatrix} \qquad (4.12)$$

and

$$\boldsymbol{R}_{M+1}(N) = \begin{bmatrix} \rho_N(0,0) & \rho_n(0,1) & \cdots & \rho_N(0, M) \\ \rho_N(1,0) & \rho_N(1,1) & \cdots & \rho_N(1, M) \\ \vdots & & & \\ \rho_N(M,0) & \rho_N(M,1) & \cdots & \rho_N(M, M) \end{bmatrix} \qquad (4.13)$$

$\boldsymbol{A}_M(N)$ is an M-dimensional forward prediction coefficient vector, $\overline{\boldsymbol{A}}_M(N)$ is an extended $(M + 1)$-dimensional forward prediction coefficient vector, $\boldsymbol{E}_M^f(N)$ is an $(M + 1)$-dimensional forward error vector, and $\boldsymbol{R}_{M+1}(N)$ is an $(M + 1)$ by $(M + 1)$-dimensional autocorrelation matrix. It should be noted that $\boldsymbol{R}_{M+1}(N)$ is not a Toeplitz matrix. Consequently, the Levinson recursion is no longer applicable for solving the prediction coefficients. Actually, $\boldsymbol{R}_{M+1}(N)$ is a Hermitian matrix which has the property $\rho_N(m, j) = \rho_N^*(j, m)$. Similarly, the backward error signal, $e_M^b(i, N)$, is defined as

$$e_M^b(i, N) = \sum_{m=0}^{M} b_M(m, N)x_{i-m}, \qquad i = 1,\ldots, N \qquad (4.14)$$

where

$$b_M(M, N) = 1 \qquad (4.15)$$

and $b_M(m, N)$ is the backward prediction coefficient. The mean square error for backward filter is defined as

$$P_M^b(N) = \sum_{i=1}^{N} \omega^{N-i} \left| e_M^b(i, N) \right|^2 \qquad (4.16)$$

Minimization of (4.16) with respect to $b_M(j, N)$ leads to

$$\sum_{j=0}^{M} b_M(j, N)\rho_N(m, j) = 0, \qquad m = 0, \ldots, M-1 \qquad (4.17)$$

and

$$r_M(N) = \sum_{j=0}^{M} b_M(j, N)\rho_N(M, j) \qquad (4.18)$$

where $r_M(N)$ is the minimum mean square error power for the backward filter. Equations (4.17) and (4.18) can be written in matrix form as

$$R_{M+1}(N)\overline{B}_M(N) = E_M^b(N) \qquad (4.19)$$

where

$$\overline{B}_M(N) = \begin{pmatrix} b_M(0, N) \\ \vdots \\ b_M(M-1, N) \\ 1 \end{pmatrix} = \begin{pmatrix} B_M(N) \\ 1 \end{pmatrix} \qquad (4.20)$$

$$B_M(N) = \begin{pmatrix} b_M(0, N) \\ \vdots \\ b_M(M-1, N) \end{pmatrix} \qquad (4.21)$$

$$E_M^b(N) = \begin{pmatrix} 0 \\ \vdots \\ 0 \\ r_M(N) \end{pmatrix} \qquad (4.22)$$

and $B_M(N)$ is the M-dimensional backward prediction coefficient vector. $B_M(N)$ is the extended $M + 1$-dimensional backward prediction coefficient vector, $E_M^b(N)$ is an $M + 1$ dimensional backward error vector and $R_{M+1}(N)$ is the autocorrelation matrix given in (4.13).

Equations (4.9) and (4.19) must now be solved for the forward and backward predictor coefficients, respectively. The solution is based on properties associated with the correlation matrix $R_{M+1}(N)$ and the autocorrelation coefficients $\rho_N(m, j)$. Note that the autocorrelation coefficient $\rho_{N+1}(m + 1, j + 1)$ can be expressed as

$$\rho_{N+1}(m + 1, j + 1) = \sum_{i=1}^{N+1} \omega^{N+1-i} x_{i-(m+1)}^* x_{i-(j+1)}$$

$$= \sum_{i=0}^{N} \omega^{N-i} x_{i-m}^* x_{i-j}$$

Since $x_{-m}^* = 0$ (i.e., for $i = 0$) as a result of data prewindowing, then

$$\rho_{N+1}(m + 1, j + 1) = \rho_N(m, j) \tag{4.23}$$

Furthermore, it is noted that

$$\rho_N(m, j) = \rho_N^*(j, m) \tag{4.24}$$

and

$$\rho_N(0, 0) = \sum_{i=1}^{N} \omega^{N-i} |x_i|^2 \tag{4.25}$$

As a result the matrix $R_{M+1}(N)$ and its inverse are found to be complex conjugate symmetric:

$$R_{M+1}^{t*}(N) = R_{M+1}(N)$$

4.1.2   Order Update for Coefficients

In this section recursive formulas for the order update of the forward coefficients $A_{M+1}(N)$ and the backward coefficients $B_{M+1}(N)$ are obtained. The solution for the forward and backward coefficients results by expressing the $(M + 1)$-th order matrix $R_{M+1}(N)$ in terms of an Mth order matrix at the prior time instant $R_M(N - 1)$. Note that

$$R_M(N - 1) = \begin{bmatrix} \rho_{N-1}(0, 0) & \cdots & \rho_{N-1}(0, M - 1) \\ \vdots & \ddots & \\ \rho_{N-1}(M - 1, 0) & \cdots & \rho_{N-1}(M - 1, M - 1) \end{bmatrix}$$

$$\tag{4.26}$$

Using the reduction formula (4.23), Eqs. (4.26) can be reexpressed as

$$R_M(N-1) = \begin{bmatrix} \rho_N(1,1) & \cdots & \rho_N(1,M) \\ \vdots & \ddots & \\ \rho_N(M,1) & \cdots & \rho_N(M,M) \end{bmatrix} \tag{4.27}$$

Now it can be seen from Eqs. (4.13), (4.24), and (4.27) that $R_{M+1}(N)$ can be written as

$$R_{M+1}(N) = \begin{bmatrix} q(N) & Q_M^{t*}(N) \\ Q_M(N) & R_M(N-1) \end{bmatrix} \tag{4.28}$$

where $q(N) = \rho_N(0,0)$ and

$$Q_M(N) = \begin{pmatrix} \rho_N(1,0) \\ \vdots \\ \rho_N(M,0) \end{pmatrix} \tag{4.29}$$

Note that $Q_M(N)$ is the lower M-element vector from the first column of $R_{M+1}(N)$. Inversion of the matrix in (4.28) is now accomplished by use of the following matrix identity[6] (see Appendix Section A.2.7):

$$\begin{bmatrix} A & B \\ B^{t*} & C \end{bmatrix}^{-1}$$

$$= \begin{bmatrix} (A - BC^{-1}B^{t*})^{-1} & -(A - BC^{-1}B^{t*})^{-1}BC^{-1} \\ -C^{-1}B^{t*}(A - BC^{-1}B^{t*})^{-1} & C^{-1}B^{t*}(A - BC^{-1}B^{t*})^{-1}BC^{-1} + C^{-1} \end{bmatrix}$$

$$\tag{4.30}$$

Therefore, the inverse of $R_{M+1}(N)$ is

$$R_{M+1}^{-1}(N)$$

$$= \begin{bmatrix} \left[q(N) - Q_M^{t*}(N)R_M^{-1}(N-1)Q_M(N)\right]^{-1} & -\left[q(N) - Q_M^{t*}(N)R_M^{-1}(N-1)Q_M(N)\right]^{-1}Q_M^{t*}(N)R_M^{-1}(N-1) \\ -R_M^{-1}(N-1)Q_M(N) & R_M^{-1}(N-1)Q_M(N)\left[q(N) - Q_M^{t*}(N)R_M^{-1}(N-1)Q_M(N)\right]^{-1} \\ \times\left[q(N) - Q_M^{t*}(N)R_M^{-1}(N-1)Q_M(N)\right]^{-1} & \times Q_M^{t*}(N)R_M^{-1}(N-1) + R_M^{-1}(N-1) \end{bmatrix}$$

$$\tag{4.31}$$

Equation (4.31) can be reexpressed in terms of the minimum mean square error power for the forward filter, $f_M(N)$ and the forward prediction coefficients. First expand Eq. (4.7):

$$f_M(N) = \sum_{j=1}^{M} a_M(j, N)\rho_N(0, j) + \rho_N(0,0)$$

$$= Q_M^{t*}(N)A_M(N) + q(N) \qquad (4.32)$$

Next, Eq. (4.5) is converted to matrix form:

$$\sum_{j=1}^{M} a_M(j, N)\rho_N(m, j) = -\rho_N(m,0), m = 1,\ldots, M$$

or equivalently

$$R_M(N - 1)A_M(N) = -Q_M(N) \qquad (4.33)$$

Using (4.33) in (4.32) results in

$$f_M(N) = q(N) - Q_M^{t*}(N)R_M^{-1}(N - 1)Q_M(N) \qquad (4.34)$$

Combining (4.31), (4.33), and (4.34) results in

$$R_{M+1}^{-1}(N) = \begin{bmatrix} \dfrac{1}{f_M(N)} & \dfrac{A_M^{t*}(N)}{f_M(N)} \\ \dfrac{A_M(N)}{f_M(N)} & R_M^{-1}(N - 1) + \dfrac{A_M(N)}{f_M(N)}A_M^{t*}(N) \end{bmatrix}$$

$$= \begin{bmatrix} 0 & 0 \\ 0 & R_M^{-1}(N - 1) \end{bmatrix} + \frac{1}{f_M(N)}\begin{bmatrix} 1 \\ A_M(N) \end{bmatrix}\begin{bmatrix} 1 & A_M^{t*}(N) \end{bmatrix}$$

$$(4.35)$$

The solutions to the forward predictor coefficients are obtained by first expressing the $M + 1$ order matrix $R_{M+1}(N)$ in terms of the M order matrix $R_M(N)$:

$$R_{M+1}(N) = \begin{bmatrix} R_M(N) & V_M(N) \\ V_M^{t*}(N) & v_M(N) \end{bmatrix} \qquad (4.36)$$

where

$$V_M(N) = \begin{pmatrix} \rho_N(0, M) \\ \vdots \\ \rho_N(M - 1, M) \end{pmatrix}$$

$$v_M(N) = \rho_N(M, M) \qquad (4.37)$$

and

$$
R_M(N) = \begin{bmatrix}
\rho_N(0,0) & \rho_N(0,1) & \cdots & \rho_N(0, M-1) \\
\rho_N(1,0) & \rho_N(1,1) & \cdots & \rho_N(1, M-1) \\
\vdots & & \ddots & \\
\rho_N(M-1,0) & \rho_N(M-1,1) & \cdots & \rho_N(M-1, M-1)
\end{bmatrix}
$$

(4.38)

Note that $V_M(N)$ is the upper M-element vector from the last column of $R_{M+1}(N)$ (see Eq. 4.13). Inversion of the matrix in Eq. (4.36) is now accomplished by use of the following matrix identity[6] (see Appendix Section A.2.7) where the transpose is replaced by the transpose conjugate

$$
\begin{bmatrix} A & B \\ B^{t*} & C \end{bmatrix}^{-1}
$$

$$
= \begin{bmatrix}
A^{-1} + A^{-1}B(C - B^{t*}A^{-1}B)^{-1}B^{t*}A^{-1} & -A^{-1}B(C - B^{t*}A^{-1}B)^{-1} \\
-(C - B^{t*}A^{-1}B)^{-1}B^{t*}A^{-1} & (C - B^{t*}A^{-1}B)^{-1}
\end{bmatrix}
$$

(4.39)

Therefore the inverse of $R_{M+1}(N)$ is

$R_{M+1}^{-1}(N)$

$$
= \begin{bmatrix}
R_M^{-1}(N) + R_M^{-1}(N)V_M(N) & -R_M^{-1}(N)V_M(N) \\
\times \left[v_M(N) - V_M^{t*}(N)R_M^{-1}(N)V_M(N) \right]^{-1} V_M^{t*}(N)R_M^{-1}(N) & \times \left[v_M(N) - V_M^{t*}(N)R_M^{-1}(N)V_M(N) \right]^{-1} \\
-\left[v_M(N) - V_M^{t*}(N)R_M^{-1}(N)V_M(N) \right]^{-1} V_M^{t*}(N)R_M^{-1}(N) & \left[v_M(N) - V_M^{t*}(N)R_M^{-1}(N)V_M(N) \right]^{-1}
\end{bmatrix}
$$

(4.40)

Equation (4.40) can be expressed in terms of the minimum mean square error for the backward filter $r_M(N)$ and the backward prediction coefficients. First expand Eq. (4.18):

$$
r_M(N) = \sum_{j=0}^{M-1} b_M(j, N)\rho_N(M, j) + \rho_N(M, M)
$$

$$
= V_M^{t*}(N)B_M(N) + v_M(N)
$$

(4.41)

Next, Eq. (4.17) is converted to matrix form:

$$
\sum_{j=0}^{M-1} b_M(j, N)\rho_N(m, j) = -\rho_N(m, M), \qquad m = 0, \ldots, M-1
$$

or equivalently

$$R_M(N)B_M(N) = -V_M(N) \tag{4.42}$$

Using (4.42) and (4.41) results in

$$r_M(N) = v_M(N) - V_M^{t*}(N)R_M^{-1}(N)V_M(N) \tag{4.43}$$

Combining (4.40), (4.42), and (4.43) results in

$$R_{M+1}^{-1}(N) = \begin{bmatrix} R_M^{-1}(N) + \dfrac{B_M(N)}{r_M(N)}B_M^{t*}(N) & \dfrac{B_M(N)}{r_M(N)} \\[3mm] \dfrac{B_M^{t*}(N)}{r_M(N)} & \dfrac{1}{r_M(N)} \end{bmatrix}$$

$$= \begin{bmatrix} R_M^{-1}(N) & 0 \\ 0 & 0 \end{bmatrix} + \frac{1}{r_M(N)}\begin{bmatrix} B_M(N) \\ 1 \end{bmatrix}\begin{bmatrix} B_M^{t*}(N) & 1 \end{bmatrix} \tag{4.44}$$

Now the solution to the forward coefficients is obtained by substituting Eq. (4.44) into

$$A_{M+1}(N) = -R_{M+1}^{-1}(N-1)Q_{M+1}(N)$$

$$= -\left\{\begin{bmatrix} R_M^{-1}(N-1) & 0 \\ 0 & 0 \end{bmatrix} + \frac{1}{r_M(N-1)}\begin{bmatrix} B_M(N-1) \\ 1 \end{bmatrix}\right.$$

$$\left. \times \begin{bmatrix} B_M^{t*}(N-1) & 1 \end{bmatrix}\right\}Q_{M+1}(N) \tag{4.45}$$

Note that $Q_{M+1}(N)$ can be expressed in terms of $Q_M(N)$:

$$Q_{M+1}(N) = \begin{pmatrix} \rho_N(1,0) \\ \vdots \\ \rho_N(M+1,0) \end{pmatrix}$$

$$= \begin{pmatrix} Q_M(N) \\ \rho_N(M+1,0) \end{pmatrix}$$

Since $-R_M^{-1}(N-1)Q_M(N) = A_M(N)$, Eq. (4.45) becomes

$$A_{M+1}(N) = \begin{pmatrix} A_M(N) \\ 0 \end{pmatrix} - \frac{k_{M+1}(N)}{r_M(N-1)}\begin{pmatrix} B_M(N-1) \\ 1 \end{pmatrix} \tag{4.46}$$

where

$$k_{M+1}(N) = \begin{bmatrix} B_M^{t*}(N-1) & 1 \end{bmatrix} Q_{M+1}(N)$$

The solution to the feedback coefficients is obtained by substituting Eq. (4.35) into

$$B_{M+1}(N) = -R_{M+1}^{-1}(N)V_{M+1}(N)$$

$$= -\left\{ \begin{bmatrix} 0 & 0 \\ 0 & R_M^{-1}(N-1) \end{bmatrix} + \frac{1}{f_M(N)} \begin{bmatrix} 1 \\ A_M(N) \end{bmatrix} \begin{bmatrix} 1 & A_M^{t*}(N) \end{bmatrix} \right\} V_{M+1}(N)$$

$$(4.47)$$

Note that $V_{M+1}(N)$ can be expressed in terms of $V_M(N-1)$:

$$V_{M+1}(N) = \begin{pmatrix} \rho_N(0, M+1) \\ \rho_N(1, M+1) \\ \vdots \\ \rho_N(M, M+1) \end{pmatrix}$$

$$= \begin{pmatrix} \rho_N(0, M+1) \\ \rho_{N-1}(0, M) \\ \vdots \\ \rho_{N-1}(M-1, M) \end{pmatrix}$$

$$= \begin{pmatrix} \rho_N(0, M+1) \\ V_M(N-1) \end{pmatrix}$$

Since $-R_M^{-1}(N-1)V_M(N-1) = B_M(N-1)$, then Eqs. (4.47) become

$$B_{M+1}(N) = \begin{pmatrix} 0 \\ B_M(N-1) \end{pmatrix} - \frac{\hat{k}_{M+1}(N)}{f_M(N)} \begin{pmatrix} 1 \\ A_M(N) \end{pmatrix} \qquad (4.48)$$

where

$$\hat{k}_{M+1}(N) = \begin{bmatrix} 1 & A_M^{t*}(N) \end{bmatrix} V_{M+1}(N)$$

It is now shown in the following equations that

$$k_{M+1}(N) = \hat{k}_{M+1}^*(N) \qquad (4.49)$$

Using the definition for the $k_{M+1}(N)$ (see Eq. 4.46) results in

$$k_{M+1}(N) = \begin{bmatrix} B_M^{t*}(N-1) & 1 \end{bmatrix} \begin{bmatrix} Q_M(N) \\ \rho_N(M+1,0) \end{bmatrix}$$

$$= B_M^{t*}(N-1)Q_M(N) + \rho_N(M+1,0)$$

Using $-R_M^{-1}(N-1)V_M(N-1) = B_M(N-1)$ in the above equation yields

$$k_{M+1}(N) = -V_M^{t*}(N-1)R_M^{-1}(N-1)Q_M(N) + \rho_N(M+1,0)$$

$$= V_M^{t*}(N-1)A_M(N) + \rho_N(M+1,0)$$

$$= \begin{bmatrix} 1 & A_M^{t*}(N) \end{bmatrix} * \begin{pmatrix} \rho_N^*(0,M+1) \\ V_M^*(N-1) \end{pmatrix}$$

$$= \hat{k}_{M+1}^*(N)$$

Equations (4.46) and (4.48) are referred to as order update equations, since the coefficients of order $M+1$ are expressed in terms of the coefficients of order M.

4.1.3   Order Update for Minimum Mean Square Error

An order update recursion for the forward minimum mean square error power can be obtained by substituting eq. (4.46) into eq. (4.32) with M replaced by $M+1$, that is,

$$f_{M+1}(N) = q(N) + A_{M+1}^{t*}(N)Q_{M+1}(N)$$

$$= q(N) + \left\{ \begin{bmatrix} A_M^{t*}(N) & 0 \end{bmatrix} - \frac{k_{M+1}^*(N)}{r_M(N-1)} \begin{bmatrix} B_M^{t*}(N-1) & 1 \end{bmatrix} \right\}$$

$$\times \begin{pmatrix} Q_M(N) \\ \rho_N(M+1,0) \end{pmatrix}$$

$$= q(N) + A_M^{t*}(N)Q_M(N)$$

$$- \frac{k_{M+1}^*(N)}{r_M(N-1)} \begin{bmatrix} B_M^{t*}(N-1)Q_M(N) + \rho_N(M+1,0) \end{bmatrix}$$

Identifying the term in brackets in the above equation as $k_{M+1}(N)$ results in

$$f_{M+1}(N) = f_M(N) - \frac{|k_{M+1}(N)|^2}{r_M(N-1)} \tag{4.50}$$

An order update for the backward minimum mean square error power can be obtained by substituting Eq. (4.48) into Eq. (4.41) with M replaced by $M + 1$:

$$r_{M+1}(N) = v_{M+1}(N) + V_{M+1}^{t*}(N) B_{M+1}(N)$$

$$= v_{M+1}(N) + \left[\rho_N^*(0, M+1) \quad V_M^{t*}(N-1) \right]$$

$$\times \left\{ \left(\begin{array}{c} 0 \\ B_M(N-1) \end{array} \right) - \frac{\hat{k}_{M+1}(N)}{f_M(N)} \left(\begin{array}{c} 1 \\ A_M(N) \end{array} \right) \right\}$$

$$= v_{M+1}(N) + V_M^{t*}(N-1) B_M(N-1)$$

$$- \frac{\hat{k}_{M+1}(N)}{f_M(N)} \left[\rho_N^*(0, M+1) + V_M^{t*}(N-1) A_M(N) \right]$$

Identifying the term in brackets in the above equation as $k_{M+1}(N)$ and using $v_{M+1}(N) = v_M(N-1)$ results in

$$r_{M+1}(N) = r_M(N-1) - \frac{|k_{M+1}(N)|^2}{f_M(N)} \tag{4.51}$$

4.1.4 Error Updates

Recursive formulas for the forward and backward errors are derived here. The forward error, $e_M^f(i, N)$, and backward error, $e_M^b(i, N)$, are given by

$$e_M^f(i, N) = \sum_{m=0}^{M} a_M(m, N) x_{i-m}, \qquad i = 1, \ldots, N \tag{4.1}$$

$$e_M^b(i, N) = \sum_{m=0}^{M} b_M(m, N) x_{i-m}, \qquad i = 1, \ldots, N \tag{4.14}$$

Equations (4.1) and (4.14) can be written in matrix form as

$$e_M^f(i, N) = X_M^t(i) \overline{A}_M(N) \tag{4.52}$$

$$e_M^b(i, N) = X_M^t(i) \overline{B}_M(N) \tag{4.53}$$

where

$$X_M^t(i) = (x_i, \ldots, x_{i-M})$$

$$\bar{A}_M(N) = \begin{pmatrix} 1 \\ A_M(N) \end{pmatrix}$$

$$\bar{B}_M(N) = \begin{pmatrix} B_M(N) \\ 1 \end{pmatrix}$$

$$A_M(N) = \begin{pmatrix} a_M(1, N) \\ \vdots \\ a_M(M, N) \end{pmatrix} \tag{4.54}$$

$$B_M(N) = \begin{pmatrix} b_M(0, N) \\ \vdots \\ b_M(M-1, N) \end{pmatrix}$$

Now multiply the order update for $\bar{A}_{M+1}(N)$

$$\bar{A}_{M+1}(N) = \begin{bmatrix} \bar{A}_M(N) \\ 0 \end{bmatrix} - \frac{k_{M+1}(N)}{r_M(N-1)} \begin{bmatrix} 0 \\ \bar{B}_M(N-1) \end{bmatrix}$$

by $X_{M+1}^t(i)$, resulting in

$$e_{M+1}^f(i, N) = (x_i, \ldots, x_{i-(M+1)}) \begin{pmatrix} 1 \\ A_M(N) \\ 0 \end{pmatrix}$$

$$- \frac{k_{M+1}(N)}{r_M(N-1)} (x_i, \ldots, x_{i-(M+1)}) \begin{pmatrix} 0 \\ B_M(N-1) \\ 1 \end{pmatrix}$$

or

$$e_{M+1}^f(i, N) = e_M^f(i, N) - \frac{k_{M+1}(N)}{r_M(N-1)} e_M^b(i-1, N-1) \tag{4.55}$$

Similarly, multiply the order update for $\bar{B}_{M+1}(N)$

$$\bar{B}_{M+1}(N) = \begin{bmatrix} 0 \\ \bar{B}_M(N-1) \end{bmatrix} - \frac{k_{M+1}^*(N)}{f_M(N)} \begin{bmatrix} \bar{A}_M(N) \\ 0 \end{bmatrix}$$

by $X_{M+1}^t(i)$, resulting in

$$e_{M+1}^b(i, N) = e_M^b(i-1, N-1) - \frac{k_{M+1}^*(N)}{f_M(N)} e_M^f(i, N) \tag{4.56}$$

Note that the error updates involve recursions in both time and order.

4.1.5 Time Update

An update equation for $k_{M+1}(N)$ will now be provided in terms of the coefficient $k_{M+1}(N-1)$ taken at the previous time instant. The recursive equation is referred to as a time update and is given by

$$k_{M+1}(N) = \omega k_{M+1}(N-1) + \frac{e_M^f(N,N)e_M^{b*}(N-1,N-1)}{1 - \gamma_M(N-1,N-1)} \quad (4.57)$$

where

$$\gamma_M(i, N-1) = X_{M-1}^t(i)R_M^{-1}(N-1)X_{M-1}^*(i) \quad (4.58)$$

An order update equation for the parameter $\gamma_M(i, N-1)$ is given by

$$\gamma_M(i, N-1) = \gamma_{M-1}(i, N-1) + \frac{|e_{M-1}^b(i, N-1)|^2}{r_{M-1}(N-1)} \quad (4.59)$$

The mathematical derivation of Eqs. (4.57) to (4.59) is tedious and is given in the appendix at the end of the chapter. It should be noted that only a single time index, N, appears in Eq. (4.57) as a result of the definition of a correlation matrix.

4.1.6 Summary of Recursive Least Squares Lattice Algorithm

The recursive least squares lattice algorithm can be summarized as follows:

$$\text{Time Index} = N$$
$$\text{Order Index} = M$$

Signal Statistics

Prewindowed Data Samples

$$x_1, \ldots, x_N$$

Data Vector, $X_m(i)$

$$X_M^t(i) = (x_i, \ldots, x_{i-M}) \quad i = 1, \ldots, N$$

Received Signal Autocorrelation, $\rho_N(m, j)$

$$\rho_N(m, j) = \sum_{i=1}^{N} \omega^{N-i} x_{i-m}^* x_{i-j}, \quad m = 1, \ldots, M; \quad j = 1, \ldots M$$

Correlation Matrix, $R_{M+1}(N)$

$$R_{M+1}(N) = \begin{bmatrix} \rho_N(0,0) & \cdots & \rho_N(0, M) \\ \vdots & & \\ \rho_N(M,0) & \cdots & \rho_N(M, M) \end{bmatrix}$$

$$R_{M+1}(N) = \omega R_{M+1}(N-1) + X_M^*(N)X_M^t(N)$$

MSE Minimization

Forward Predictor Coefficients
$$a_M(m, N)$$
$$\{a_M(0, N), \ldots, a_M(M, N)\},$$
$$a_M(0, N) = 1$$

Backward Predictor Coefficients
$$b_M(m, N)$$
$$\{b_M(0, N), \ldots, b_M(M, N)\},$$
$$b_M(M, N) = 1$$

Forward Coefficient Vector
$$A_M(N)$$

$$\overline{A}_M(N) = \begin{pmatrix} 1 \\ A_M(N) \end{pmatrix}$$

$$= \begin{pmatrix} 1 \\ a_M(1, N) \\ \vdots \\ a_M(M, N) \end{pmatrix}$$

Backward Coefficient Vector
$$B_M(N)$$

$$\overline{B}_M(N) = \begin{pmatrix} B_M(N) \\ 1 \end{pmatrix}$$

$$= \begin{pmatrix} b_M(0, N) \\ \vdots \\ b_M(M - 1, N) \\ 1 \end{pmatrix}$$

Time-Varying Forward Error
$$e_M^f(i, N)$$

$$e_M^f(i, N) = \sum_{m=0}^{M} a_M(m, N)x_{i-m}$$
$$= X_M^t(i)\overline{A}_M(N)$$

Time-Varying Backward Error
$$e_M^b(i, N)$$

$$e_M^b(i, N) = \sum_{m=0}^{M} b_M(m, N)x_{i-m}$$
$$= X_M^t(i)\overline{B}_M(N)$$

Forward MSE
$$P_M^f(N)$$

$$P_M^f(N) = \sum_{i=1}^{N} \omega^{N-i}|e_M^f(i, N)|^2$$

Backward MSE
$$P_M^b(N)$$

$$P_M^b(N) = \sum_{i=1}^{N} \omega^{N-i}|e_M^b(i, N)|^2$$

Minimization of Forward MSE (with respect to Forward Coefficients) yields
$$R_{M+1}(N)\overline{A}_M(N) = E_M^f(N)$$

$$E_M^f(N) = \begin{pmatrix} f_M(N) \\ 0 \\ \vdots \\ 0 \end{pmatrix}$$

Minimization of Backward MSE (with respect to Backward Coefficients) yields
$$R_{M+1}(N)\overline{B}_M(N) = E_M^b(N)$$

$$E_M^b(N) = \begin{pmatrix} 0 \\ \vdots \\ 0 \\ r_M(N) \end{pmatrix}$$

Minimum Forward Error Power,
$$f_M(N)$$

$$f_M(N) = \sum_{j=0}^{M} a_M(j, N)\rho_N(0, j)$$

Minimum Backward Error Power,
$$r_M(N)$$

$$r_M(N) = \sum_{j=0}^{M} b_M(j, N)\rho_N(M, j)$$

Update Equations

Time-varying Forward Error Update

$$e^f_{M+1}(i, N) = e^f_M(i, N)$$
$$- \frac{k_{M+1}(N)}{r_M(N - 1)} e^b_M(i - 1, N - 1)$$

Time-varying Backward Error Update

$$e^b_{M+1}(i, N) = e^b_M(i - 1, N - 1)$$
$$- \frac{k^*_{M+1}(N)}{f_M(N)} e^f_M(i, N)$$

Minimum Forward Error Power Update

$$f_{M+1}(N) = f_M(N) - \frac{|k_{M+1}(N)|^2}{r_M(N - 1)}$$

Minimum Backward Error Power Update

$$r_{M+1}(N) = r_M(N - 1) - \frac{|k_{M+1}(N)|^2}{f_M(N)}$$

Forward Coefficient Order Update
$M \to M + 1$

$$\overline{A}_{M+1}(N) = \begin{pmatrix} \overline{A}_M(N) \\ 0 \end{pmatrix}$$
$$- \frac{k_{M+1}(N)}{r_M(N - 1)} \begin{pmatrix} 0 \\ \overline{B}_M(N - 1) \end{pmatrix}$$

Backward Coefficient Order Update
$M \to M + 1$

$$\overline{B}_{M+1}(N) = \begin{pmatrix} 0 \\ \overline{B}_M(N - 1) \end{pmatrix}$$
$$- \frac{k^*_{M+1}(N)}{f_M(N)} \begin{pmatrix} \overline{A}_M(N) \\ 0 \end{pmatrix}$$

Time Update for Coefficient $k_M(N)$, that is, $N - 1 \to N$

$$k_{M+1}(N) = \omega k_{M+1}(N - 1) + \frac{e^f_M(N, N) e^{b*}_M(N - 1, N - 1)}{1 - \gamma_M(N - 1, N - 1)}$$

Order Update for Parameter $\gamma_M(i, N)$

$$\gamma_M(i, N - 1) = \gamma_{M-1}(i, N - 1) + \frac{|e^b_{M-1}(i, N - 1)|^2}{r_{M-1}(N - 1)}$$

4.1.7 Lattice Structure Development

In this section a lattice structure implementation is developed for the recursive least square algorithms summarized in Section 4.1.6. It should be noted that the final solution minimizing the forward and backward mean square errors involves a single time index, N, as a result of the definition of a correlation matrix. Therefore the index i is replaced by N and then one index N is dropped. The forward error, $e^f_M(i, N)$, the backward error, $e^b_M(i, N)$, and the quantity $\gamma_M(i, N)$ are now expressed as $e^f_M(N)$, $e^b_M(N)$ and $\gamma_M(N)$, respectively. To complete the algorithm implementation several initial conditions, listed below, are required:

$$
\begin{array}{ll}
e_0^f(l) = e_0^b(l) = x_l, & l = 1, 2, \ldots \\
r_l(0) = \varepsilon, & l = 0, 1, 2, \ldots \\
r_{-1}(l) = \varepsilon, & l = 0, 1, 2, \ldots \\
e_{-1}^b(l) = 0, & l = 0, 1, 2, \ldots \\
e_l^b(0) = 0, & l = 0, 1, 2, \ldots \\
k_l(0) = 0, & l = 1, 2, \ldots \\
\gamma_{-1}(l) = 0, & l = 0, 1, 2, \ldots \\
f_0(l) = r_0(l) = \omega f_0(l - 1) + |e_0^f(l)|^2, & l = 1, 2, \ldots \\
f_0(0) = \varepsilon &
\end{array}
$$

where ε is a small positive quantity.

The lattice structure implementation will now be developed in a step-by-step fashion utilizing the above equations and the results in Section 4.1.6. For $M = 0$ and $N = 1$ the equations in Section 4.1.6 become

$$
e_1^f(1) = e_0^f(1) - \frac{k_1(1)}{r_0(0)} e_0^b(0) \quad = x_1
$$

$$
e_1^b(1) = e_0^b(0) - \frac{k_1^*(1)}{f_0(1)} e_0^f(1) \quad = 0
$$

$$
k_1(1) = \omega k_1(0) + \frac{e_0^f(1) e_0^{b*}(0)}{1 - \gamma_0(0)} \quad = 0
$$

$$
f_1(1) = f_0(1) - \frac{|k_1(1)|^2}{r_0(0)} \quad = |x_1|^2 + \omega \varepsilon
$$

$$
r_1(1) = r_0(0) - \frac{|k_1(1)|^2}{f_0(1)} \quad = \varepsilon
$$

$$
\gamma_0(0) = \gamma_{-1}(0) + \frac{|e_{-1}^b(0)|^2}{r_{-1}(0)} \quad = 0
$$

$$
r_0(1) = f_0(1) \quad = \omega f_0(0) + |e_0^f(1)|^2 = \omega \varepsilon + |x_1|^2
$$

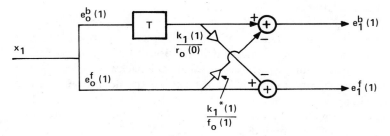

Figure 4-1. Single stage lattice.

Figure 4-1 depicts the single stage structure where T denotes a sample delay. In the next iteration $M = 0$ and $N = 2$ so that

$$e_1^f(2) = e_0^f(2) - \frac{k_1(2)e_0^b(1)}{r_0(1)} \qquad = x_2 - \frac{x_2|x_1|^2}{\omega\varepsilon + |x_1|^2}$$

$$e_1^b(2) = e_0^b(1) - \frac{k_1^*(2)e_0^f(2)}{f_0(2)} \qquad = x_1 - \frac{x_1|x_2|^2}{\omega(\omega\varepsilon + |x_1|^2) + |x_2|^2}$$

$$f_1(2) = f_0(2) - \frac{|k_1(2)|^2}{r_0(1)} \qquad = \omega(\omega\varepsilon + |x_1|^2) + |x_2|^2 - \frac{|x_2 x_1^*|^2}{\omega\varepsilon + |x_1|^2}$$

$$r_1(2) = r_0(1) - \frac{|k_1(2)|^2}{f_0(2)} \qquad = \omega\varepsilon + |x_1|^2 - \frac{|x_2 x_1^*|^2}{\omega(\omega\varepsilon + |x_1|^2) + |x_2|^2}$$

$$k_1(2) = \omega k_1(1) + \frac{e_0^f(2)e_0^{b*}(1)}{1 - \gamma_0(1)} \qquad = x_2 x_1^*$$

$$\gamma_0(1) = \gamma_{-1}(1) + \frac{|e_{-1}^b(1)|^2}{r_{-1}(1)} \qquad = 0$$

$$r_0(2) = f_0(2) = \omega f_0(1) + |e_0^f(2)|^2 = \omega(\omega\varepsilon + |x_1|^2) + |x_2|^2$$

If the order remained fixed at zero, then the single stage lattice depicted in Figure 4-1 would have x_2 as its input and $e_1^f(2)$ and $e_1^b(2)$ as its outputs. Increasing the order index by 1, that is, $M = 1$ and $N = 2$, results in

$$e_2^f(2) = e_1^f(2) - \frac{k_2(2)}{r_1(1)}e_1^b(1) \qquad = x_2 - \frac{x_2|x_1|^2}{\omega\varepsilon + |x_1|^2}$$

$$e_2^b(2) = e_1^b(1) - \frac{k_2^*(2)}{f_1(2)}e_1^f(2) \qquad = 0$$

$$f_2(2) = f_1(2) - \frac{|k_2(2)|^2}{r_1(1)} \qquad = \omega(\omega\varepsilon + |x_1|^2) + |x_2|^2 - \frac{|x_2 x_1^*|^2}{\omega\varepsilon + |x_1|^2}$$

$$r_2(2) = r_1(1) - \frac{|k_2(2)|^2}{f_1(2)} \qquad = \varepsilon$$

$$k_2(2) = \omega k_2(1) + \frac{e_1^f(2)e_1^{b*}(1)}{1 - \gamma_1(1)} = 0$$

$$\gamma_1(1) = \gamma_0(1) + \frac{|e_0^b(1)|^2}{r_0(1)} \qquad = \frac{|x_1|^2}{\omega\varepsilon + |x_1|^2}$$

$$k_2(1) = \omega k_2(0) + \frac{e_1^f(1)e_1^{b*}(0)}{1 - \gamma_1(0)} = 0$$

$$\gamma_1(0) = \gamma_0(0) + \frac{|e_0^b(0)|^2}{r_0(0)} \qquad = 0$$

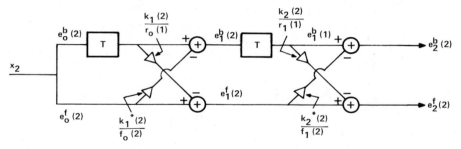

Figure 4-2. Two-stage lattice.

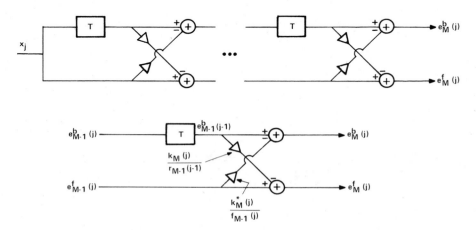

Figure 4-3. Mth stage lattice.

These quantities can be computed from the initial conditions and the prior iteration results. Figure 4-2 depicts the two-stage lattice structure. In general, the equations in Section 4.1.6 can be used recursively with the initial conditions. The general Mth-stage lattice is depicted in Figure 4-3. This lattice filter is built up from the previous lattice filter by adding another stage. When sufficient stages are established, the minimum mean square error quantities are small and no further stages are required. In this case the error quantities can still be updated in accordance with the equations in Section 4.1.6 for the fixed order.

4.1.8 Specialization of Lattice Structure to Burg Algorithm

In this section the assumptions required to reduce the recursive least square lattice algorithm to the Burg algorithm are given. The forward and backward update equations of the recursive least square lattice algorithm can be written using the notation provided in Section 4.1.7:

$$e_M^f(N) = e_{M-1}^f(N) - \frac{k_M(N)}{r_{M-1}(N-1)} e_{M-1}^b(N-1) \qquad (4.60)$$

$$e_M^b(N) = e_{M-1}^b(N-1) - \frac{k_M^*(N)}{f_{M-1}(N)} e_{M-1}^f(N) \qquad (4.61)$$

For convenience, the corresponding equations for the Burg algorithm given by Eqs. (3.145) and (3.146) are repeated here with i replaced by N:

$$e_{N,M}^f = e_{N,M-1}^f + a_{M,M} e_{N-1,M-1}^b \qquad (3.145)$$

$$e_{N,M}^b = e_{N-1,M-1}^b + a_{M,M}^* e_{N,M-1}^f \qquad (3.146)$$

where

$$a_{M,M} = - \frac{\sum_{j=M+1}^{N} e_{j-1,M-1}^{b*} e_{j,M-1}^f}{\sum_{j=M+1}^{N}\left[|e_{j-1,M-1}^b|^2 + |e_{j,M-1}^f|^2\right]/2} \qquad (3.155)$$

It is now shown that under certain assumptions Eqs. (4.60) and (4.61) reduce to (3.145) and (3.146), respectively. Since the forward and backward minimum error powers are real, the equivalence of these equations requires that

$$a_{M,M} \cong - \frac{k_M(N)}{r_{M-1}(N-1)} \qquad (4.62)$$

and

$$r_{M-1}(N-1) = f_{M-1}(N) \qquad (4.63)$$

An important assumption required in the development of the Burg algorithm is that the received signal is stationary. In this case it will now be shown that the forward lattice coefficients are time reversed complex conjugates of the backward coefficients:

$$a_M(j, N) = b_M^*(M - j, N - 1) \qquad (4.64)$$

Proof of Eq. (4.64) is similar to that followed in Section 3.5.1 and requires Eqs. (4.5) and (4.17)

$$\sum_{j=0}^{M} a_M(j, N)\rho_N(m, j) = 0, \qquad m = 1, \dots, M \qquad (4.5)$$

$$\sum_{j=0}^{M} b_M(j, N)\rho_N(m, j) = 0, \qquad m = 0, \dots, M - 1 \qquad (4.17)$$

Using the identity (4.23) and replacing N with $N - 1$ in (4.17) yields

$$\sum_{j=0}^{M} b_M(j, N - 1)\rho_N(m + 1, j + 1) = 0, \qquad m = 0, \ldots, M - 1 \quad (4.65)$$

Taking the complex conjugate of (4.65) and applying the identity given by (4.24) results in

$$\sum_{j=0}^{M} b_M^*(j, N - 1)\rho_N(j + 1, m + 1) = 0, \qquad m = 0, \ldots, M - 1 \quad (4.66)$$

In (4.66), j is now replaced by $M - j$ and m is replaced by $m - 1$

$$\sum_{j=0}^{M} b_M^*(M - j, N - 1)\rho_N(M - j + 1, m) = 0, \qquad m = 1, \ldots, M$$

$$(4.67)$$

Assuming that the received signal is stationary, $\rho_N(m, j) = \rho_N(m - j)$, so that Eqs. (4.5) and (4.67) can be respectively written as

$$\sum_{j=0}^{M} a_M(j, N)\rho_N(m - j) = 0, \qquad m = 1, \ldots, M \quad (4.68)$$

and

$$\sum_{j=0}^{M} b_M^*(M - j, N - 1)\rho_N(M - j + 1 - m) = 0, \qquad m = 1, \ldots, M \quad (4.69)$$

Expressing (4.68) and (4.69) in matrix form yields

$$\begin{pmatrix} \rho_N(1) & \cdots & \rho_N(1 - M) \\ \vdots & \ddots & \vdots \\ \rho_N(M) & \cdots & \rho_N(0) \end{pmatrix} \begin{pmatrix} a_M(0, N) \\ \vdots \\ a_M(M, N) \end{pmatrix} = 0 \qquad (4.70)$$

and

$$\begin{pmatrix} \rho_N(1) & \cdots & \rho_N(1 - M) \\ \vdots & \ddots & \vdots \\ \rho_N(M) & \cdots & \rho_N(0) \end{pmatrix} \begin{pmatrix} b_M^*(M, N - 1) \\ \vdots \\ b_M^*(0, N - 1) \end{pmatrix} = 0 \qquad (4.71)$$

Since (4.70) and (4.71) are identical, the proof of (4.64) is complete.

The proof of (4.63) proceeds by substituting (4.64) into the minimum forward error power given by Eq. (4.7):

$$f_M(N) = \sum_{j=0}^{M} a_M(j, N)\rho_N(0, j)$$

$$= \sum_{j=0}^{M} b_M^*(M - j, N - 1)\rho_N(-j) \tag{4.7}$$

Letting $l = M - j$ in the above equation and noting that $f_M(N)$ is real leads to

$$f_M(N) = \sum_{l=0}^{M} b_M(l, N - 1)\rho_N(M - l) \tag{4.72}$$

The minimum backward error power is given by Eq. (4.18):

$$r_M(N) = \sum_{j=0}^{M} b_M(j, N)\rho_N(M, j) \tag{4.18}$$

Using the identity (4.23) in the stationary case with N replaced by $N - 1$ results in

$$r_M(N - 1) = \sum_{j=0}^{M} b_M(j, N - 1)\rho_N(M - j) \tag{4.73}$$

Since (4.72) and (4.73) are identical, Eq. (4.63) has now been proved. Proof of (4.62) begins by expressing (4.57) in nonrecursive form. First it is assumed that $\omega = 1$ and that (4.57) is written using the single index notation presented in this section with $M + 1$ replaced by M:

$$k_M(N) = k_M(N - 1) + \frac{e_{M-1}^f(N)e_{M-1}^{b*}(N - 1)}{1 - \gamma_{M-1}(N - 1)} \tag{4.74}$$

For the stationary case $R_{M-1}^{-1}(N - 1)$ represents the error covariance matrix and approaches zero with coefficient convergence. Therefore $\gamma_{M-1}(N - 1)$ approaches zero so that (4.74) can be expressed in nonrecursive form as

$$k_M(N) = \sum_{i=1}^{N} e_{M-1}^f(i)e_{M-1}^{b*}(i - 1) \tag{4.75}$$

where

$$k_M(0) = 0$$

From equations (4.60) and (4.61) the minimum backward error power is expressed as

$$r_{M-1}(N-1) = \sum_{i=1}^{N} |e_{M-1}^{b}(i-1)|^2 \tag{4.76}$$

and the minimum forward error power is expressed as

$$f_{M-1}(N) = \sum_{i=1}^{N} |e_{M-1}^{f}(i)|^2 \tag{4.77}$$

Combining (4.63), (4.75), (4.76) and (4.77) results in

$$\frac{-k_M(N)}{r_{M-1}(N-1)} = \frac{-\sum_{i=1}^{N} e_{M-1}^{f}(i) e_{M-1}^{b*}(i-1)}{\frac{1}{2}\left[\sum_{i=1}^{N} |e_{M-1}^{b}(i-1)|^2 + \sum_{i=1}^{N} |e_{M-1}^{f}(i)|^2\right]} \tag{4.78}$$

Except for the summation range in (4.78) it can be seen that (4.78) and (3.155) are identical in form. This difference in the range results from the correlations coefficient estimation using the prewindowed method in the recursive least squares lattice algorithm. The Burg algorithm uses only the available data and does not require correlation coefficient estimation. Thus, the equivalence given by (4.62) is approximately true.

APPENDIX: DERIVATION OF THE TIME UPDATE RECURSION

The time update recursion given in Section 4.1.5 is derived here. First the recursion equation for the parameter $\gamma_M(i, N-1)$ expressed in Eq. (4.59) is obtained. This derivation begins with the definition of an intermediate variable, $D_M(i, N-1)$, given by

$$D_M(i, N-1) = R_M^{-1}(N-1) X_{M-1}^{*}(i) \tag{4A.1}$$

Thus, the variable $\gamma_M(i, N-1)$ can be expressed as

$$\gamma_M(i, N-1) = X_{M-1}^{t}(i) R_M^{-1}(N-1) X_{M-1}^{*}(i)$$

$$= X_{M-1}^{t}(i) D_M(i, N-1) \tag{4A.2}$$

A recursion for $\gamma_M(i, N-1)$ is obtained by developing a recursion for $D_M(i, N-1)$. For consistency with previous equations, the recursion equation for $D_{M+1}(i, N)$ will be derived. The proof will be accomplished by assuming that the recursion equation for $D_{M+1}(i, N)$ is given. Subsequently, the equation is proved by suitable manipulation. The recursion for $D_{M+1}(i, N)$ is expressed by

$$D_{M+1}(i, N) = \begin{pmatrix} D_M(i, N) \\ 0 \end{pmatrix} + \frac{e_M^{b*}(i, N)}{r_M(N)} \bar{B}_M(N) \qquad (4A.3)$$

The proof of Eq. (4A.3) begins by multiplying this equation by $R_{M+1}(N)$, resulting in

$$X_M^*(i) = R_{M+1}(N) \left\{ \begin{pmatrix} D_M(i, N) \\ 0 \end{pmatrix} + \frac{e_M^{b*}(i, N)}{r_M(N)} \bar{B}_M(N) \right\} \qquad (4A.4)$$

If the right hand side of Eq. (4A.4) can be shown to be $X_M^*(i)$, then the recursion given by eq. (4A.3) will have been proved. Since

$$R_{M+1}(N) \bar{B}_M(N) = E_M^b(N) = \begin{pmatrix} 0 \\ \vdots \\ 0 \\ r_M(N) \end{pmatrix}$$

and

$$R_{M+1}(N) = \begin{bmatrix} R_M(N) & V_M(N) \\ V_M^{t*}(N) & v_M(N) \end{bmatrix}$$

Eq. (4A.4) can be written as

$$X_M^*(i) = \begin{pmatrix} X_{M-1}^*(i) \\ V_M^{t*}(N) D_M(i, N) \end{pmatrix} + \frac{e_M^{b*}(i, N)}{r_M(N)} \begin{pmatrix} 0 \\ \vdots \\ 0 \\ r_M(N) \end{pmatrix}$$

$$= \begin{pmatrix} X_{M-1}^*(i) \\ V_M^{t*}(N) D_M(i, N) + e_M^{b*}(i, N) \end{pmatrix} \qquad (4A.5)$$

An expression for the backward prediction error is now derived and used in Eq. (4A.5):

$$e_M^b(i, N) = X_M^t(i) \bar{B}_M(N)$$

$$= (x_i, \ldots, x_{i-M}) \begin{pmatrix} B_M(N) \\ 1 \end{pmatrix}$$

$$= X_{M-1}^t(i) B_M(N) + x_{i-M}$$

$$= B_M^t(N) X_{M-1}(i) + x_{i-M} \qquad (4A.6)$$

Since $R_M(N)B_M(N) = -V_M(N)$ and $R_M(N) = R_M^{*t}(N)$, the complex conjugate of Eq. (4A.6) can be written as

$$e_M^{b*}(i, N) = -V_M^{*t}(N)R_M^{-1}(N)X_{M-1}^*(i) + x_{i-M}^* \qquad (4A.7)$$

When Eq. (4A.7) is substituted into eq. (4A.5), the right hand side of Eq. (4A.5) reduces to $X_M^*(i)$, thereby proving the relationship given by Eq. (4A.3).

If Eq. (4A.3), with time and order indices reduced by 1, is substituted into Eq. (4A.2), then Eq. (4.59) is demonstrated:

$$\gamma_M(i, N-1) = X_{M-1}^t(i)\left[\left(\begin{array}{c} D_{M-1}(i, N-1) \\ 0 \end{array}\right)\right.$$

$$\left. + \frac{e_{M-1}^{b*}(i, N-1)}{r_{M-1}(N-1)}\overline{B}_{M-1}(N-1)\right]$$

$$= \left[\begin{array}{cc} X_{M-2}^t(i) & x_{i-(M-1)} \end{array}\right]\left(\begin{array}{c} D_{M-1}(i, N-1) \\ 0 \end{array}\right)$$

$$+ \frac{e_{M-1}^{b*}(i, N-1)}{r_{M-1}(N-1)}X_{M-1}^t(i)\overline{B}_{M-1}(N-1)$$

$$= X_{M-2}^t(i)D_{M-1}(i, N-1) + \frac{e_{M-1}^{b*}(i, N-1)}{r_{M-1}(N-1)}e_{M-1}^b(i, N-1)$$

$$= \gamma_{M-1}(i, N-1) + \frac{|e_{M-1}^b(i, N-1)|^2}{r_{M-1}(N-1)} \qquad (4.59)$$

Derivation of the time update for $k_{M+1}(N)$ begins with the definition

$$k_{M+1}(N) = \left[\begin{array}{cc} 1 & A_M^t(N) \end{array}\right]V_{M+1}^*(N)$$

$$= V_{M+1}^{t*}(N)\overline{A}_M(N) \qquad (4A.8)$$

To develop the recursion for $k_{M+1}(N)$, recursions for $V_{M+1}(N)$ and $\overline{A}_M(N)$ are needed. To focus attention on the recursion for $k_{M+1}(N)$, the recursions for $V_{M+1}(N)$ and $\overline{A}_M(N)$ are now presented and will be derived subsequently:

$$V_{M+1}(N) = \omega V_{M+1}(N-1) + x_{N-(M+1)}X_M^*(N) \qquad (4A.9)$$

$$\overline{A}_M(N) = \overline{A}_M(N-1) - \frac{e_M^f(N, N)}{1 - \gamma_M(N-1, N-1)}$$

$$\times \left(\begin{array}{c} 0 \\ D_M(N-1, N-1) \end{array}\right) \qquad (4A.10)$$

Substituting (4A.9) and (4A.10) into (4A.8) and using (4A.7) results in

$$k_{M+1}(N) = \omega k_{M+1}(N-1) + x^*_{N-(M+1)} X^t_M(N)$$

$$\times \left[\bar{A}_M(N) + \frac{e^f_M(N,N)}{1 - \gamma_M(N-1,N-1)} \left(\begin{matrix} 0 \\ D_M(N-1,N-1) \end{matrix} \right) \right]$$

$$- \frac{e^f_M(N,N)}{1 - \gamma_M(N-1,N-1)} \left[x^*_{N-(M+1)} - e^{b*}_M(N-1,N-1) \right.$$

$$\left. - x^*_{N-(M+1)} \gamma_M(N-1,N-1) \right]$$

$$- x^*_{N-(M+1)} \frac{e^f_M(N,N)}{1 - \gamma_M(N-1,N-1)} \gamma_M(N-1,N-1)$$

$$(4A.11)$$

Since $e^f_M(N,N) = X^t_M(N)\bar{A}_M(N)$, Eq. (4A.11) becomes

$$k_{M+1}(N) = \omega k_{M+1}(N-1) + x^*_{N-(M+1)}$$

$$\times \left[e^f_M(N,N) + \frac{e^f_M(N,N)\gamma_M(N-1,N-1)}{1 - \gamma_M(N-1,N-1)} \right]$$

$$- \frac{e^f_M(N,N)}{1 - \gamma_M(N-1,N-1)} \left[x^*_{N-(M+1)} - e^{b*}_M(N-1,N-1) \right]$$

$$= \omega k_{M+1}(N-1) + \frac{e^f_M(N,N)e^{b*}_M(N-1,N-1)}{1 - \gamma_M(N-1,N-1)} \qquad (4.57)$$

Now that the recursion for $k_{M+1}(N)$ has been developed, Eqs. (4A.9) and (4A.10), given previously without proof, are derived. To prove (4A.9) the definition of $V_{M+1}(N)$ obtained from Eq. (4.37) is needed:

$$V_{M+1}(N) = \left(\begin{matrix} \rho_N(0, M+1) \\ \vdots \\ \rho_N(M, M+1) \end{matrix} \right) \qquad (4A.12)$$

Since

$$\rho_N(m, j) = \sum_{i=1}^N \omega^{N-i} x^*_{i-m} x_{i-j} \qquad (4A.13)$$

then

$$\rho_N(m, M+1) = \omega\rho_{N-1}(m, M+1) + x^*_{N-m}x_{N-(M+1)}, \qquad m = 0, \ldots, M$$

$$(4A.14)$$

Therefore

$$V_{M+1}(N) = \omega V_{M+1}(N-1) + x_{N-(M+1)}\begin{pmatrix} x^*_N \\ \vdots \\ x^*_{N-M} \end{pmatrix}$$

$$= \omega V_{M+1}(N-1) + x_{N-(M+1)}X^*_M(N) \qquad (4A.9)$$

Recursions similar to Eq. (4A.9) for $\boldsymbol{R}_{M+1}(N)$ and $\boldsymbol{Q}_{M+1}(N)$ are required for the proof of Eq. (4A.10). The definition of $\boldsymbol{Q}_{M+1}(N)$ obtained from Eq. (4.29) is given by

$$\boldsymbol{Q}_{M+1}(N) = \begin{pmatrix} \rho_N(1,0) \\ \vdots \\ \rho_N(M+1,0) \end{pmatrix} \qquad (4A.15)$$

Since

$$\rho_{N+1}(m,0) = \omega\rho_N(m,0) + x_{N+1}x^*_{N+1-m}, \qquad m = 1, \ldots, M+1$$

then

$$\boldsymbol{Q}_{M+1}(N+1) = \omega\boldsymbol{Q}_{M+1}(N) + x_{N+1}X^*_M(N) \qquad (4A.16)$$

The definition for $\boldsymbol{R}_{M+1}(N)$ is given by

$$\boldsymbol{R}_{M+1}(N) = \begin{bmatrix} \rho_N(0,0) & \cdots & \rho_N(0,M) \\ \vdots & \ddots & \vdots \\ \rho_N(M,0) & \cdots & \rho_N(M,M) \end{bmatrix}$$

Since

$$\rho_{N+1}(m,j) = \omega\rho_N(m,j) + x^*_{N+1-m}x_{N+1-j}, \qquad m = 1, \ldots, M+1,$$

$$j = 1, \ldots, M+1$$

and

$$X^*_M(N)X^t_M(N) = \begin{pmatrix} |x_N|^2 & \cdots & x^*_N x_{N-M} \\ \vdots & \ddots & \vdots \\ x^*_{N-M}x_N & \cdots & |x_{N-M}|^2 \end{pmatrix}$$

then

$$R_{M+1}(N) = \omega R_{M+1}(N-1) + X_M^*(N) X_M^t(N) \qquad (4A.17)$$

To derive Eq. (4A.10), a recursion for the inverse correlation matrix is obtained by use of the following matrix identity[6] (see Appendix Section A.2.7):

$$(C - B^{t*} A^{-1} B)^{-1} = C^{-1} - C^{-1} B^{t*} (BC^{-1} B^{t*} - A)^{-1} BC^{-1}$$

Letting $C = R_{M+1}(N)$, $A = 1$ and $B = X_M^t(N)$ in this identity results in

$$\omega^{-1} R_{M+1}^{-1}(N-1) = R_{M+1}^{-1}(N) - \frac{R_{M+1}^{-1}(N) X_M^*(N) X_M^t(N) R_{M+1}^{-1}(N)}{X_M^t(N) R_{M+1}^{-1}(N) X_M^*(N) - 1}$$

$$(4A.18)$$

From Eq. (4A.2) it can be seen that Eq. (4A.18) can be reexpressed as

$$R_{M+1}^{-1}(N-1) = \left[R_{M+1}^{-1}(N) + \frac{R_{M+1}^{-1}(N) X_M^*(N) X_M^t(N) R_{M+1}^{-1}(N)}{1 - \gamma_{M+1}(N, N)} \right] \omega$$

$$(4A.19)$$

In Eq. (4A.19), it should be noted that there is now a single time index N as a result of the definition of the correlation matrix. Using Eq. (4.33), a recursion for $A_{M+1}(N)$ can be developed by multiplying Eq. (4A.19) by $Q_{M+1}(N)$. Since

$$A_{M+1}(N) = -R_{M+1}^{-1}(N-1) Q_{M+1}(N)$$

use of Eq. (4A.16) results in

$$-A_{M+1}(N) = \left[R_{M+1}^{-1}(N) + \frac{R_{M+1}^{-1}(N) X_M^*(N) X_M^t(N) R_{M+1}^{-1}(N)}{1 - \gamma_{M+1}(N, N)} \right]$$

$$\times \left[Q_{M+1}(N+1) - x_{N+1} X_M^*(N) \right]$$

$$= -A_{M+1}(N+1) - \frac{R_{M+1}^{-1}(N) X_M^*(N) X_M^t(N)}{1 - \gamma_{M+1}(N, N)} A_{M+1}(N+1)$$

$$- x_{N+1} R_{M+1}^{-1}(N) X_M^*(N)$$

$$- \frac{x_{N+1} R_{M+1}^{-1}(N) X_M^*(N) X_M^t(N) R_{M+1}^{-1}(N) X_M^*(N)}{1 - \gamma_{M+1}(N, N)}$$

Rearranging the terms in the above equation yields

$$A_{M+1}(N) = A_{M+1}(N+1)$$

$$+ \frac{R_{M+1}^{-1}(N) X_M^*(N)}{1 - \gamma_{M+1}(N,N)} \left[x_{N+1} + X_M^t(N) A_{M+1}(N+1) \right]$$

$$(4A.20)$$

Another form of Eq. (4A.20) can be obtained by using Eq. (4.52):

$$e_M^f(N,N) = X_M^t(N) \bar{A}_M(N)$$

$$= (x_N, \ldots, x_{N-M}) \begin{pmatrix} 1 \\ A_M(N) \end{pmatrix}$$

$$= x_N + X_{M-1}^t(N-1) A_M(N) \qquad (4A.21)$$

Using Eqs. (4A.1) and (4A.21) in (4A.20) results in

$$A_{M+1}(N) = A_{M+1}(N+1) + \frac{D_{M+1}(N,N)}{1 - \gamma_{M+1}(N,N)} e_{M+1}^f(N+1, N+1)$$

$$(4A.22)$$

Reducing the indices on M and N by one and expressing Eq. (4A.22) in terms of $\bar{A}_M(N)$ yields Eq. (4A.10).

REFERENCES

1. J. D. Pack, and E. H. Satorius, Least Squares Adaptive Lattice Algorithms, Naval Ocean Systems Center, San Diego, Ca., Tech. Rep. TR423, April 1979.

2. M. Morf, D. Lee, J. Nickolls, and A. Vierira, A Classification of Algorithms for ARMA Models and Ladder Realizations, *Proc. IEEE Int. Conf. on Acoustics, Speech and Signal Processing*, Hartford, Conn., May 1977, pp. 13–19.

3. M. Morf, A. Vierira, and D. Lee, Ladder Forms for Identification and Speech Processing, In *Proc. IEEE Conf. Decision and Control* 1977, New Orleans, La., Dec. 1977, pp. 1074–1078.

4. M. Morf, Ladder Forms in Estimation and System Identification, 11th Ann. Asilomar Conf. on Circuits, Systems and Computers, Monterey Ca., Nov. 7–9, 1977, pp. 424–429.

5. M. Morf, and D. Lee, Recursive Least Squares Ladder Forms for Fast Parameter Tracking, In *Proc. 1978 IEEE Conf. on Decision and Control*, San Diego Ca., Jan. 10–12, 1979, pp. 1362–1367.

6. P. B. Liebelt, *An Introduction to Optimal Estimation*, Addison-Wesley, Reading, Mass., 1967.

chapter

5

LEAST SQUARE ESTIMATION
ALGORITHMS—PART 3

The Kalman recursive least square estimation algorithm is described in this section. To provide a better understanding of the Kalman algorithm the methods of simple least squares and weighted least squares are treated first. Simple least squares techniques are applicable to deterministic and random variables. The weighted least squares techniques to be presented consider only the case of minimum variance estimation of random variables. The Kalman recursive estimation algorithm is then developed by extending the minimum variance weighted least squares methods to the case where new measurement data are added to an existing measurement data base and prior minimum variance weighted least squares estimates.

5.1 SIMPLE LEAST SQUARES ESTIMATION

Consider the model described in Section 3.2.1, where the Levinson recursive algorithm was developed. The problem was to estimate the transmitted signal s_i from a linear combination of previous real samples x_{i-m} with $m \geq 1$. The error signal, ε_i, is

$$\varepsilon_i = s_i - \sum_{m=1}^{M} b_m x_{i-m}, \qquad i = 1, \ldots, N \tag{5.1}$$

An error vector E can be defined in terms of its components $\varepsilon_1, \ldots, \varepsilon_N$ according to

$$E = \begin{pmatrix} \varepsilon_1 \\ \vdots \\ \varepsilon_N \end{pmatrix} \tag{5.2}$$

The mean square error is given by

$$P_M = \mathscr{E}\left[\varepsilon_i^2\right] \tag{5.3}$$

where $\mathscr{E}$ denotes an ensemble average. In Section 3.4 it was shown that an estimate of the mean square error can be obtained by forming a time average, L_f, over the N samples according to

$$L_f = \sum_{i=1}^{N} |\varepsilon_i|^2 \tag{5.4}$$

$$= E^{t*}E \tag{5.5}$$

L_f is often termed a loss function and is the quantity to be minimized by selecting the coefficients b_m for $m = 1, \ldots, M$. As a result of Eq. (5.4), the least squares solution could apply to either deterministic or random variables.

Using the orthogonality principle or setting the derivatives $\partial L_f / \partial b_j = 0$ for $j = 1, \ldots, M$ yields the least square error. From Section 3.23 with time averages replacing ensemble averages the optimum coefficients, $\hat{b}_m$, are the solutions of

$$\sum_{i=1}^{N} x_{i-j}^{*} s_i = \sum_{m=1}^{M} \hat{b}_m \rho_N(j, m), \qquad j = 1, \ldots M \tag{5.6}$$

where

$$\rho_N(m, j) = \sum_{i=1}^{N} x_{i-j} x_{i-m}^{*}$$

Note that $\rho_N^{*}(m, j) = \rho_N(j, m)$ and it is assumed that $\omega = 1$ (see Eq. (4.6)).

Equation (5.6) can be put in matrix form using the following definitions:

$$
X = \begin{pmatrix}
x_0 & x_{-1} & \cdots & x_{-(M-1)} \\
x_1 & x_0 & & \\
\vdots & & \ddots & \vdots \\
\vdots & & & x_0 \\
\vdots & & & \vdots \\
x_{N-1} & \cdots & & x_{N-M}
\end{pmatrix}
\tag{5.7}
$$

$$
S = \begin{pmatrix} s_1 \\ \vdots \\ s_N \end{pmatrix} \qquad
B = \begin{pmatrix} b_1 \\ \vdots \\ b_M \end{pmatrix} \qquad
\hat{B} = \begin{pmatrix} \hat{b}_1 \\ \vdots \\ \hat{b}_M \end{pmatrix}
$$

Note that the covariance matrix $R_M(N-1) = X^{t*}X$ can now be expressed as (See eq. (4.27))

$$
X^{t*}X = \begin{pmatrix}
\rho_N(1,1) & \cdots & \rho_N(1,M) \\
\vdots & \ddots & \vdots \\
\rho_N(M,1) & \cdots & \rho_N(M,M)
\end{pmatrix}
\tag{5.8}
$$

Equation (5.6) can now be expressed in matrix form:

$$
X^{t*}S = X^{t*}X\hat{B}
\tag{5.9}
$$

The best estimate of B is then given by $\hat{B}$:

$$
\hat{B} = (X^{t*}X)^{-1}X^{t*}S
\tag{5.10}
$$

Equation (5.10) is recognized as the least squares solution for an overdetermined set of linear equations (case 2) and was presented in Section 2.3.3.[1] The quantity $(X^{t*}X)^{-1}X^{t*}$ is termed a pseudoinverse operator since it provides a solution to the original set of equations $S = XB$ in a least squares sense. Note that in this presentation X is an $N \times M$ matrix and B is an M-dimensional column vector. (In Section 2.3.3 A was the $m \times n$ matrix and X was an n-dimensional column vector.)

5.2 MINIMUM VARIANCE WEIGHTED LEAST SQUARES (GAUSS-MARKOV)

A generalization of the problem described in Section 5.1 can be obtained by computing a pseudoinverse that is appropriate for the Hilbert space of random vectors. In this instance the inner product is a generalization of the form given by Eq. (2.24). Using the vector notation introduced in Section 5.1, the problem can be stated as follows. The transmitted signal vector, S, is estimated from a set of measurements of the received signal matrix X with an error vector E according to

$$S = XB + E \tag{5.11}$$

Note that X is assumed to be known once it is received and is not random in computing the estimate of the coefficient B.

Let us assume that the error vector is zero mean with a covariance, R_ε, given by

$$R_\varepsilon = \mathscr{E}\left[EE^{t*}\right] \tag{5.12}$$

The best estimate, $\hat{B}$, which provides a minimum variance estimate of B, is obtained by minimizing $\mathscr{E}[\|B - \hat{B}\|^2]$ (see Appendix Section A.1.16). This is accomplished by requiring that the estimate $\hat{B}$ be unbiased. Assume that $\hat{B}$ is obtained by weighting Eq. (5.11) by a constant $M \times N$ matrix K:

$$\hat{B} = KS$$

$$= KXB + KE \tag{5.13}$$

The average value of the estimate $\hat{B}$ is then

$$\mathscr{E}(\hat{B}) = KX\mathscr{E}(B)$$

since $\mathscr{E}[E] = 0$. For the estimate $\hat{B}$ to be unbiased, then

$$KX = I \tag{5.14}$$

This result implies that minimizing $\mathscr{E}[\|B - \hat{B}\|^2]$ is the same as minimizing the mean square error $\mathscr{E}[\|KE\|^2]$. An equivalent minimization problem is to minimize $\mathscr{E}[\|E\|^2]$ since K is a constant.

In order to proceed with the minimization an estimate of the mean square error is obtained by forming a time average (see Section 5.1). An inner product definition which is a generalization of that given by (2.24) and leads to a Gauss Markov solution[1,2] is then

$$(S, X) = S^{t*}R_\varepsilon^{-1}X \tag{5.15}$$

The orthogonality principle requires that

$$(S - X\hat{B}, X) = 0$$

or

$$(S, X) = (X\hat{B}, X) \tag{5.16}$$

Using the inner product definition given by (5.15),

$$S'^*R_\varepsilon^{-1}X = (X\hat{B})'^* R_\varepsilon^{-1}X \tag{5.17}$$

Taking the conjugate transpose of (5.17) yields

$$X'^*(R_\varepsilon^{-1})'^* S = X'^*(R_\varepsilon^{-1})'^* X\hat{B} \tag{5.18}$$

Since R_ε is a covariance matrix, it can be represented by $R_\varepsilon = VV'^*$ (see Appendix Section A.2.6) where V is an intermediate variable. Then $(R_\varepsilon^{-1})'^* = [(V'^*)^{-1}V^{-1}]'^* = (V^{-1})'^*[(V'^*)^{-1}]'^* = (V'^*)^{-1}V^{-1} = (VV'^*)^{-1} = R_\varepsilon^{-1}$. Using this identity, (5.18) becomes

$$\hat{B} = (X'^*R_\varepsilon^{-1}X)^{-1}X'^*R_\varepsilon^{-1}S \tag{5.19}$$

The estimate in (5.19) is referred to as a Gauss-Markov estimate. From (5.19) the constant K can be identified as

$$K = (X'^*R_\varepsilon^{-1}X)^{-1}X'^*R_\varepsilon^{-1} \tag{5.20}$$

From Eq. (5.20) it can be seen that $KX = I$ which implies that $\hat{B}$ is an unbiased estimate of B. Notice that if $R_\varepsilon = I$, the identity matrix, then (5.19), corresponding to the unbiased minimum variance weighted least squares case, reduces to the simple least squares case obtained in Section 5.1.

To determine the minimum mean square error for the estimate $\hat{B}$, the error covariance matrix is computed according to

$$\mathscr{E}[(\hat{B} - B)(\hat{B} - B)'^*] = \mathscr{E}[KE(KE)'^*] \tag{5.21}$$

$$= KR_\varepsilon K'^* \tag{5.22}$$

Since $X'^*R_\varepsilon^{-1}X$ is a quadratic form (see Appendix Sections A.2.5 and A.2.6), then

$$[(X'^*R_\varepsilon^{-1}X)^{-1}]'^* = (X'^*R_\varepsilon^{-1}X)^{-1}$$

Substituting Eq. (5.20) into (5.22) then results in

$$\mathscr{E}[(\hat{B} - B)(\hat{B} - B)'^*] = (X'^*R_\varepsilon^{-1}X)^{-1} \tag{5.23}$$

The minimum mean square error is then given by

$$\|\hat{B} - B\|^2 = \mathrm{Trace}\{\mathscr{E}[(\hat{B} - B)(\hat{B} - B)'^*]\} = \mathrm{Trace}\{(X'^*R_\varepsilon^{-1}X)^{-1}\}.$$

(see Appendix Section A.2.3).

5.3 MINIMUM VARIANCE LEAST SQUARES (KALMAN)

A generalization of the problem described in Section 5.2 can be obtained by assuming that the coefficient vector B is random. The received signal matrix X is again assumed to be known once it is received. The current transmitted signal, denoted by S_c, is given by

$$S_c = XB_c + E \qquad (5.24)$$

where B_c is the current coefficient vector to be estimated. The estimate of B_c is termed $\hat{B}_c$ and is referred to as the current Kalman coefficient estimate. The previous coefficient vector and its estimate are termed B_p and $\hat{B}_p$, respectively. The error in the transmitted signal is defined as $\tilde{S}$ where

$$\tilde{S} = S_c - X\hat{B}_p \qquad (5.25)$$

and $X\hat{B}_p$ is the current estimate of S_c. Let $B_e = B_c - \hat{B}_p$ and denote the estimate of B_c by $\hat{B}_c$. Note that

$$\tilde{S} = XB_e + E \qquad (5.26)$$

Defining $\hat{B}_e = K\tilde{S}$, it can be seen that $\hat{B}_e$ is unbiased if $KX = I$ and $\mathscr{E}[E] = 0$, that is,

$$\mathscr{E}(\hat{B}_e) = \mathscr{E}(KXB_e) = \mathscr{E}(B_e)$$

To develop the Kalman recursive algorithm, it is assumed the estimate $\hat{B}_e$ of B_e is orthogonal to the coefficient error $B_e - \hat{B}_e$ (see Figure 5-1). An equivalent orthogonality condition is that $\hat{B}_e = K\tilde{S}$ is orthogonal to KE. The proof uses $KX = I$ and proceeds as follows: $\hat{B}_e - B_e = K\tilde{S} - B_e = K(XB_e + E) - B_e = KE$. Since K is a constant matrix, for $K\tilde{S}$ to be orthogonal to KE, then $\tilde{S}$ is orthogonal to E. Therefore,

$$(\tilde{S} - XB_e, \tilde{S}) = 0 \qquad (5.27)$$

or

$$(\tilde{S}, \tilde{S}) = (XB_e, \tilde{S}) \qquad (5.28)$$

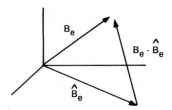

Figure 5-1. Estimation of B_e.

The inner product, which leads to the Kalman least square algorithm, is now defined to be

$$(\tilde{S}, X) = \mathscr{E}(\tilde{S}X^{t*}) \tag{5.29}$$

Therefore, (5.28) becomes

$$\mathscr{E}(\tilde{S}\tilde{S}^{t*}) = \mathscr{E}\left(XB_e\tilde{S}^{t*}\right) \tag{5.30}$$

Multiplying both sides by K with $KX = I$ leads to

$$K = \mathscr{E}\left(B_e\tilde{S}^{t*}\right)\left[\mathscr{E}(\tilde{S}\tilde{S}^{t*})\right]^{-1} \tag{5.31}$$

Since $\hat{B}_e = K\tilde{S}$, (5.31) can be reexpressed as

$$\hat{B}_e = \mathscr{E}\left(B_e\tilde{S}^{t*}\right)\left[\mathscr{E}(\tilde{S}\tilde{S}^{t*})\right]^{-1}\tilde{S} \tag{5.32}$$

Letting $\hat{B}_e = \hat{B}_c - \hat{B}_p$ be the estimate of B_e yields

$$\hat{B}_c = \hat{B}_p + \mathscr{E}\left(B_e\tilde{S}^{t*}\right)\left[\mathscr{E}(\tilde{S}\tilde{S}^{t*})\right]^{-1}\tilde{S} \tag{5.33}$$

Equation (5.33) can be expressed in a more familiar form by computing the indicated averages.

Assume that the error, E, is uncorrelated with the error in the coefficient vector:

$$\mathscr{E}\left(EB_e^{t*}\right) = 0 \tag{5.34}$$

Also define the covariances for B_e and the error E according to

$$\mathscr{E}\left(B_eB_e^{t*}\right) = P_{c,p} \tag{5.35}$$

$$\mathscr{E}\left(EE^{t*}\right) = R_\varepsilon \tag{5.36}$$

Note that the covariance $P_{c,p}$ represents a predicted error covariance between the new coefficient vector B_c and the old estimate $\hat{B}_p$. Therefore, from (5.35) and (5.36), we obtain

$$\mathscr{E}(\tilde{S}\tilde{S}^{t*}) = \mathscr{E}\left[(XB_e + E)(XB_e + E)^{t*}\right]$$

$$= XP_{c,p}X^{t*} + R_\varepsilon \tag{5.37}$$

$$\mathscr{E}\left(B_e\tilde{S}^{t*}\right) = \mathscr{E}\left[B_e(XB_e + E)^{t*}\right] = P_{c,p}X^{t*} \tag{5.38}$$

Using (5.37), (5.38), and (5.25) in (5.33) results in

$$\hat{B}_c = \hat{B}_p + P_{c,p}X^{t*}\left[XP_{c,p}X^{t*} + R_\varepsilon\right]^{-1}\left(S_c - X\hat{B}_p\right) \qquad (5.39)$$

Equation (5.39) is the recursive minimum least squares solution for the coefficients and is referred to as the Kalman algorithm.[1,2,3,4,5] The quantity, G_c, embedded in Eq. (5.39), is now defined as

$$G_c = P_{c,p}X^{t*}\left[XP_{c,p}X^{t*} + R_\varepsilon\right]^{-1} \qquad (5.40)$$

G_c, defined by Eq. (5.40), is known as the current Kalman gain.

The error covariance matrix, P_c, for the current coefficient vector, B_c, is defined by

$$P_c = \mathscr{E}\left[(B_c - \hat{B}_c)(B_c - \hat{B}_c)^{t*}\right]$$

It can now be shown that the estimate for the current coefficient vector $\hat{B}_c$ is unbiased and that the error covariance matrix can be recursively computed. Since $KX = I$, then

$$\hat{B}_c = \hat{B}_p + \hat{B}_e$$

$$= \hat{B}_p + K\tilde{S}$$

$$= \hat{B}_p + K(XB_e + E)$$

$$= \hat{B}_p + B_e + KE$$

From the above and the definition $B_c = \hat{B}_p + B_e$, it can be seen that $\hat{B}_c$ is unbiased, that is,

$$\mathscr{E}(\hat{B}_c) = \mathscr{E}(B_c)$$

To compute the error covariance matrix, P_c, the quantity $B_c - \hat{B}_c$ is formed using Eqs. (5.25), (5.39), (5.40), and the definition for $B_e = B_c - \hat{B}_p$:

$$B_c - \hat{B}_c = B_e - G_c\tilde{S}$$

The error covariance matrix can now be expressed as

$$P_c = \mathscr{E}\left[(B_e - G_c\tilde{S})(B_e - G_c\tilde{S})^{t*}\right]$$

$$= \mathscr{E}\left[B_eB_e^{t*}\right] - G_c\mathscr{E}(\tilde{S}B_e^{t*}) - \mathscr{E}(B_e\tilde{S}^{t*})G_c^{t*} + G_c\mathscr{E}(\tilde{S}\tilde{S}^{t*})G_c^{t*}$$

Using (5.35), (5.37), (5.38), and (5.40) and the property $P_{c,p}^{t^*} = P_{c,p}$ leads to

$$P_c = P_{c,p} - G_c X P_{c,p} \tag{5.41}$$

Equation (5.41) represents the recursive computation for the error covariance matrix.

5.3.1   Kalman Recursive Estimation Algorithm

Equations (5.39), (5.40), and (5.41) collectively are referred to as the Kalman recursive estimation algorithm.[5] A more explicit form of this algorithm is obtained by replacing indices c and p. Indices p and c refer to the previous and current time indices, respectively. Using an index k to represent the current sample, $k - 1$ then represents the previous sample. Table 5-1 shows the correspondence between the quantities due to the index changes. For the Kalman algorithm the received signal samples under consideration at time sample index k correspond to the following $N + M - 1$ samples:

$$x_{-(M-1)}, \ldots, x_{-1}, x_0, x_1, \ldots, x_{N-1}$$

In general, a larger set of received samples is assumed to be available. For example, at time sample index $k + 1$ the samples correspond to the following $N + M - 1$ set:

$$x_{-(M-2)}, \ldots, x_{-1}, x_0, x_1, \ldots, x_N$$

Therefore the received signal matrix is an $(N \times M)$ matrix and takes the form

$$X_k = \begin{pmatrix} X_k^t(1) \\ \vdots \\ X_k^t(N) \end{pmatrix}$$

where $X_k^t(i) = (x_{i,k}, \ldots, x_{i-(M-1),k})$ for $i = 1, \ldots, N$.

Similarly the desired N-dimensional signal vector, S_k, the N-dimensional error signal vector, E_k, and the M-dimensional Kalman coefficient vector, B_k, are, respectively, given by

$$S_k = \begin{pmatrix} s_{1,k} \\ \vdots \\ s_{N,k} \end{pmatrix}$$

$$E_k = \begin{pmatrix} \varepsilon_{1,k} \\ \vdots \\ \varepsilon_{N,k} \end{pmatrix}$$

TABLE 5-1. CORRESPONDENCE IN KALMAN ALGORITHM QUANTITIES DUE TO INDEX CHANGE

Quantity	Kalman Algorithm in General Recursive Form Given in Section 5.3	Kalman Algorithm with Time Sample Index k
Received Signal	X	$X_k = \begin{pmatrix} X_k^t(1) \\ \vdots \\ X_k^t(N) \end{pmatrix} = \begin{pmatrix} x_{1,k}, & \cdots & x_{1-(M-1),k} \\ \cdots & & \cdots \\ x_{N,k}, & \cdots & x_{N-(M-1),k} \end{pmatrix}$
Desired Signal	S_c	$S_k = \begin{pmatrix} s_{1,k} \\ \vdots \\ s_{N,k} \end{pmatrix}$
Current Kalman Coefficient	B_c	$B_k = \begin{pmatrix} b_{1,k} \\ \vdots \\ b_{M,k} \end{pmatrix}$
Error	$E = S_c - XB_c$	$E_k = S_k - X_k B_k$
Covariance	$R_\varepsilon = \mathscr{E}[EE^*]$	$R_{k\varepsilon} = \mathscr{E}[E_k E_k^*]$
Previous Kalman Coefficient Estimate	$\hat{B}_p$	$\hat{B}_{k-1}$
Current Kalman Coefficient Estimate	$\hat{B}_c = \hat{B}_p + G_c(S_c - X\hat{B}_p)$	$\hat{B}_k = \hat{B}_{k-1} + G_k(S_k - X_k \hat{B}_{k-1})$
Kalman Gain	$G_c = P_{c,p} X^{t*}[XP_{c,p}X^{t*} + R_\varepsilon]^{-1}$	$G_k = P_{k,k-1}X_k^*[X_k P_{k,k-1}X_k^* + R_{k\varepsilon}]^{-1}$
Error Covariance Matrix	$P_c = \mathscr{E}[(B_c - \hat{B}_c)(B_c - \hat{B}_c)^t{}^*]$ $= P_{c,p} - G_c X P_{c,p}$	$P_k = \mathscr{E}[(B_k - \hat{B}_k)(B_k - \hat{B}_k)^t{}^*]$ $= P_{k,k-1} - G_k X_k P_{k,k-1}$

and

$$B_k = \begin{pmatrix} b_{1,k} \\ \vdots \\ b_{M,k} \end{pmatrix}$$

The covariance matrix, $R_{k\varepsilon}$, is an $N \times N$-dimensional matrix defined by

$$R_{k\varepsilon} = \mathscr{E}\left[E_k E_k^{t*} \right]$$

The M-dimensional vectors $\hat{B}_k$ and $\hat{B}_{k-1}$ denote the current and previous Kalman coefficient estimates, respectively. The error covariance matrix, P_k, is an $M \times M$ matrix defined by

$$P_k = \mathscr{E}\left\{ \left[B_k - \hat{B}_k \right]\left[B_k - \hat{B}_k \right]^{t*} \right\}$$

The Kalman gain metrix is an $M \times N$ matrix denoted by G_k.

Often a system dynamic model for the Kalman coefficient is assumed. The form of this model is given by

$$B_k = \Phi(k, k-1) B_{k-1} + \delta_k \tag{5.42}$$

where $\Phi(k, k-1)$ is an $M \times M$ transition matrix used to determine the new Kalman coefficient from the previous Kalman coefficient, and δ_k is a zero mean random vector with an $M \times M$ covariance given by

$$Q_k = \mathscr{E}\left[\delta_k \delta_k^{t*} \right] \tag{5.43}$$

The random vector δ_k is assumed to be uncorrelated with $B_{k-1} - \hat{B}_{k-1}$ and the error E_k:

$$\mathscr{E}\left[\delta_k E_k^{t*} \right] = 0 \tag{5.44}$$

$$\mathscr{E}\left[\delta_k (B_{k-1} - \hat{B}_{k-1})^{t*} \right] = 0$$

The predicted value, $\hat{B}_{k,k-1}$, of the current Kalman coefficient estimate $\hat{B}_k$ is given by

$$\hat{B}_{k,k-1} = \Phi(k, k-1)\hat{B}_{k-1}. \tag{5.45}$$

The predicted error covariance is an $M \times M$ matrix defined by

$$P_{k,k-1} = \mathscr{E}\left\{ \left[B_k - \hat{B}_{k,k-1} \right]\left[B_k - \hat{B}_{k,k-1} \right]^{t*} \right\} \tag{5.46}$$

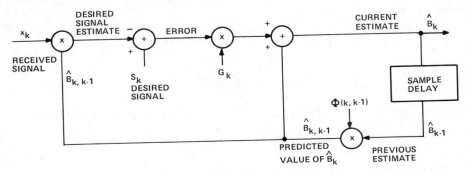

Figure 5-2.   Discrete Kalman filter.

Substituting (5.42) to (5.45) into (5.46) using the definition for the error covariance matrix, P_k, in Table 5-1 yields

$$P_{k,k-1} = \Phi(k, k-1) P_{k-1} \Phi^{t^*}(k, k-1) + Q_k \qquad (5.47)$$

Figure 5-2 illustrates the form of the discrete Kalman filter based on the above relationship. A summary of the Kalman algorithm using a dynamic model is shown in Table 5-2.

5.3.2   Derivation of Kalman Recursive Estimation Algorithm Using the Loss Function

The Kalman recursive estimation algorithm can also be developed using a loss function, L_f, that is a generalization of (5.5):

$$L_f = E_k^{t^*} R_{k\varepsilon}^{-1} E_k + W_k^{t^*} P_{k,k-1}^{-1} W_k \qquad (5.48)$$

TABLE 5-2. SUMMARY OF KALMAN ALGORITHM USING DYNAMIC MODEL

System Dynamic Model	$B_k = \Phi(k, k-1) B_{k-1} + \delta_k$
System Dynamic Model Covariance	$Q_k = \mathcal{E}[\delta_k \delta_k^{t^*}]$
Error	$E_k = S_k - X_k B_k$
Error covariance	$R_{k\varepsilon} = \mathcal{E}[E_k E_k^{t^*}]$
Current Kalman Coefficient Estimate	$\hat{B}_k = \hat{B}_{k,k-1} + G_k(S_k - X_k \hat{B}_{k,k-1})$
Prediction of Current Kalman Coefficient Estimate	$\hat{B}_{k,k-1} = \Phi(k, k-1) \hat{B}_{k-1}$
Kalman Gain	$G_k = P_{k,k-1} X_k^{t^*} [X_k P_{k,k-1} X_k^{t^*} + R_{k\varepsilon}]^{-1}$
Error Covariance Matrix	$P_k = P_{k,k-1} - G_k X_k P_{k,k-1}$
Predicted Error Covariance Matrix	$P_{k,k-1} = \Phi(k, k-1) P_{k-1} \Phi^{t^*}(k, k-1) + Q_k$

where the new quantity $W_k \equiv B_k - \hat{B}_{k,k-1}$. It is further assumed that

$$\mathscr{E}\left[W_k E_k^{t*}\right] = 0$$

$$\mathscr{E}[W_k] = 0 \tag{5.49}$$

Minimization of the loss function requires that

$$\frac{\partial L_f}{\partial B_k} = 0 \tag{5.50}$$

Computing the derivative in (5.50) using the definition in (5.48) yields

$$\frac{\partial L_f}{\partial B_k} = E_k^{t*} R_{k\varepsilon}^{-1} \frac{\partial E_k}{\partial B_k} + W_k^{t*} P_{k,k-1}^{-1} \frac{\partial W_k}{\partial B^k}$$

Since

$$\frac{\partial E_k}{\partial B_k} = -X_k \qquad \text{and} \qquad \frac{\partial W_k}{\partial B_k} = I$$

then

$$\frac{\partial L_f}{\partial B_k} = -E_k^{t*} R_{k\varepsilon}^{-1} X_k + W_k^{t*} P_{k,k-1}^{-1} \tag{5.51}$$

When (5.50) is set to zero, B_k is replaced by $\hat{B}_k$. Since

$$\left(R_{k\varepsilon}^{-1}\right)^{t*} = R_{k\varepsilon}^{-1} \qquad \text{and} \qquad \left(P_{k,k-1}^{-1}\right)^{t*} = P_{k,k-1}^{-1}$$

Eq. (5.51) becomes

$$X_k^{t*} R_{k\varepsilon}^{-1}\left(S_k - X_k \hat{B}_k\right) = P_{k,k-1}^{-1}\left(\hat{B}_k - \hat{B}_{k,k-1}\right) \tag{5.52}$$

Expanding (5.52) and solving for $\hat{B}_k$ results in

$$\hat{B}_k = \left(P_{k,k-1}^{-1} + X_k^{t*} R_{k\varepsilon}^{-1} X_k\right)^{-1}\left[X_k^{t*} R_{k\varepsilon}^{-1} S_k + P_{k,k-1}^{-1} \hat{B}_{k,k-1}\right] \tag{5.53}$$

Equation (5.53) can be rewritten using the following identity (see Appendix Section A.2.7):

$$\left(P_{k,k-1}^{-1} + X_k^{t*} R_{k\varepsilon}^{-1} X_k\right)^{-1} = P_{k,k-1} - P_{k,k-1} X_k^{t*}\left(X_k P_{k,k-1} X_k^{t*} + R_{k\varepsilon}\right)^{-1}$$

$$\times X_k P_{k,k-1} \tag{5.54}$$

A few algebraic steps are required to demonstrate the validity of (5.54) and are now outlined. Form the product

$$
\left(P_{k,k-1}^{-1} + X_k^{t*}R_{k\varepsilon}^{-1}X_k \right)\left[P_{k,k-1} - P_{k,k-1}X_k^{t*} \right.
$$

$$
\left. \times \left(X_k P_{k,k-1}X_k^{t*} + R_{k\varepsilon} \right)^{-1} X_k P_{k,k-1} \right]
$$

$$
= I + X_k^{t*}R_{k\varepsilon}^{-1}X_k P_{k,k-1} - X_k^{t*}\left(X_k P_{k,k-1}X_k^{t*} + R_{k\varepsilon} \right)^{-1}X_k P_{k,k-1}
$$

$$
- X_k^{t*}R_{k\varepsilon}^{-1}X_k P_{k,k-1}X_k^{t*}\left(X_k P_{k,k-1}X_k^{t*} + R_{k\varepsilon} \right)^{-1}X_k P_{k,k-1}
$$

$$
= I + X_k^{t*}\left[R_{k\varepsilon}^{-1} - \left(X_k P_{k,k-1}X_k^{t*} + R_{k\varepsilon} \right)^{-1} \right.
$$

$$
\left. - R_{k\varepsilon}^{-1}X_k P_{k,k-1}X_k^{t*}\left(X_k P_{k,k-1}X_k^{t*} + R_{k\varepsilon} \right)^{-1} \right] X_k P_{k,k-1}
$$

$$
= I + X_k^{t*}\left[R_{k\varepsilon}^{-1} - \left(I + R_{k\varepsilon}^{-1}X_k P_{k,k-1}X_k^{t*} \right) \right.
$$

$$
\left. \times \left(X_k P_{k,k-1}X_k^{t*} + R_{k\varepsilon} \right)^{-1} \right] X_k P_{k,k-1}
$$

$$
= I + X_k^{t*}\left[R_{k\varepsilon}^{-1} - R_{k\varepsilon}^{-1}\left(R_{k\varepsilon} + X_k P_{k,k-1}X_k^{t*} \right) \right.
$$

$$
\left. \times \left(X_k P_{k,k-1}X_k^{t*} + R_{k\varepsilon} \right)^{-1} \right] X_k P_{k,k-1}
$$

$$
= I
$$

Thus (5.54) is proved.

Using (5.54) in (5.53) and identifying the quantity

$$
G_k = P_{k,k-1}X_k^{t*}\left[X_k P_{k,k-1}X_k^{t*} + R_{k\varepsilon} \right]^{-1}
$$

as the Kalman gain results in

$$
\hat{B}_k = \left(P_{k,k-1} - G_k X_k P_{k,k-1} \right)\left(X_k^{t*}R_{k\varepsilon}^{-1}S_k + P_{k,k-1}^{-1}\hat{B}_{k,k-1} \right)
$$

$$
= \hat{B}_{k,k-1} + P_{k,k-1}X_k^{t*}R_{k\varepsilon}^{-1}S_k - G_k X_k P_{k,k-1}X_k^{t*}R_{k\varepsilon}^{-1}S_k
$$

$$
- G_k X_k P_{k,k-1}P_{k,k-1}^{-1}\hat{B}_{k,k-1} \tag{5.55}
$$

$$
= \hat{B}_{k,k-1} + \left[P_{k,k-1}X_k^{t*} - P_{k,k-1}X_k^{t*}\left(X_k P_{k,k-1}X_k^{t*} + R_{k\varepsilon} \right)^{-1} \right.
$$

$$
\left. \times X_k P_{k,k-1}X_k^{t*} \right] R_{k\varepsilon}^{-1}S_k - G_k X_k \hat{B}_{k,k-1}
$$

$$
= \hat{B}_{k,k-1} + G_k\left[S_k - X_k\hat{B}_{k,k-1} \right] \tag{5.56}
$$

The expression for the error covariance matrix

$$P_k = \mathscr{E}\left[(B_k - \hat{B}_k)(B_k - \hat{B}_k)^{t*}\right]$$

can be developed as follows. Note that from (5.56)

$$
\begin{aligned}
B_k - \hat{B}_k &= B_k - \hat{B}_{k,k-1} - G_k[S_k - X_k\hat{B}_{k,k-1}] \\
&= W_k - G_k[S_k - X_kB_k - X_k\hat{B}_{k,k-1} + X_kB_k] \\
&= W_k - G_k[E_k + X_kW_k]
\end{aligned}
$$

(5.57)

Therefore, substituting (5.57) in the definition for P_k yields

$$
\begin{aligned}
P_k &= \mathscr{E}\left\{\left[W_k - G_k(E_k + X_kW_k)\right]\left[W_k^{t*} - (E_k + X_kW_k)^{t*}G_k^{t*}\right]\right\} \\
&= \mathscr{E}\left[W_kW_k^{t*}\right] - G_k\mathscr{E}\left[(E_k + X_kW_k)W_k^{t*}\right] \\
&\quad - \mathscr{E}\left[W_k(E_k + X_kW_k)^{t*}\right]G_k^{t*} \\
&\quad + G_k\mathscr{E}\left[(E_k + X_kW_k)(E_k + X_kW_k)^{t*}\right]G_k^{t*} \\
&= P_{k,k-1} - G_kX_kP_{k,k-1} - P_{k,k-1}X_k^{t*}G_k^{t*} + G_kR_{k\epsilon}G_k^{t*} \\
&\quad + G_kX_kP_{k,k-1}X_k^{t*}G_k^{t*}
\end{aligned}
$$

(5.58)

Using the expression for G_k in (5.58) results in

$$P_k = P_{k,k-1} - G_kX_kP_{k,k-1}$$

(5.59)

EXAMPLE

Consider the simple communication channel with a transmitted signal, S_k, a constant channel attenuation, C, and an additive noise term n_k shown in Figure 5-3. The received signal, X_k, is given by

$$X_k = CS_k + n_k$$

(5.60)

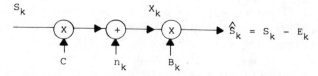

Figure 5-3. Kalman Recursive Estimation of a Constant Channel

The Kalman coefficient, B_k, weights the received signal to produce a least mean square estimate, $\hat{S}_k$, of the transmitted signal with an error, E_k, so that

$$S_k = X_k B_k + E_k \tag{5.61}$$

with covariance

$$R_{k\varepsilon} = \mathscr{E}\left[E_k E_k^{t*}\right]$$

In this case the system dynamic model is assumed to be

$$B_k = B_{k-1} + \delta_k \tag{5.62}$$

where the transition matrix $\Phi(k, k-1)$ is the identity matrix and the prediction of the current Kalman coefficient estimate is

$$\hat{B}_{k,k-1} = \hat{B}_{k-1}.$$

The predicted error covariance is then given by

$$P_{k,k-1} = P_{k-1} + Q_k$$

where $Q_k = \mathscr{E}[\delta_k \delta_k^{t*}]$. The following three equations are then used in the recursive computation:

Current Kalman coefficient estimate

$$\hat{B}_k = \hat{B}_{k-1} + G_k\left(S_k - X_k \hat{B}_{k-1}\right) \tag{5.63}$$

Kalman gain

$$G_k = (P_{k-1} + Q_k)X_k^{t*}\left[X_k(P_{k-1} + Q_k)X_k^{t*} + R_{k\varepsilon}\right]^{-1} \tag{5.64}$$

and Error Covariance Matrix

$$P_k = P_{k-1} + Q_k - G_k X_k(P_{k-1} + Q_k) \tag{5.65}$$

The signal S_k is assumed to take the values ± 1 with a specific sequence $\{1, 1, -1, -1, -1, -1, 1, \ldots\}$. The signal to noise ratio (SNR) is assumed to be infinite (i.e., the noise is negligible). For $C = 0.4$, the received signal, X_k, has a corresponding sequence given by $\{0.4, 0.4, -0.4, -0.4, -0.4, -0.4, 0.4, \ldots\}$. The initial error covariance is assumed to be $P_0 = 10$. Two cases, corresponding to slow and fast convergence, are considered. In the slow convergence case it is assumed that $R_{k\varepsilon} = Q_k = 1$, and in the fast convergence

**TABLE 5-3. NUMERICAL RESULTS FOR KALMAN RECURSIVE
ESTIMATION OF A CONSTANT CHANNEL**

Iteration (k)	$R_{k\varepsilon} = Q_k = 1$			$R_{k\varepsilon} = Q_k = 0.1$		
	G_k	P_k	$\hat{B}_k$	G_k	P_k	$\hat{B}_k$
1	1.594	3.986	1.594	2.354	0.5886	2.354
2	1.109	2.773	1.996	1.311	0.3276	2.431
3	−0.9411	2.353	2.186	−1.016	0.2539	2.459
4	−0.8729	2.182	2.296	−0.9038	0.2260	2.474
5	−0.8434	2.109	2.365			
6	−0.8304	2.076	2.410			
7	0.8246	2.061	2.439			

case $R_{k\varepsilon} = Q_k = 0.1$. For $R_{k\varepsilon} = Q_k = 1$

$$G_1 = (P_0 + Q_1)X_1[X_1(P_0 + Q_1)X_1 + R_{1\varepsilon}]^{-1}$$

$$= 11(0.4)[0.4^2(11) + 1]^{-1} = 1.5942$$

$$P_1 = P_0 + Q_1 - G_1 X_1(P_0 + Q_1)$$

$$= 11 - (1.5942)0.4(11) = 3.9855$$

$$\hat{B}_1 = \hat{B}_0 + G_1(S_1 - X_1\hat{B}_0)$$

$$= 0 + 1.5942(1 - 0.4(0)) = 1.5942$$

Table 5-3 provides the numerical recursive solution.

It is useful to compare the Kalman recursive solution given above to the least mean square solution given by (3.53). The correlation terms are given by

$$\mathcal{E}(S_k^2) = 1$$

$$\mathcal{E}(S_k X_k) = \mathcal{E}(S_k(CS_k + n_k)) = C$$

$$\mathcal{E}(n_k X_k) = \mathcal{E}(n_k(CS_k + n_k)) = \sigma_n^2$$

where σ_n^2 is the noise variance. Equation (3.53) for $M = 1$ is then given by

$$C = C^2\hat{B}_k + \sigma_n^2\hat{B}_k$$

or

$$\hat{B}_k = \frac{C}{C^2 + \sigma_n^2}$$

for infinite SNR, σ_n^2 approaches 0 and $\hat{B}_k = 1/C = 2.5$. From Table 5-3 it can be seen that the Kalman coefficient estimate $\hat{B}_k$ converges rapidly for $R_{k\varepsilon} = Q_k = 0.1$, and is at about 97 percent of its final value in two iterations.

5.4   SQUARE ROOT KALMAN ALGORITHM

The Kalman algorithm described above is sensitive to computer roundoff errors and can lead to numerical results that are inaccurate or even meaningless. However, a square root formulation of the Kalman algorithm has inherently better stability and numerical accuracy than the conventional Kalman algorithm.[6-15] The improved numerical behavior of the square root algorithm is due in large part to a reduction in the numerical ranges of the variables. Loosely speaking, one can say that computations which involve numbers ranging between 10^{-N} and 10^N are reduced to ranges between $10^{-N/2}$ and $10^{N/2}$. Thus the square root algorithm achieves accuracies that are comparable to a conventional Kalman algorithm that uses twice the numerical precision. There exist many square root Kalman filters.[6] In this section, an upper triangular diagonal (U-D) covariance factorization filter is discussed.[7,8,14,15] This factorization is a generalization of the Cholesky decomposition algorithm described in Section 3.3.

5.4.1   U-D Factorization

The covariance update formula (5.59) for the Kalman filter is numerically unstable. The main reason for the instability is that P_k is computed as a difference of two positive semidefinite matrices. Numerical accuracy is reduced in every iteration. Moreover, numerical deterioration is also associated with the high order dependence of G_k and P_k on X_k and with the accumulated effects of roundoff error. The accuracy degeneracy of the classic Kalman filter may result in a P_k matrix which is indefinite (having negative eigenvalues). The U-D covariance factorization filter involves an upper triangular factorization of the filter error covariance matrix, that is, $P_k = U_k^* D_k U_k^t$ (see Appendix Section A.2.6). Efficient and stable measurement updating recursions are achieved for the upper triangular matrix, U_k, and the diagonal matrix, D_k. In this form the algorithm can be referred to as square root Kalman filtering although the computation of square roots is not required. The Cholesky decomposition algorithm in Section 3.3 does require square root computations resulting in excessive computation. The U-D factorization procedure guarantees non-negativity of the computed covariance matrix.

5.4.2   Mathematical Development of U-D Kalman Algorithm

In the square root Kalman formulation several simplifying assumptions are made. To begin with, let $M = N$ and assume that the received signal, $\bar{X}_k$, is an

N-dimensional vector given by

$$\overline{X}_k^t = (x_{1,k}, x_{2,k}, \ldots, x_{N,k}) \tag{5.66}$$

The Kalman coefficient vector B_k is an N-dimensional vector denoted by

$$B_k^t = (b_{1,k}, b_{2,k}, \ldots, b_{N,k}) \tag{5.67}$$

Now let the state transition matrix $\Phi(k, k - 1)$ be the identity matrix and let the covariance matrix Q_k be zero. Then the error covariance matrix is an $N \times N$ matrix satisfying the relationship $P_{k-1} = P_{k,k-1}$. The previous Kalman coefficient estimate satisfies $\hat{B}_{k-1} = \hat{B}_{k,k-1}$. The Kalman gain matrix, G_k, then reduces to an N-dimensional Kalman gain vector, K_k, denoted by

$$K_k^t = (K_{1,k}, K_{2,k}, \ldots, K_{N,k}) \tag{5.68}$$

The measurement error, ε_k, can then be expressed as

$$\varepsilon_k = s_k - \overline{X}_k^t B_k \tag{5.69}$$

Then the Kalman filter equations become

$$\hat{B}_k = \hat{B}_{k-1} + K_k\left[s_k - \overline{X}_k^t \hat{B}_{k-1}\right] \tag{5.70}$$

$$K_k = P_{k-1}\overline{X}_k^* \alpha^{-1} \tag{5.71}$$

$$P_k = P_{k-1} - K_k\overline{X}_k^t P_{k-1} \tag{5.72}$$

where

$$\alpha = \left[\overline{X}_k^t P_{k-1}\overline{X}_k^* + R_{k\varepsilon}\right] \tag{5.72a}$$

The variables α and $R_{k\varepsilon}$ are scalars due to the definition given above.

Suppose that the error covariance matrix, P_k, is expressed in factored form as

$$P_k = U_k^* D_k U_k^t \tag{5.73}$$

where U_k is an upper triangular matrix with unit diagonal elements and nonequal off-diagonal elements $u_{i,j}$, $i = 1, 2, \ldots, N - 1$; $j = i + 1$, $i + 2, \ldots, N$, and D_k is a diagonal matrix with real-valued diagonal elements $(d_1(k), d_2(k), \ldots, d_N(k))$. Also let the vectors F and V be defined by

$$F_{k-1} = U_{k-1}^t \overline{X}_k^* \tag{5.74}$$

$$V_{k-1} = D_{k-1}F_{k-1} \tag{5.75}$$

Then Eq. (5.72) can be written as

$$U_k^* D_k U_k^t = U_{k-1}^* \big[D_{k-1} - \alpha^{-1} V_{k-1} V_{k-1}^{*t} \big] U_{k-1}^t \qquad (5.76)$$

by substituting Eqs. (5.71), (5.73), (5.74), and (5.75) into (5.72). If we define $\overline{U}_{k-1}$ and $\overline{D}_{k-1}$ to be the U-D factors of the terms inside the brackets of Eq. (5.76) with elements $\overline{u}_{i,j}(k-1)$ and $\overline{d}_i$, respectively, then

$$\overline{U}_{k-1}^* \overline{D}_{k-1} \overline{U}_{k-1}^t = D_{k-1} - \alpha^{-1} V_{k-1} V_{k-1}^{*t} \qquad (5.77)$$

Equation (5.76) may then be written as

$$U_k^* D_k U_k^t = (U_{k-1} \overline{U}_{k-1})^* \overline{D}_{k-1} (U_{k-1} \overline{U}_{k-1})^t \qquad (5.78)$$

From Eq. (5.78) we can identify

$$U_k = U_{k-1} \overline{U}_{k-1} \qquad (5.79)$$

$$D_k = \overline{D}_{k-1} \qquad (5.80)$$

Thus the updated U-D factors are determined in terms of the U-D factors of $D_{k-1} - \alpha^{-1} V_{k-1} V_{k-1}^{*t}$ and U_{k-1}. Let the vectors F_{k-1} and V_{k-1} be denoted as

$$F_{k-1}^t = \big(f_1(k-1), f_2(k-1), \ldots, f_N(k-1) \big) \qquad (5.81)$$

$$V_{k-1}^t = \big(v_1(k-1), v_2(k-1), \ldots, v_N(k-1) \big) \qquad (5.82)$$

Note that the upper triangular matrix U_k and the diagonal matrix D_k can be expressed as

$$U_k = \begin{pmatrix} 1 & u_{12}(k) & \cdots & & u_{1N}(k) \\ & 1 & u_{23}(k) & & \cdots \\ & \phi & & \ddots & u_{N-1,N}(k) \\ & & & & 1 \end{pmatrix} \qquad (5.83)$$

$$D_k = \begin{pmatrix} d_1(k) & & & \phi \\ & d_2(k) & & \\ & & \ddots & \\ \phi & & & d_N(k) \end{pmatrix} \qquad (5.84)$$

Denote

$$\overline{P}_{k-1} = \overline{U}_{k-1}^* \overline{D}_{k-1} \overline{U}_{k-1}^t$$

and form a quadratic expression (see Appendix Section A.2.5)

$$\overline{X}_k^t \overline{P}_{k-1} \overline{X}_k^* = \overline{X}_k^t D_{k-1} \overline{X}_k^* - \alpha_N^{-1} \overline{X}_k^t V_{k-1} V_{k-1}^{t*} \overline{X}_k^*$$

$$= \sum_{j=1}^{N} d_j(k-1)|x_{j,k}|^2 - \alpha_N^{-1} \left| \sum_{j=1}^{N} v_j(k-1)x_{j,k} \right|^2$$

$$= \sum_{j=1}^{N-1} d_j(k-1)|x_{j,k}|^2 + d_N(k-1)|x_{N,k}|^2$$

$$- \alpha_N^{-1} \left\{ \left| \sum_{j=1}^{N-1} v_j(k-1)x_{j,k} \right|^2 + |v_N(k-1)x_{N,k}|^2 \right.$$

$$+ v_N^*(k-1)x_{N,k}^* \sum_{j=1}^{N-1} v_j(k-1)x_{j,k}$$

$$\left. + v_N(k-1)x_{N,k} \sum_{j=1}^{N-1} v_j^*(k-1)x_{j,k}^* \right\}$$

where $\alpha_N = \alpha$. The subscript N was appended to α in order to recursively compute this parameter. Assume

$$\bar{d}_N(k-1) = d_N(k-1) - \alpha_N^{-1}|v_N(k-1)|^2$$

Then, the quadratic form becomes

$$\overline{X}_k^t \overline{P}_{k-1} \overline{X}_k^* = \sum_{j=1}^{N-1} d_j(k-1)|x_{j,k}|^2$$

$$+ \bar{d}_N(k-1) \left| - \sum_{j=1}^{N-1} \frac{v_N(k-1)}{\alpha_N \bar{d}_N(k-1)} v_j^*(k-1)x_{j,k}^* + x_{N,K}^* \right|^2$$

$$- \frac{d_N(k-1)}{\alpha_N \bar{d}_N(k-1)} \left| \sum_{j=1}^{N-1} v_j(k-1)x_{j,k} \right|^2$$

Let

$$\bar{d}_N(k-1) = d_N(k-1) \frac{\alpha_{N-1}}{\alpha_N}$$

and

$$\bar{u}_{j,N}(k-1) = -\frac{v_j^*(k-1)v_N(k-1)}{\alpha_N \bar{d}_N(k-1)}$$

The quadratic form can then be written as

$$\bar{X}_k^t \bar{P}_{k-1} \bar{X}_k^* = \bar{d}_N(k-1)\left| \sum_{j=1}^{N-1} \bar{u}_{jN}(k-1)x_{j,k}^* + x_{N,k}^* \right|^2$$

$$+ \sum_{j=1}^{N-1} d_j(k-1)|x_{j,k}|^2 - \alpha_{N-1}^{-1}\left| \sum_{j=1}^{N-1} v_j(k-1)x_{j,k} \right|^2$$

Since

$$\bar{F}_{k-1} = \bar{U}_{k-1}^t \bar{X}_k^*$$

$$= \begin{pmatrix} 1 & & & \\ \bar{u}_{12}(k-1) & 1 & \phi & \\ \vdots & & \ddots & \\ \bar{u}_{1N}(k-1) & \cdots & \bar{u}_{N-1,N}(k-1) & 1 \end{pmatrix} \begin{pmatrix} x_{1k}^* \\ \vdots \\ x_{N,k}^* \end{pmatrix}$$

and

$$\bar{f}_N(k-1) = \sum_{j=1}^{N-1} \bar{u}_{jN}(k-1)x_{j,k}^* + x_{N,k}^*$$

where $\{\bar{f}_i(k-1)\}$ are elements of $\bar{F}_{k-1}$, then the quadratic form after one step becomes

$$\bar{X}_k^t \bar{P}_{k-1} \bar{X}_k^* = \bar{d}_N(k-1)|\bar{f}_N(k-1)|^2$$

$$+ \sum_{j=1}^{N-1} d_j(k-1)|x_{j,k}|^2 - \alpha_{N-1}^{-1}\left| \sum_{j=1}^{N-1} v_j(k-1)x_{j,k} \right|^2$$

After $N-1$ additional steps have been taken, the quadratic form may be written as

$$\bar{X}_k^t \bar{P}_{k-1} \bar{X}_k^* = \sum_{j=2}^{N} \bar{d}_j(k-1)|\bar{f}_j(k-1)|^2$$

$$+ d_1(k-1)|x_{1,k}|^2 - \alpha_1^{-1}|v_1(k-1)x_{1,k}|^2$$

$$= \sum_{j=1}^{N} \bar{d}_j(k-1)|\bar{f}_j(k-1)|^2$$

where

$$d_1(k-1)|x_{1,k}|^2 - \alpha_1^{-1}|v_1(k-1)x_{1,k}|^2 = |\bar{f}_1(k-1)|^2\bar{d}_1(k-1)$$

$$\bar{d}_j(k-1) = d_j(k-1)\frac{\alpha_{j-1}}{\alpha_j}, \qquad\qquad j = 1,\ldots,N \qquad (5.85)$$

$$\bar{d}_j(k-1) = d_j(k-1) - \alpha_j^{-1}|v_j(k-1)|^2, \qquad j = 1,\ldots,N \qquad (5.86)$$

$$\bar{f}_1(k-1) = x_{1,k}^* \qquad\qquad (5.87)$$

$$\bar{f}_j(k-1) = \sum_{i=1}^{j-1} \bar{u}_{i,j}(k-1)x_{i,k}^* + x_{j,k}^*, \qquad j = 2,\ldots,N \qquad (5.88)$$

$$\bar{u}_{i,j}(k-1) = \frac{-v_i^*(k-1)v_j(k-1)}{\alpha_j\bar{d}_j(k-1)}, \qquad i = 1,\ldots,j-1;$$

$$j = 1,\ldots,N \qquad (5.89)$$

Since

$$v_j(k-1) = d_j(k-1)f_j(k-1), \qquad j = 1,\ldots,N \qquad (5.90)$$

Eq. (5.85) can be written as

$$d_j(k-1)\frac{\alpha_{j-1}}{\alpha_j} = d_j(k-1) - \frac{1}{\alpha_j}|d_j(k-1)|^2|f_j(k-1)|^2$$

or

$$\alpha_j = \alpha_{j-1} + v_j(k-1)f_j^*(k-1), \qquad j = 1,\ldots,N \qquad (5.91)$$

For $j = N$, α_N can be expressed as

$$\alpha_N = \alpha = \sum_{j=1}^{N} v_j(k-1)f_j^*(k-1) + \alpha_0$$

Note that

$$\sum_{j=1}^{N} v_j(k-1)f_j^*(k-1) = \sum_{j=1}^{N} d_j(k-1)|f_j(k-1)|^2$$

$$= \bar{X}_k^t P_{k-1}\bar{X}_k^*$$

As a result of the above relationships and (5.72a),

$$\alpha_0 = R_{k\varepsilon} = \xi \tag{5.92}$$

where ξ is a scalar. Equation (5.80) leads to

$$d_j(k) = \bar{d}_j(k-1), \qquad j = 1, \dots, N \tag{5.93}$$

Equation (5.79) can be written as

$$U_k = U_{k-1}\bar{U}_{k-1}$$

$$= U_{k-1} \begin{pmatrix} 1 & \bar{u}_{12}(k-1) & \cdots & \bar{u}_{1N}(k-1) \\ & 1 & \ddots & \vdots \\ & \phi & & \bar{u}_{N-1,N}(k-1) \\ & & & 1 \end{pmatrix}$$

Let

$$W_j^t(k-1) \equiv \left(v_1(k-1), v_2(k-1), \dots, v_{j-1}(k-1), 0, \dots, 0 \right)$$

$$\overset{\uparrow}{\times\, N\text{th element}}$$

and

$$\lambda_j \equiv -\frac{v_j(k-1)}{\alpha_j \bar{d}_j(k-1)} = -\frac{f_j(k-1)}{\alpha_{j-1}}, \qquad j = 2, 3, \dots, N \tag{5.94}$$

Then

$$U_k = U_{k-1} + \left[0, \lambda_2 U_{k-1} W_2^*(k-1), \right.$$

$$\left. \lambda_3 U_{k-1} W_3^*(k-1), \dots, \lambda_N U_{k-1} W_N^*(k-1) \right]$$

The quantity in brackets depicts the updating of the N columns of the upper triangular matrix and

$$\lambda_2 U_{k-1} W_2^*(k-1) = \frac{-v_2(k-1)}{\alpha_2 \bar{d}_2(k-1)} U_{k-1} \begin{pmatrix} v_1^*(k-1) \\ 0 \\ \vdots \\ 0 \end{pmatrix}$$

$$\vdots$$

$$\lambda_N U_{k-1} W_N^*(k-1) = \frac{-v_N(k-1)}{\alpha_N \bar{d}_N(k-1)} U_{k-1} \begin{pmatrix} v_1^*(k-1) \\ \vdots \\ v_{N-1}^*(k-1) \\ 0 \end{pmatrix}$$

Note that

$$K_k = P_{k-1}\bar{X}_k^*\alpha^{-1}$$

$$= U_{k-1}^* D_{k-1} U_{k-1}^t \bar{X}_k^* \alpha^{-1}$$

$$= \alpha^{-1} U_{k-1}^* V_{k-1}$$

Let

$$\gamma_j^* = U_{k-1} W_j^*(k-1) \qquad j = 2, \ldots, N$$

Then

$$\gamma_{N+1}^* = K_k^* \alpha$$

by noting $W_{N+1}(k-1) = V_{k-1}$. If γ_j^t is defined by

$$\gamma_j^t = (g_{1,j}, \ldots, g_{j-1,j}, 0, \ldots, 0) \qquad j = 2, \ldots, N+1$$

$$\uparrow$$

$$\times N\text{th element}$$

Then

$$\gamma_j = \begin{pmatrix} g_{1,j} \\ g_{2,j} \\ \vdots \\ g_{j-1,j} \\ 0 \\ \vdots \\ 0 \end{pmatrix} = \begin{pmatrix} 1 & u_{1,2}^*(k-1) & \cdots & u_{1,N}^*(k-1) \\ & 1 & & \vdots \\ & \phi & \ddots & \vdots \\ & & & 1 \end{pmatrix} \begin{pmatrix} v_1(k-1) \\ v_2(k-1) \\ \vdots \\ v_{j-1}(k-1) \\ 0 \\ \vdots \\ 0 \end{pmatrix}$$

or

$$g_{i,j} = v_i(k-1) + \sum_{l=i+1}^{j-1} u_{i,l}^*(k-1) v_l(k-1), \qquad (5.94a)$$

$$j = 2, \ldots, N; \quad i = 1, \ldots, j-1$$

Now replacing j with $j+1$ yields

$$g_{i,j+1} = v_i(k-1) + \sum_{l=i+1}^{j} u_{i,l}^*(k-1) v_l(k-1)$$

$$j = 2, \ldots, N-1; \quad i = 1, \ldots, j-1$$

$$= g_{i,j} + u_{i,j}^*(k-1) v_j(k-1) \qquad (5.95)$$

and

$$U_k = U_{k-1} + \left[0, \lambda_2 \gamma_2^*, \lambda_3 \gamma_3^*, \ldots, \lambda_N \gamma_N^*\right]$$

$$= U_{k-1} + \begin{pmatrix} 0 & \lambda_2 g_{12}^* & \lambda_3 g_{13}^* & \cdots & \lambda_N g_{1N}^* \\ 0 & 0 & \lambda_3 g_{23}^* & & \lambda_N g_{2N}^* \\ \vdots & \vdots & \vdots & & \vdots \\ & & & & \lambda_N g_{N-1,N}^* \\ 0 & 0 & 0 & & 0 \end{pmatrix}$$

Note that (5.94a) is used to compute $g_{12}, g_{23}, \cdots g_{N-1,N}$ and all other g_{ij}'s are computed recursively from (5.95). The elements of U_k are

$$u_{i,j}(k) = u_{i,j}(k-1) + g_{i,j}^* \lambda_j, \qquad i = 1, \ldots, j-1; \quad j = 2, \ldots, N$$

(5.96)

The initial values of $\{g_{i,j+1}\}$ are given by

$$(g_{1,2}, g_{2,3}, \ldots, g_{N-1,N}, g_{N,N+1}) = (v_1, v_2, \ldots, v_{N-1}, v_N) \qquad (5.97)$$

It can also be seen that

$$\gamma_N^t = (g_{1,N}, \ldots, g_{N-1,N}, 0) \qquad (5.98)$$

$$\gamma_{N+1}^t = (g_{1,N+1}, \ldots, g_{N,N+1}) \qquad (5.99)$$

Since

$$K_k = \alpha^{-1} U_{k-1}^* V_{k-1}$$

$$= \alpha^{-1} U_{k-1}^* \left(W_N(k-1) + \begin{pmatrix} 0 \\ \vdots \\ 0 \\ v_N(k-1) \end{pmatrix} \right)$$

and noting that

$$\gamma_N = U_{k-1}^* W_N(k-1)$$

then

$$K_k = \alpha^{-1} \gamma_N + \alpha^{-1} \begin{pmatrix} u_{1,N}^*(k-1) \\ u_{2,N}^*(k-1) \\ \vdots \\ u_{N-1,N}^*(k-1) \\ 1 \end{pmatrix} v_N(k-1)$$

or

$$\gamma_{N+1} = \gamma_N + \begin{pmatrix} u^*_{1,N}(k-1) \\ \vdots \\ u^*_{N-1,N}(k-1) \\ 1 \end{pmatrix} v_N(k-1)$$

$$= K_k \alpha = K_k \alpha_N \tag{5.100}$$

5.4.3   Summary of U-D Kalman Algorithm

From the mathematical development given in Section 5.4.2, the U-D Kalman filter can be summarized as follows:

$$K_k^t = \alpha_N^{-1}(g_{1,N+1}, g_{2,N+1}, \ldots, g_{N,N+1}) \tag{5.101}$$

where

$$f_1(k-1) = x^*_{1,k} \tag{5.102}$$

$$f_j(k-1) = \sum_{i=1}^{j-1} u_{i,j}(k-1)x^*_{i,k} + x^*_{j,k}, \qquad j = 2,3,\ldots,N \tag{5.103}$$

$$g_{j,j+1} = v_j(k-1) = d_j(k-1)f_j(k-1), \quad j = 1,2,\ldots,N \tag{5.104}$$

$$\alpha_1 = \xi + g_{1,2}f_1^*(k-1) \tag{5.105}$$

$$\alpha_j = \alpha_{j-1} + g_{j,j+1}f_j^*(k-1), \qquad\qquad j = 2,3,\ldots,N \tag{5.106}$$

$$\lambda_j = -\frac{f_j(k-1)}{\alpha_{j-1}}, \qquad\qquad j = 2,3,\ldots,N \tag{5.107}$$

$$d_j(k) = d_j(k-1)\frac{\alpha_{j-1}}{\alpha_j}, \qquad\qquad j = 1,2,\ldots,N \tag{5.108}$$

$$u_{i,j}(k) = u_{i,j}(k-1) + g^*_{i,j}\lambda_j, \qquad\qquad i = 1,\ldots,j-1;$$

$$j = 2,\ldots,N \tag{5.109}$$

$$g_{i,j+1} = g_{i,j} + v_j(k-1)u^*_{i,j}(k-1), \qquad i = 1,\ldots,j-1;$$

$$j = 2,\ldots,N-1 \tag{5.110}$$

All constants except ξ, α_j, $j = 1, 2, \ldots, N$, d_j, $j = 1, 2, \ldots, N$ defined in (5.101)–(5.110) are complex quantities. The quantities defined in (5.102)–(5.108) are computed first using the initial conditions $u_{i,j}(k - 1) = 0$ and $d_j(k - 1) = 1$. Equations (5.109) and (5.110) are evaluated recursively for $j = 2, 3 \ldots$, and $i = 1, 2, \ldots, j - 1$. Finally, the Kalman gains are computed according to Eq. (5.101).

REFERENCES

1. D. G. Luenberger, *Optimization by Vector Space Methods*, Wiley, New York, 1969.

2. H. W. Sorenson, Least-Squares Estimation from Gauss to Kalman, *IEEE Spectrum*, July 1970, pp. 63–68.

3. R. E. Kalman and R. S. Bucy, New Results in Linear Filtering and Prediction Theory, *J. Basic Engineering 83D*, 1961, pp. 95–108.

4. R. E. Kalman, A New Approach to Linear Filtering and Prediction Problems, *J. Basic Engineering*, Trans. ASME Series D, Vol. 82, 1960, pp. 35–45.

5. A. P. Sage and J. L. Melsa, *Estimation Theory with Applications to Communication and Control*, McGraw-Hill, New York, 1971.

6. F. M. Hsu, Square Root Kalman Filtering for High Speed Data Received Over Fading Dispersive HF Channels, *IEEE Trans. Information Theory IT-28*, 5 Sept. 1982.

7. G. J. Bierman, *Factorization Methods for Discrete Sequential Estimation*, Academic, New York, 1977.

8. N. A. Carlson, Fast Triangular Formulation of the Square Root Filter, *AIAA J. 11*, 9 Sept. 1973.

9. N. A. Carlson and A. F. Culmone, Efficient Algorithms for On-Board Array Processing, Int Conf. on Communications, Boston Mass., June 10–14, 1979, pp. 58.1.1-58.1.5.

10. A. Andrews, A Square Root Formulation of the Kalman Covariance Equations, *AIAA J. 6*, 6 July 1967.

11. J. F. Bellantoni and K. W. Dodge, A Square Root Formulation of the Kalman-Schmidt Filter, *AIAA J. 5*, 7 July 1967.

12. R. H. Batten, *Astronomical Guidance*, McGraw-Hill, New York, 1964, pp. 338–339.

13. S. F. Schmidt, Computational Techniques in Kalman Filtering, In *Theory and Applications of Kalman Filtering*, (C. T. Leondes Ed.), Advisory Group for Aerospace Res Develop., AGARD graph 139, 1970.

14. G. T. Bierman, Measurement Updating Using the U-D Factorization, In *Proc. IEEE Conf. Decision Control*, Houston Tx., 1975.

15. W. M. Gentleman, Least Squares Computations by Givens Transformations Without Square Roots, *J. Inst. Math. Appl. 12*, 1973.

part
———————
2

Applications of Least Square Analysis to Digital Signal Processing

At this juncture important least square algorithms have been formally derived and are mathematically well defined. Applications of these algorithms to a variety of digital signal processing problems are presented in the succeeding chapters. Often, an appropriate least square algorithm requires modification to fit the specific application. Thus, versions of these algorithms are numerous.

The applications presented are representative rather than exhaustive, even when the subject treated is restricted to digital signal processing. However, these applications often uncover algorithm weaknesses, such as complexity or numerical instabilities, that are not obvious from a conceptual viewpoint. Hence, the marriage of formal least square algorithms and their applications provides important insight to the intended users.

chapter

6

EQUALIZATION

This chapter describes equalization algorithms for mitigating intersymbol interference and/or multipath introduced by a dispersive channel. First a discrete model of a continuous-time communication system is presented. Subsequently, a variety of dispersive channel models are presented for both wireline and radio channels. Then equalization algorithms based on the least squares analysis techniques presented in Part 1 are described. An excellent companion reference to this chapter is Proakis.[1]

6.1 DISCRETE MODEL OF A CONTINUOUS-TIME COMMUNICATION SYSTEM

Digital data transmission involves the transmission of an information sequence consisting of discrete symbols through a bandpass channel. The channel has some nominal bandwidth, B. It is mathematically convenient to represent a channel with a bandpass frequency response characteristic by an equivalent lowpass or baseband form. Moreover, a bandpass signal which is to be transmitted over a bandpass channel has a lowpass equivalent representation that is appropriate for transmission over the lowpass equivalent channel.[1] Thus, unless otherwise specified, all signals and filter response functions are written for convenience in complex-valued, lowpass equivalent form, rather than in the form used for real-valued bandpass signals and filters that exist in the physical system.

A mathematical model used to represent the digital communication system to be considered is shown in Figure 6-1. An information sequence $\{I_k\}$ modulates a basic transmitting filter pulse $p(t)$ at a rate $1/T$. The total transmitted signal is

$$s(t) = \sum_{k=0}^{\infty} I_k p(t - kT), \qquad (6.1)$$

where T is the duration of the signaling interval or symbol interval and I_k is the time-independent information symbol that is transmitted in the interval $kT < t < (k + 1)T$. The information symbols in (6.1) may be complex valued in general. It is convenient to assume that the symbols $\{I_k\}$ are statistically independent, with zero mean and unit variance.

Three types of modulations fit the mathematical formulation presented above. First, in the special case where I_k can take one of M_a real values, equally spaced about zero, the signal in (6.1) is a pulse-amplitude-modulated (PAM) signal. Second, an M_a-phase modulated (PM) or phase-shift-keying (PSK) signal is obtained if one allows I_k to take any one of the M_a values $\exp[2\pi i(m - 1)/M_a]$, where $m = 1, 2, \ldots, M_a$. Finally, hybrid PAM-PSK modulation also fits the formulation in (6.1).

The channel through which the signal is transmitted is characterized in general by a time-variant impulse response. Often a time-variant channel has time variations that are much slower than the duration of the signaling interval. Then, the channel can be considered as being essentially time invariant over a large number of signaling intervals. This assumption is realistic not only for telephone channels but also for some radio multipath channels such as tropospheric scatter channels. In this case a variety of adaptive techniques such as steepest descent algorithms can follow the time variations in the channel impulse response since these variations are slow relative to the duration of the signaling interval. For other radio multipath channels, such as high frequency (HF), faster tracking adaptive techniques such as Kalman tracking algorithms are needed to follow the time variations in the channel. Both of these types of algorithms are described in subsequent sections.

Noise, $\eta(t)$, is added to the signal at the output of the channel. The noise process is assumed to be stationary, zero mean, white, and Gaussian. Its autocorrelation function is $N_0\delta(t)$.

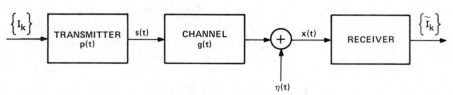

Figure 6-1. Model of digital communication system.

The input to the receiver can be represented as

$$x(t) = \sum_{k=0}^{\infty} I_k h(t - kT) + \eta(t) \qquad (6.2)$$

where

$$h(t) = \int_{-\infty}^{\infty} g(\tau) p(t - \tau) \, d\tau \qquad (6.3)$$

and $g(t)$ is the impulse response of the channel. It is assumed that $h(t)$ is square integrable and finite in duration so that $h(t) = 0$, $t \geq (L + 1)T$.

The problem at the receiver is to detect the sequence $\{I_k\}$ of information symbols from observation of the received signal. The detected sequence is denoted as $\{\tilde{I}_k\}$ and differs from the transmitted sequence as a result of transmission errors. In a detection problem of this type it is well known that the output of a filter matched to the received signal and sampled every T seconds (once every signaling interval) constitutes a set of sufficient statistics for the detection of the sequence $\{I_k\}$ (see Appendix Section A.1.16.4).[2] However, the noise sequence at the sampler output is now a colored Gaussian noise sequence. Since it is more convenient to deal with a white noise sequence when calculating the error rate performance, the sampler output sequence is passed through a digital noise whitening filter. A detection algorithm such as decision feedback equalization is then used to reduce the deleterious effects of intersymbol interference and additive noise.

Since the transmitter and receiver operate with discrete-time symbols at a rate $1/T$ symbols per second, it seems reasonable to develop an equivalent

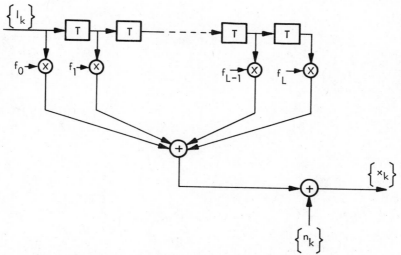

Figure 6-2. Equivalent discrete-time channel model with white noise.

discrete-time model of the communication system[3] (see Appendix Section A.1.18). The cascade of the transmitting filter, the channel, the matched filter, the sampler and the discrete-time noise whitening filter can then be represented as a discrete-time transversal filter having $(L + 1)$ tap coefficients $\{f_k\}$. The additive noise sequence $\{n_k\}$ corrupting the output of the discrete-time transversal filter is a white Gaussian noise sequence having zero mean and variance σ_n^2. The output sequence $\{x_k\}$ can then be expressed as

$$x_k = \sum_{m=0}^{L} f_m I_{k-m} + n_k, \qquad k = 0, 1, \ldots \tag{6.4}$$

Figure 6-2 illustrates the model of the equivalent discrete-time channel model with white noise.

For the case in which the channel impulse response is changing slowly with time, the matched filter becomes a time-variable filter and the time variations of the channel/matched filter pair give rise to a discrete-time filter with time-variable coefficients. The time-variant transfer characteristic of the fading multipath channel then manifests itself through time variations in the tap coefficients denoted by $\{f_k(t)\}$. A widely accepted model describing the statistics of a fading multipath channel such as troposcatter or HF assumes a Rayleigh fading distribution for the channel transfer characteristic as a function of frequency. Thus $f_k(t)$ is assumed to be a narrowband zero-mean Gauss random process with bandwidth B_0, and a spectral density which characterizes

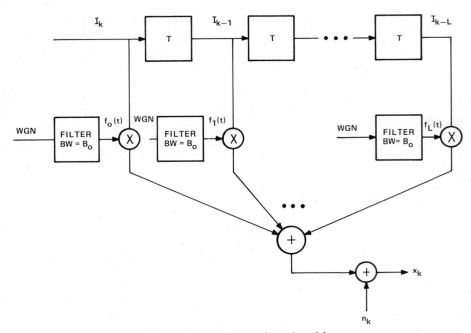

Figure 6-3. A discrete channel model.

the Doppler spreading. The coefficients $\{f_k(t)\}$ are then modeled by passing white Gaussian noise through a filter with bandwidth B_0. The resulting time-varying discrete channel model is depicted in Figure 6-3. More detailed descriptions of practical wireline and radio channels are presented in the following section.

6.2 WIRELINE AND RADIO CHANNELS

A problem encountered in high-speed serial data transmission over telephone channels and multipath radio channels is intersymbol interference. In tele-

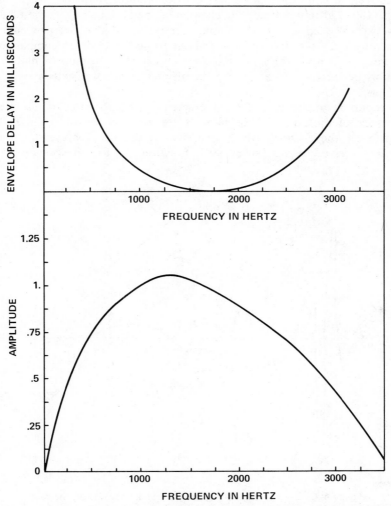

Figure 6-4. Average amplitude and delay characteristics of a medium-range telephone channel.

phone channels intersymbol interference is a consequence of the imperfect frequency response characteristics of the channel, which are usually expressed in terms of attenuation and envelope delay as functions of frequency. When the attenuation and envelope delay of the channel are not constant across the band of frequencies occupied by the transmitted signal, the signal is said to undergo amplitude distortion and delay distortion, respectively. In data transmission this distortion takes the form of a time dispersion of the transmitted symbols resulting in intersymbol interference. On radio channels such as HF and tropospheric scatter channels, intersymbol interference is the result of multiple propagation modes or paths with different path delays.

The nature of the intersymbol interference can be appreciated by observation of some channel response characteristics. Figure 6-4 illustrates the measured average amplitude and delay as a function of frequency for a medium range (180–725 miles) telephone channel of the switched telecommunications network.[4] The corresponding impulse response of this average channel is shown in Figure 6-5. Its duration is about 10 milliseconds. In comparison, the transmitted symbol rates on such a channel may be approximately 2400 symbols per second. Hence, intersymbol interference will extend over 20–30 symbols.

In contrast to the method for characterizing telephone channels, the characteristics of multipath radio channels are given in terms of the scattering function of the channel, which is a two-dimensional representation of the average received signal power as a function of time delay and Doppler

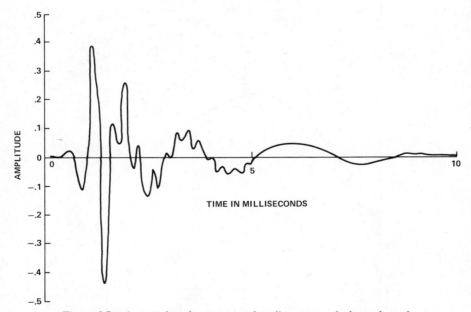

Figure 6-5. Average impulse response of medium-range telephone channel.

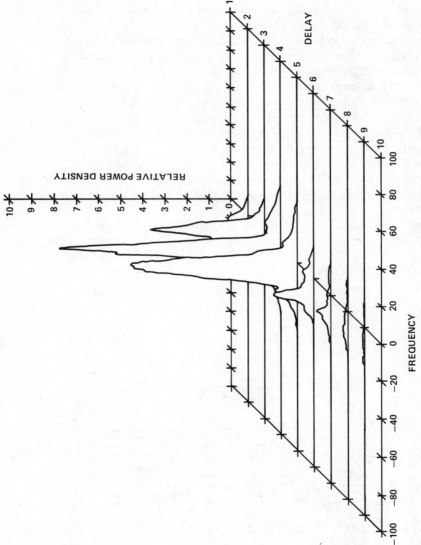

Figure 6-6. Scattering function of a medium range tropospheric scatter channel.

145

frequency. The total time duration of the channel response representing the difference in time between the latest and earliest arrivals from a single transmission is known as the multipath spread. The root mean square (RMS) bandwidth of the fading process along the Doppler frequency axis of the scattering function is known as the Doppler spread. For illustrative purposes, a scattering function measured on a medium range (150 miles) tropospheric scatter channel is shown in Figure 6-6.[1,5] The total time duration of the channel response is approximately 0.7 microseconds on the average, and the spread between "half power points" in the Doppler frequency is a little less than one hertz on the strongest path and somewhat larger on the other paths. In the troposcatter channel high-speed transmission at the rate of 10^7 symbols per second will result in intersymbol interference that spans at least seven symbols for a multipath spread of 0.7 microseconds.

The HF channel is a fading dispersive channel with representative multipath intensity profiles, shown in Figure 6-7.[6] Typical Doppler spreads for this channel commonly range from 0.1 Hz to 1 Hz for midlatitude paths with auroral and transequatorial paths exhibiting spreads in the 1–10 Hz range. Multipath spreads on HF links are about 1–5 ms with most links in the 1 to 2 ms range.

One important characteristic of the channels under consideration is that their impulse responses or frequency responses are unknown to the receiver

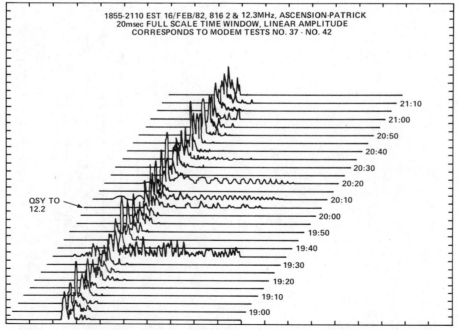

Figure 6-7. Typical multipath intensity profiles (© 1982 IEEE Anderson, et al., reference 6-6).

a priori. Furthermore, the response characteristics vary with time, especially in the case of radio channels, as illustrated by the Doppler frequency spread in Figure 6-6. Telephone channels exhibit very small and relatively slow time variations in their characteristics. In any case, the removal of the intersymbol interference, which is most conveniently performed at the receiving terminal, implies that the signal processing technique must incorporate some means of measuring and, usually, tracking the channel response, that is, the receiver must be adaptive.

6.3 EQUALIZATION TECHNIQUES

Mitigation of intersymbol interference or multipath can be accomplished by a variety of linear and nonlinear adaptive filtering techniques referred to as equalization. Linear filtering techniques for reducing intersymbol interference have been widely used successfully in high-speed data transmitted over telephone channels. This filtering technique has been called linear equalization.[1,7,8] However, it has proved inadequate in handling the severe intersymbol interference encountered on fading multipath channels. Consequently, various nonlinear techniques known as decision-feedback equalization (DFE)[9-11] and maximum likelihood sequence estimation (MLSE) using the Viterbi algorithm (VA) have been developed.[12-14] MLSE techniques are not least squares algorithms but do yield optimum performance. MLSE performance data will thus be used as a benchmark but will not be explicitly described. It should be noted that the performance improvement is often small compared to certain least squares algorithms (such as decision feedback) and is usually attained with increased computational complexity. It should also be noted that channel

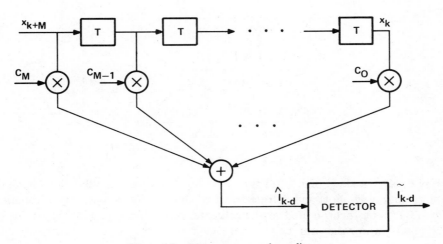

Figure 6-8.   Linear transversal equalizer.

estimation required for MLSE operation is typically accomplished via least squares algorithms.

6.3.1   Linear Equalization Using Transversal Filters

The linear filter most often used for equalization is the transversal filter shown in Figure 6-8. Its input is the sequence $\{x_k\}$ of samples corrupted by intersymbol interference and noise and its output consists of estimates, $\{\tilde{I}_k\}$, of the information sequence. At the kth signaling interval the output is denoted as

$$\hat{I}_{k-d} = \sum_{j=0}^{M} c_j x_{k+j}, \qquad k = 0, 1, \ldots \tag{6.5}$$

where $\{c_j\}$ are the tap coefficients of the equalizer, d is a delay parameter that controls symbol synchronization, $\hat{I}_k$ denotes the estimate of the kth information symbol, and $M + 1$ is the total number of equalizer taps. $\hat{I}_k$ is quantized to the nearest information symbol (nearest in amplitude for PAM, in phase for PSK and in amplitude and phase for combined PAM/PSK) to form the decision $\tilde{I}_k$. If $\tilde{I}_k$ is not identical to I_k, an error has been committed.

Note that equalization accomplished by transversal filters allows coefficient weighting of the received signal samples that contain future, current and previous information symbols. To emphasize this point. let $d = 0$ and expand Eq. (6.5) using Eq. (6.4) as follows:

$$\hat{I}_k = \sum_{j=0}^{M} c_j \left[\sum_{m=0}^{L} f_m I_{k+j-m} + n_{k+j} \right], \qquad k = 0, 1, \ldots$$

$$= \underbrace{\cdots + (c_0 f_1 + c_1 f_2 + \cdots) I_{k-1}}_{\text{previous symbols}}$$

$$+ \underbrace{(c_0 f_0 + c_1 f_1 + \cdots) I_k}_{\text{current symbol}}$$

$$+ \underbrace{(c_1 f_0 + c_2 f_1 + \cdots) I_{k+1} + \cdots}_{\text{future symbols}}$$

$$+ \text{noise terms}$$

Use of x_{k-j} instead of x_{k+j} in Eq. (6.5) with $d = 0$ would only allow previous information symbols and a fraction of the energy in the current information symbol to be incorporated in the estimate. Assuming a nonzero value for d, time synchronization can be controlled and either x_{k-j} or x_{k+j} can be used in the estimate.

The equalizer coefficients are obtained by minimizing the mean square error (MSE) defined as

$$E(c) = \mathscr{E}\left(|\varepsilon_{k-d}|^2\right) \tag{6.6}$$

where $\varepsilon_k = I_k - \hat{I}_k$, and the expectation is over the signal and noise statistics. Using a modified version[†] of the orthogonality principle given in Eq. (3.54):

$$\mathscr{E}\left(\varepsilon_{k-d} x^*_{k+l}\right) = 0, \qquad l = 0,\dots, M$$

results in the normal equations (see 3.5.3).

$$g_l = \sum_{j=0}^{M} c_j r_{j-l} \tag{6.7}$$

where

$$g_l \equiv \mathscr{E}\left\{ I_{k-d} x^*_{k+l} \right\}$$

and

$$r_{j-l} \equiv \mathscr{E}\left\{ x_{k+j} x^*_{k+l} \right\}$$

Evaluating the correlation coefficients utilizing Eq. (6.4) results in

$$g_l = \mathscr{E}\left\{ I_{k-d} \left[\sum_{m=0}^{L} f_m I_{k+l-m} + n_{k+l} \right]^* \right\}, \qquad l = 0,\dots, M \tag{6.8}$$

The symbol sequence is assumed to be uncorrelated with zero mean and unit variance:

$$\mathscr{E}\left\{ I_k I_j^* \right\} = \begin{cases} 0, & k \neq j \\ 1, & k = j \end{cases} \tag{6.9}$$

Since the signal and noise terms are uncorrelated, Eq. (6.8) becomes

$$g_l = f^*_{l+d} \tag{6.10}$$

The autocorrelation coefficients can be similarly derived:

$$r_{j-l} = \mathscr{E}\left\{ \left[\sum_{s=0}^{L} f_s I_{k+l-s} + n_{k+l} \right]^* \left[\sum_{\nu=0}^{L} f_\nu I_{k+j-\nu} + n_{k+j} \right] \right\}$$

$$= \sum_{s=0}^{L} \sum_{\nu=0}^{L} f_s^* f_\nu \mathscr{E}\left\{ I^*_{k+l-s} I_{k+j-\nu} \right\} + \mathscr{E}\left\{ n^*_{k+l} n_{k+j} \right\}$$

$$r_{j-l} = \sum_{s=0}^{L} f_s^* f_{j-l+s} + \phi_{j-l} \tag{6.11}$$

[†]Several equations are equivalent in form to those developed in Chapter 3 but are modified here for convenience in notation.

where $\phi_{j-l} = \mathscr{E}\{n^*_{k+l}n_{k+j}\}$ and is the covariance of the additive noise in the received signal.

In summary, the tap coefficients are the solution of the following set of linear equations

$$\sum_{j=0}^{M} c_j r_{j-l} = f^*_{l+d}, \qquad l = 0, \ldots, M \tag{6.12}$$

where r_{j-l} is given by Eq. (6.11).

The minimum MSE can now be obtained (see Eq. (3.56c)):

$$P_{M_{\min}} = 1 - \sum_{j=0}^{M} c_j f_{j+d}$$

6.3.2 Decision-Feedback Equalization

The decision-feedback equalizer (DFE) consists of two sections, a feedforward section and a feedback section, as shown in Figure 6-9. The feedforward

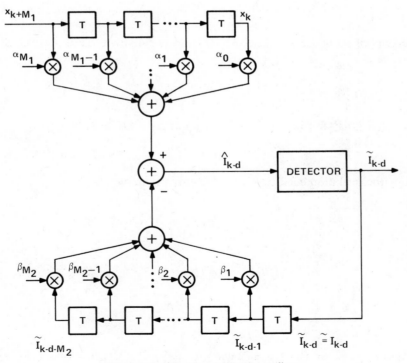

Figure 6-9. Decision feedback equalizer.

section is a transversal filter with taps spaced at the symbol interval T, and has as an input the sequence $\{x_k\}$ of received signal-plus-noise samples. In this respect its structure is identical to the linear transversal equalizer discussed above. The feedback section is also a transversal filter with taps spaced at the symbol interval. Its inputs are decisions based on previously detected symbols. Functionally, the feedback section is used to remove that portion of the intersymbol interference caused by previously detected symbols from the estimates of future symbols. The combination of the feedforward and feedback transversal filters used in this manner has been called a decision-feedback equalizer.

From the description given above, the equalizer output can be expressed as

$$\hat{I}_{k-d} = \sum_{j=0}^{M_1} \alpha_j x_{k+j} - \sum_{j=1}^{M_2} \beta_j \tilde{I}_{k-d-j} \qquad (6.13)$$

where $\hat{I}_k$ is an estimate of the kth information symbol, d is a delay parameter that controls symbol synchronization, $\{\alpha_j\}$ are the tap gain coefficients of the feedforward section, $\{\beta_j\}$ are the tap gain coefficients of the feedback section and $\{\tilde{I}_{k-1}, \ldots, \tilde{I}_{k-M_2}\}$ are previously detected symbols. The decision $\tilde{I}_k$ is formed by quantizing the estimate $\hat{I}_k$ to the nearest information symbol.

The minimum MSE criterion can be used to compute the tap coefficients $\{\alpha_j\}$ and $\{\beta_j\}$. Based on the assumption that previously detected symbols in the feedback section are correct, the minimization of the MSE

$$E(\alpha, \beta) = \mathscr{E}|I_{k-d} - \hat{I}_{k-d}|^2$$

can be accomplished. In this case the orthogonality principle requires that the error signal be orthogonal to the received signal samples in the forward filter and to the decisions in the feedback filter:

$$\mathscr{E}\{x_{k+l}^* \varepsilon_{k-d}\} = 0, \qquad l = 0, \ldots, M_1 \qquad (6.14)$$

$$\mathscr{E}\{I_{k-d-m}^* \varepsilon_{k-d}\} = 0, \qquad m = 1, \ldots, M_2 \qquad (6.15)$$

Substituting (6.13) into (6.15) leads to

$$\mathscr{E}\left\{I_{k-d-m}^*\left[I_{k-d} - \sum_{j=0}^{M_1} \alpha_j x_{k+j} + \sum_{j=1}^{M_2} \beta_j I_{k-d-j}\right]\right\} = 0 \qquad (6.16)$$

where the decisions fed back are assumed correct ie $\tilde{I}_{k-d-j} = I_{k-d-j}$ for $j = 1, \ldots, M_2$. Simplifying Eq. (6.16) results in

$$\beta_m = \sum_{j=0}^{M_1} \alpha_j f_{d+m+j}, \qquad m = 1, \ldots, M_2 \qquad (6.17)$$

Substituting (6.13) into (6.14) leads to

$$
\mathscr{E}\left[x_{k+l}^{*}I_{k-d}\right] = \sum_{j=0}^{M_1} \alpha_j \mathscr{E}\left[x_{k+j}x_{k+l}^{*}\right] - \sum_{j=1}^{M_2} \beta_j \mathscr{E}\left[I_{k-d-j}x_{k+l}^{*}\right] \quad (6.18)
$$

Simplifying Eq. (6.18) results in

$$
\sum_{j=0}^{M_1} \alpha_j r_{j-l} - \sum_{j=1}^{M_2} \beta_j f_{l+d+j}^{*} = f_{l+d}^{*}, \qquad l = 0, \ldots, M_1 \quad (6.19)
$$

Equations (6.19) reduce to those of a linear equalizer (see Eq. (6.12)) if the feedback coefficients are set to zero.

Substituting Eq. (6.17), which relates the feedback and feedforward coefficients into Eq. (6.19), results in an explicit expression for the feedforward coefficients:

$$
\sum_{j=0}^{M_1} \alpha_j \left[\sum_{s=0}^{l+d} f_s^{*} f_{s+j-l} + \phi_{j-l} \right] = f_{l+d}^{*}, \qquad l = 0, \ldots, M_1 \quad (6.20)
$$

In the derivation of (6.20), it is assumed that the delay, d, plus the number of feedback tap coefficients, M_2, exceeds the number of channel taps, that is,

$$
d + M_2 \geq L + 1
$$

In summary, the tap coefficient for the decision feedback equalizer are given by

$$
\sum_{j=0}^{M_1} \alpha_j \left[\sum_{s=0}^{l+d} f_s^{*} f_{s+j-l} + \phi_{j-l} \right] = f_{l+d}^{*}, \qquad l = 0, \ldots, M_1
$$

and

$$
\beta_l = \sum_{j=0}^{M_1} \alpha_j f_{d+l+j}, \qquad l = 1, \ldots, M_2; \; d + M_2 \geq L + 1 \quad (6.21)
$$

It can be shown that these values of the feedback coefficients result in complete elimination of intersymbol interference from previously detected symbols, provided that previous decisions are correct.

6.3.3   Adaptive Equalization Techniques

The solution for optimum equalizer tap coefficients involves the inversion of a matrix that depends on the signal and noise statistics as well as the characteristics of the channel. The simultaneous solution of Eq. (6.12) is required for the linear equalizer, and Eqs. (6.20) and (6.21) are required for the decision feedback equalizer. Generally, the required statistics and channel characteristics are unknown to the receiver, so that iterative procedures can be used to determine the tap coefficients. In these methods, the equalizer tap coefficients are derived directly from the received signal samples and can be self-adapting to changes in channel characteristics during the transmission. Several adaptive algorithms are described in the following sections.

6.3.3.1   Steepest Descent Method

Let $\tilde{C}_{k-1}^{t}$, the vector of equalizer tap coefficients for the decision feedback equalizer at time instant $k - 1$, be denoted by

$$\tilde{C}_{k-1}^{t} = \left(\alpha_0, \alpha_1, \ldots, \alpha_{M_1}, \beta_1, \beta_2, \ldots, \beta_{M_2} \right) \tag{6.22}$$

Similarly, $C_{k-1}^{t} = (c_0, \ldots, c_M)$ is the linear equalizer tap coefficient vector. In the DFE case at time instant k the received signal inputs located in the forward filter are given by $(x_k, \ldots, x_{k+M_1})$, and the decisions in the feedback filter are denoted by $(\tilde{I}_{k-1}, \ldots, \tilde{I}_{k-M_2})$ for $d = 0$. The signals in the DFE and linear cases can be represented, respectively, by single vectors denoted by

$$\tilde{X}_{k}^{t} = \left(x_k, \ldots, x_{k+M_1}, \tilde{I}_{k-1}, \ldots, \tilde{I}_{k-M_2} \right) \tag{6.22a}$$

$$X_{k}^{\prime t} = \left(x_k, \ldots, x_{k+M} \right)$$

Using this formulation the linear equalizer is a special case in which the feedback coefficients are set to zero. Thus in the remainder of this section only the DFE case is described. The optimum tap coefficient vector is then denoted by $\tilde{C}_{k\,\text{opt}}$ and is obtained by minimizing the mean square error $E(\tilde{C}_k)$. In the development the minimization was accomplished by use of the orthogonality principle. It could also have been accomplished by forming the gradient components

$$\frac{\partial E(\tilde{C}_k)}{\partial \alpha_i} \quad \text{and} \quad \frac{\partial E(\tilde{C}_k)}{\partial \beta_j}$$

where $i = 0, 1, \ldots, M_1$, and $j = 1, \ldots, M_2$.

The mean square error can be geometrically viewed as a quadratic surface in a space whose dimension is the number of tap coefficients, $M_2 + M_1 + 1$. An initial value of the tap coefficients corresponds to some point on the quadratic surface. The gradient vector, Γ_k, with components

$$\frac{\partial E(\tilde{C}_k)}{\partial \alpha_i}, \frac{\partial E(\tilde{C}_k)}{\partial \beta_j}$$

can be computed at this point on the surface. To determine the minimum mean square error, each tap coefficient must be changed in proportion to the size of each gradient component in a direction opposite to this component. The minimum mean square is then achieved when the gradient vector is zero. An iterative procedure then requires that

$$\tilde{C}_k = \tilde{C}_{k-1} - \Delta\Gamma_k, \qquad k = 1, 2, \ldots \tag{6.23}$$

where Δ is a positive number chosen small enough to guarantee convergence. Note that the index k now refers to both time and iteration number.

Since the gradient is difficult to obtain, an estimate of the gradient can be used. This estimate is based on the orthogonality principle which forces the gradient Γ_k to be proportional to $\mathscr{E}(\varepsilon_k \tilde{X}_k^*)$ where ε_k denotes the error signal at the kth iteration. Note that ε_k corresponds to $I_k - \hat{I}_k$ where I_k and $\hat{I}_k$ are the symbol and its estimate, respectively, at the kth iteration. A noisy unbiased estimate of the gradient is thus given by[7]

$$\Gamma_k = -\varepsilon_k \tilde{X}_k^* \tag{6.24}$$

The basic algorithm given by combining (6.23) and (6.24), and variations of it, have been incorporated in commercial and military adaptive equalizers. One variation of the basic algorithm uses only sign information contained in the error signal. This particular algorithm is easy to implement but reduces the rate of convergence of the coefficients.

In the above discussion, it was assumed that the receiver had knowledge of the transmitted information sequence in forming the error signal between the desired symbol and its estimate. Such knowledge can be made available during a short training period in which a signal with a known information sequence is transmitted to the receiver for initial adjustment of the tap gains. One practical scheme for continuous adjustment of the tap gains uses a decision-directed mode of operation in which decisions on the information symbols are assumed to be correct and used in place of the transmitted symbols in forming the error signal ε_k. Another method uses a known pseudorandom probe sequence which is inserted in the information-bearing signal either additively or by interleaving in time with tap gains adjusted by comparing the received probe symbols with the known transmitted probe symbols. In the decision-directed mode of operation the error signal becomes $\varepsilon_k = \tilde{I}_k - \hat{I}_k$, where $\tilde{I}_k$ is the decision of the receiver based on the estimate $\hat{I}_k$. As long as the receiver is operating at low error rates, an occasional error will have a negligible effect on the convergence of the algorithm.

6.3.3.2 *Lattice Structure Method for Linear Equalization*

The lattice structure developed in Chapter 4 can be utilized for adaptive equalization. Both linear lattice structures[15,16] as well as decision feedback lattice structures[17,18,19] have been developed. In this section the linear lattice structure for adaptive equalization is presented.

The notation used in this section is similar to that used in Chapter 4, with the order index M and the time index N. The equalizer is assumed to have $M + 1$ tap coefficients denoted by $c_M(0, N), \ldots, c_M(M, N)$ and an $(M + 1)$-dimensional vector of the equalizer coefficients is denoted by $C_M^t(N) = (c_M(0, N), \ldots, c_M(M, N))$. The estimate of the kth information symbol, I_k, is denoted by $\hat{I}_k$ and is formed from a weighted sum of the received signal samples with the delay parameter, d, set to zero according to

$$\hat{I}_k = \sum_{j=0}^{M} c_M(j, N) x_{k-j} \tag{6.25}$$

Since the vector of received signal samples is given by

$$X_M^t(k) = (x_k, \ldots, x_{k-M}),$$

then (6.25) can be rewritten as

$$\hat{I}_k = C_M^t(N) X_M(k)$$

$$= X_M^t(k) C_M(N) \tag{6.26}$$

An error signal for the kth symbol at sample time N, denoted by $e_M(k, N)$, can now be formed:

$$e_M(k, N) = I_k - \hat{I}_k \tag{6.27}$$

If the mean square error, $P_M(N)$, is defined by

$$P_M(N) = \sum_{k=1}^{N} \omega^{N-k} |e_M(k, N)|^2$$

for the received signal samples $x_1, \ldots, x_N$, then the orthogonality principle can be utilized to derive the normal equations:

$$\sum_{k=1}^{N} \omega^{N-k} e_M(k, N) x_{k-m}^* = 0, \qquad m = 0, \ldots, M \tag{6.28}$$

Substituting (6.25) and (6.26) into (6.28) yields

$$\sum_{k=1}^{N} \omega^{N-k} \left(I_k - \sum_{j=0}^{M} c_M(j, N) x_{k-j} \right) x_{k-m}^* = 0, \qquad m = 0, \ldots, M$$

or

$$\sum_{j=0}^{M} c_M(j, N) \rho_N(m, j) = w_m, \qquad m = 0, \ldots, M \tag{6.29}$$

where

$$w_m = \sum_{k=1}^{N} \omega^{N-k} I_k x_{k-m}^*$$

and

$$\rho_N(m, j) = \sum_{k=1}^{N} \omega^{N-k} x_{k-j} x_{k-m}^*$$

Note that $\rho_N(m, j)$ is the received signal autocorrelation identified in Chapter 4. By defining the vector, $W_M^t(N) = (w_0, \ldots, w_M)$ equation (6.29) can be put in matrix form:

$$R_{M+1}(N) C_M(N) = W_M(N) \tag{6.30}$$

An order-recursive equation for the equalizer coefficients can be obtained by using the backward form of the matrix inverse from the lattice formulation (see Eq. (4.44)):

$$R_{M+1}^{-1}(N) = \begin{bmatrix} R_M^{-1}(N) & 0 \\ 0 & 0 \end{bmatrix} + \frac{1}{r_M(N)} \begin{bmatrix} B_M(N) \\ 1 \end{bmatrix} \begin{bmatrix} B_M^{t*}(N) & 1 \end{bmatrix} \tag{6.31}$$

Using (6.31) in (6.30) with

$$W_M(N) = \begin{pmatrix} W_{M-1}(N) \\ w_M \end{pmatrix}$$

yields

$$C_M(N) = \left\{ \begin{bmatrix} R_M^{-1}(N) & 0 \\ 0 & 0 \end{bmatrix} + \frac{1}{r_M(N)} \begin{bmatrix} B_M(N) \\ 1 \end{bmatrix} \begin{bmatrix} B_M^{t*}(N) & 1 \end{bmatrix} \right\}$$

$$\times \begin{pmatrix} W_{M-1}(N) \\ w_M \end{pmatrix}$$

$$= \begin{bmatrix} C_{M-1}(N) \\ 0 \end{bmatrix} + \frac{\mu_M(N)}{r_M(N)} \begin{bmatrix} B_M(N) \\ 1 \end{bmatrix} \tag{6.32}$$

where

$$\mu_M(N) = \begin{bmatrix} B_M^{t*}(N) & 1 \end{bmatrix} W_M(N)$$

An order recursion for the error signal can be obtained by substituting (6.32) into the combination (6.26) and (6.27):

$$e_M(k, N) = I_k - C_M^t(N)X_M(k)$$

$$= I_k - \left\{ [C_{M-1}^t(N) \quad 0] + \frac{\mu_M(N)}{r_M(N)} [B_M^t(N) \quad 1] \right\} \begin{pmatrix} X_{M-1}(k) \\ x_{k-M} \end{pmatrix}$$

$$= e_{M-1}(k, N) - \frac{\mu_M(N)}{r_M(N)} [B_M^t(N)X_{M-1}(k) + x_{k-M}] \qquad (6.33)$$

An alternate form for (6.33) can be obtained by using the backward error given by (4.14) with $b_M(M, N) = 1$:

$$e_M^b(k, N) = \sum_{m=0}^{M} b_M(m, N)x_{k-m}$$

$$= (b_M(0, N), \ldots, b_M(M - 1, N)) \begin{pmatrix} x_k \\ \vdots \\ x_{k-(M-1)} \end{pmatrix} + x_{k-M}$$

$$= B_M^t(N)X_{M-1}(k) + x_{k-M} \qquad (6.34)$$

Combining (6.33) and (6.34) results in

$$e_M(k, N) = e_{M-1}(k, N) - \frac{\mu_M(N)}{r_M(N)} e_M^b(k, N) \qquad (6.35)$$

Time recursions for the backward predictor coefficients and the crosscorrelation $W_M(N)$ are required to obtain a recursion for the scalar parameter $\mu_M(N)$, defined in (6.32). To derive the recursive form of the backward predictor coefficients, Eqs. (4.42), (4A.19), (4A.9), and (4A.2) are required using different time and order indices:

$$-R_M^{-1}(N)V_M(N) = B_M(N) \qquad (6.36)$$

$$\omega^{-1}R_M^{-1}(N) = R_M^{-1}(N + 1)$$

$$+ \frac{R_M^{-1}(N + 1)X_{M-1}^*(N + 1)X_{M-1}^t(N + 1)R_M^{-1}(N + 1)}{1 - \gamma_M(N + 1, N + 1)}$$

$$(6.37)$$

$$\omega V_M(N) = V_M(N + 1) - x_{N-(M-1)}X_{M-1}^*(N + 1) \qquad (6.38)$$

$$\gamma_M(N + 1, N + 1) = X_{M-1}^t(N + 1)R_M^{-1}(N + 1)X_{M-1}^*(N + 1) \qquad (6.39)$$

Using (6.37) and (6.38) in (6.36) with (6.39) yields

$$
\boldsymbol{B}_M(N) = -\Bigg[\boldsymbol{R}_M^{-1}(N+1)
$$

$$
+ \frac{\boldsymbol{R}_M^{-1}(N+1)\boldsymbol{X}_{M-1}^*(N+1)\boldsymbol{X}_{M-1}^t(N+1)\boldsymbol{R}_M^{-1}(N+1)}{1-\gamma_M(N+1,N+1)} \Bigg]
$$

$$
\times \Big[\boldsymbol{V}_M(N+1) - x_{N-(M-1)}\boldsymbol{X}_{M-1}^*(N+1) \Big]
$$

$$
= \boldsymbol{B}_M(N+1) + \frac{\boldsymbol{R}_M^{-1}(N+1)\boldsymbol{X}_{M-1}^*(N+1)}{1-\gamma_M(N+1,N+1)}
$$

$$
\times \Big[\boldsymbol{X}_{M-1}^t(N+1)\boldsymbol{B}_M(N+1) + x_{N-(M-1)}\gamma_M(N+1,N+1)
$$

$$
+ x_{N-(M-1)}\big(1-\gamma_M(N+1,N+1)\big) \Big]
$$

The bracket in the above equation can be replaced by use of (6.34) so that

$$
\boldsymbol{B}_M(N) = \boldsymbol{B}_M(N+1) + \frac{\boldsymbol{R}_M^{-1}(N+1)\boldsymbol{X}_{M-1}^*(N+1)}{1-\gamma_M(N+1,N+1)} e_M^b(N+1,N+1)
$$

$$
(6.40)
$$

The recursion for the crosscorrelation, $\boldsymbol{W}_M(N)$ is obtained from its definition, that is,

$$
\boldsymbol{W}_M(N) = \begin{pmatrix} \displaystyle\sum_{k=1}^{N} \omega^{N-k}I_k x_k^* \\ \vdots \\ \displaystyle\sum_{k=1}^{N} \omega^{N-k}I_k x_{k-M}^* \end{pmatrix}
$$

$$
= I_N \begin{pmatrix} x_N^* \\ \vdots \\ x_{N-M}^* \end{pmatrix} + \begin{pmatrix} \displaystyle\sum_{k=1}^{N-1} \omega^{N-k}I_k x_k^* \\ \vdots \\ \displaystyle\sum_{k=1}^{N-1} \omega^{N-k}I_k x_{k-M}^* \end{pmatrix}
$$

$$
= \omega \boldsymbol{W}_M(N-1) + I_N \boldsymbol{X}_M^*(N) \qquad (6.41)
$$

Using (6.41) and (6.40) leads to a recursive form for the scalar parameter $\mu_M(N)$:

$$\mu_M(N) = \left\{\left[B_M(N-1) - \frac{R_M^{-1}(N)X_{M-1}^*(N)}{1 - \gamma_M(N, N)}e_M^b(N, N)\right]^{t^*} 1\right\}W_M(N)$$

$$= \left[B_M^{t^*}(N-1) \quad 1\right]W_M(N)$$

$$- \left[\frac{X_{M-1}^t(N)R_M^{-1}(N)e_M^{b^*}(N, N)}{1 - \gamma_M(N, N)} \quad 0\right]W_M(N)$$

$$= \left[B_M^{t^*}(N-1) \quad 1\right]\left[\omega W_M(N-1) + I_N X_M^*(N)\right]$$

$$- \frac{X_{M-1}^t(N)R_M^{-1}(N)e_M^{b^*}(N, N)}{1 - \gamma_M(N, N)}W_{M-1}(N)$$

$$= \omega\mu_M(N-1) + \left[B_M^{t^*}(N-1) \quad 1\right]I_N X_M^*(N)$$

$$- \frac{X_{M-1}^t(N)R_M^{-1}(N)e_M^{b^*}(N, N)}{1 - \gamma_M(N, N)}W_{M-1}(N)$$

Replacing M by $M-1$ in (6.30) and using the result in the above equation together with a recursion for $B_M(N-1)$ obtained from (6.40) yields

$$\mu_M(N) = \omega\mu_M(N-1)$$

$$+ I_N\left[B_M^{t^*}(N) + \frac{X_{M-1}^t(N)R_M^{-1}(N)e_M^{b^*}(N, N)}{1 - \gamma_M(N, N)} \quad 1\right]X_M^*(N)$$

$$- \frac{X_{M-1}^t(N)e_M^{b^*}(N, N)}{1 - \gamma_M(N, N)}C_{M-1}(N)$$

Further simplification occurs by use of (6.34), (6.39) and the combination of (6.26) and (6.27) with

$$X_M(N) = \begin{pmatrix} X_{M-1}(N) \\ x_{N-M} \end{pmatrix}$$

that is,

$$\mu_M(N) = \omega\mu_M(N-1) + I_N e_M^{b^*}(N, N) + \frac{I_N\gamma_M(N, N)e_M^{b^*}(N, N)}{1 - \gamma_M(N, N)}$$

$$- \frac{\left[I_N - e_{M-1}(N, N)\right]e_M^{b^*}(N, N)}{1 - \gamma_M(N, N)}$$

or

$$\mu_M(N) = \omega\mu_M(N-1) + \frac{e_{M-1}(N,N)e_M^{b*}(N,N)}{1 - \gamma_M(N,N)} \tag{6.42}$$

The estimate of the Nth information symbol can be expressed in terms of the lattice structure parameters and the backward error using (6.26) and the recursion for the equalizer coefficients in (6.32):

$$\hat{I}_N = C_M^t(N)X_M(N)$$

$$= \left\{ \begin{bmatrix} C_{M-1}^t(N) & 0 \end{bmatrix} + \frac{\mu_M(N)}{r_M(N)} \begin{bmatrix} B_M^t(N) & 1 \end{bmatrix} \right\} \begin{pmatrix} X_{M-1}(N) \\ x_{N-M} \end{pmatrix}$$

The above equation can be rewritten using (6.34):

$$\hat{I}_N = C_{M-1}^t(N)X_{M-1}(N) + \frac{\mu_M(N)}{r_M(N)}e_M^b(N,N) \tag{6.43}$$

Following the notation used in Section 4.1.7 the backward error $e_M^b(N,N)$ is replaced by $e_M^b(N)$. Repeated application of the recursion for the equalizer coefficients in (6.32) results in

$$\hat{I}_N = \sum_{l=0}^{M} \frac{\mu_l(N)}{r_l(N)}e_l^b(N) \tag{6.44}$$

Equation (6.44) can be used to implement the lattice equalizer structure shown in Figure 6-10. The backward error is obtained from the general lattice form shown in Figure 4-3. In Figure 6-10 the error signal in (6.35) is specialized for $N = k$ and $e_M(N,N)$ is replaced by $e_M(N)$:

$$e_M(N) = e_{M-1}(N) - \frac{\mu_M(N)}{r_M(N)}e_M^b(N)$$

Initial conditions required to complete the estimation of the Nth symbol include

$$f_0(N) = r_0(N) = \omega f_0(N-1) + |x_N|^2$$

$$e_0^b(N) = e_0^f(N) = x_N$$

$$r_M(0) = f_0(0) = r_{-1}(N-1) = \varepsilon$$

$$\mu_M(0) = e_M^b(0) = k_M(0) = 0$$

$$\gamma_{-1}(N-1) = 0$$

$$e_{-1}^b(N-1) = 0$$

$$e_{-1}(N) = I_N$$

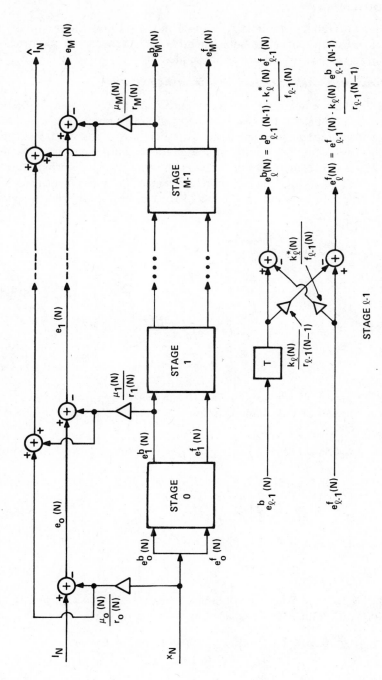

Figure 6-10. Lattice equalizer structure.

$$e_\ell^b(N) = e_{\ell-1}^b(N-1) - k_\ell^*(N) \frac{e_{\ell-1}^f(N)}{f_{\ell-1}(N)}$$

$$e_\ell^f(N) = e_{\ell-1}^f(N) - k_\ell(N) \frac{e_{\ell-1}^b(N-1)}{r_{\ell-1}(N-1)}$$

STAGE ℓ-1

Note that the training symbol, I_N, is used in the training mode initially to allow the equalizer to converge. Subsequently, I_N is replaced by the decision symbol, $\tilde{I}_N$, in the data (or information) mode.

6.3.3.3   Gradient Lattice Structure for Linear Equalization

The lattice structure method for linear equalization developed in Section 6.3.3.2 can be simplified in the case of time-invariant channels. The simplification results from a gradient lattice structure which has a substantial reduction in computational complexity with a slower convergence rate. The gradient lattice structure is presented in this section following a brief reformulation of the linear lattice equalizer, which is more convenient for subsequent use.

The order update equation (4.55) and (4.56) can be reexpressed in special forms, that is,

$$e_M^f(N, N-1) = e_{M-1}^f(N, N-1) - \frac{k_M(N-1)e_{M-1}^b(N-1, N-2)}{r_{M-1}(N-2)}$$

$$(6.45)$$

$$e_M^b(N, N-1) = e_{M-1}^b(N-1, N-2) - \frac{k_M^*(N-1)e_{M-1}^f(N, N-1)}{f_{M-1}(N-1)}$$

$$(6.46)$$

with $M+1$, i and N replaced by M, N and $N-1$, respectively. In the appendix at the end of this chapter, time update error recursions are given as follows:

$$e_M(N+1, N+1) = \alpha_{M+1}(N+1)e_M(N+1, N) \qquad (6A.1)$$

$$e_M^f(N, N) = \alpha_M(N-1)e_M^f(N, N-1) \qquad (6A.2)$$

$$e_M^b(N, N) = \alpha_M(N)e_M^b(N, N-1) \qquad (6A.3)$$

where the parameter $\alpha_M(N)$ is defined by

$$\alpha_M(N) = 1 - \gamma_M(N, N) \qquad (6A.4)$$

A time update for $k_M(N)$ can be obtained by substituting (6A.2), (6A.3), and (6A.4) into (4.57), with $M+1$ replaced by M, yielding

$$k_M(N) = \omega k_M(N-1) + \alpha_{M-1}(N-1)e_{M-1}^f(N, N-1)$$

$$\times e_{M-1}^{b*}(N-1, N-2) \qquad (6.47)$$

The order updates for $f_M(N)$ and $r_M(N)$ can be obtained from (4.50) and

(4.51), respectively, by replacing $M + 1$ by M:

$$f_M(N) = f_{M-1}(N) - \frac{|k_M(N)|^2}{r_{M-1}(N-1)} \qquad (6.48)$$

$$r_M(N) = r_{M-1}(N-1) - \frac{|k_M(N)|^2}{f_{M-1}(N)} \qquad (6.49)$$

An alternate definition for the error signal for kth symbol, $e_M(k, N)$, is defined in the chapter appendix as

$$e_M(k, N) = I_k - \hat{I}_M(k, N) \qquad (6A.18)$$

where $\hat{I}_M(k, N)$ is given by

$$\hat{I}_M(k, N) = X_M^t(k) C_M(N) \qquad (6A.17)$$

Using the definition given in (6A.18) and (6A.17), further recursions required for the lattice structure for linear equalization can be written as follows:

$$\hat{I}_M(N, N-1) = \hat{I}_{M-1}(N, N-1) + \frac{\mu_M(N-1) e_M^b(N, N-1)}{r_M(N-1)} \qquad (6.50)$$

$$\alpha_M(N) = \alpha_{M-1}(N) - \frac{|e_{M-1}^b(N, N-1)|^2 \alpha_{M-1}^2(N)}{r_{M-1}(N)} \qquad (6.51)$$

$$e_M(N, N-1) = I_N - \hat{I}_M(N, N-1) = I_N - C_M^t(N-1) X_M(N) \qquad (6.52)$$

$$\mu_M(N) = \omega\mu_M(N-1) + \alpha_M(N) e_{M-1}(N, N-1) e_M^{b*}(N, N-1) \qquad (6.53)$$

$$e_M(N, N-1) = e_{M-1}(N, N-1) - \frac{\mu_M(N-1) e_M^b(N, N-1)}{r_M(N-1)} \qquad (6.54)$$

Equations (6.50)–(6.54) are proved in the chapter appendix. The order updates for $f_M(N)$ and $r_M(N)$ can also be expressed as time updates:

$$f_M(N) = \omega f_M(N-1) + \alpha_M(N-1)|e_M^f(N, N-1)|^2 \qquad (6.55)$$

$$r_M(N) = \omega r_M(N-1) + \alpha_M(N)|e_M^b(N, N-1)|^2 \qquad (6.56)$$

Proof of (6.55) and (6.56) are also given in the chapter appendix. Equations (6.45)–(6.56) constitute an alternate form of the lattice structure presented in

Section 6.3.3.2 for linear equalization. If $e_M^b(N, N-1)$ is very small, then $\gamma_M(N, N)$ in (4.59) is also very small:

$$\gamma_M(N, N) = \gamma_{M-1}(N, N) + \frac{|e_{M-1}^b(N, N)|^2}{r_{M-1}(N)} \tag{4.59}$$

Thus, in this case $\alpha_M(N)$ in (6A.4) may be set to 1. When $\alpha_M(N)$ is set to 1, Eqs. (6.45)–(6.56) are referred to as the gradient lattice structure for linear equalization.

6.3.3.4 Lattice Structure for Decision Feedback Equalization

The least square decision feedback lattice structures have been developed in the literature for real variables.[17,18,19] Here, the DFE lattice structure is extended to include the use of complex variables. The lattice structures for both the linear and decision feedback equalizers have been shown to have many advantages, including fast convergence, reduced sensitivity to round-off noise, efficiency in computation, and flexibility for increasing or decreasing the numbers of stages.

The DFE lattice structure is assumed to have $M_1 + 1$ feedforward tap coefficients associated with the received signals and M_2 feedback tap coefficients associated with the information symbol decisions denoted as

$$C_{M_T}^t(N) = \left[c_{M_T}(0, N), c_{M_T}(1, N), \ldots, c_{M_T}(M_T, N) \right] \tag{6.57}$$

where $M_T = M_1 + M_2 + 1$ is the total number of tap coefficients. The estimate of the kth information symbol, I_k, is denoted by $I_{M_T}(k, N)$, and is formed with the delay parameter, d, set to zero from a weighted sum of the received signals, x_{k-j}, and previously detected signals, $\tilde{I}_{k-j}$, according to

$$\hat{I}_{M_T}(k, N) = \sum_{j=0}^{M_1} c_{M_T}(j, N) x_{k-j} + \sum_{j=1}^{M_2} c_{M_T}(M_1 + j, N) \tilde{I}_{k-j}$$

$$\equiv C_{M_T}^t(N) X_{M_T}(k) \tag{6.58}$$

where

$$X_{M_T}^t(k) = \left[x_k, x_{k-1}, \ldots, x_{k-M_1}, \tilde{I}_{k-1}, \tilde{I}_{k-2}, \ldots, \tilde{I}_{k-M_2} \right] \tag{6.59}$$

The error signal can then be defined as

$$e_{M_T}(k, N) = I_k - C_{M_T}^t(N) X_{M_T}(k) \tag{6.59a}$$

In order to solve the least squares problem for the symbol estimate $\hat{I}_{M_T}(k, N)$

in a recursive manner, a set of data vectors are defined as follows:

$$X_M^t(k) = [x_k, \ldots, x_{k-M}] \text{ for } M = 0 \text{ to } L \tag{6.60}$$

$$X_M^t(k) = [x_k, \ldots, x_{k-M}, I_{k-1}, \ldots, I_{k-M+L}] \text{ for } M = L + 1 \text{ to } M_1, \tag{6.61}$$

where $L \equiv M_1 - M_2$ and M_1 is assumed to be greater than M_2.

Three regions for the DFE are now identified as the linear region, the transition region and the two-dimensional region. For $M = 0$ to L the signal elements given in (6.60) are equivalent to those of the linear structure defined in Section 6.3.3.3. At the transition region the data vector is

$$X_M^t(k) = [x_k, \ldots, x_{k-M}, I_{k-1}]$$

and $M = L + 1$. For $M > L + 1$ a two-dimensional signal formulation is used in order to account for both the received signal sequence and the information symbol sequence defined by the data vector (6.61).

In the linear region equations (6.45) to (6.56) are used directly to obtain a recursive solution for $M = 0$ to L. For $M > L$ the estimation problem involves both $\{x_k\}$ and $\{\tilde{I}_k\}$. In this case two permutation matrices, T_{M+1}, and S_{M+1}, associated with the minimization of the feedforward and feedback error prediction respectively are defined as

$$T_{M+1}X_{M+1}(k) = [x_k, \tilde{I}_{k-1}, X_M^t(k-1)]^t \tag{6.62}$$

$$S_{M+1}X_{M+1}(k) = [X_M^t(k), x_{k-M-1}, \tilde{I}_{k-M-1+L}]^t \tag{6.63}$$

Using an analogy with the linear case, DFE forward and backward prediction error signals, defined by $\underline{e}_{M+1}^f(k, N)$ and $\underline{e}_{M+1}^b(k, N)$, respectively, can be expressed as

$$\underline{e}_{M+1}^f(k, N) = [e_{1,M+1}^f(k, N), e_{2,M+1}^f(k, N)]^t \tag{6.64}$$

$$\underline{e}_{M+1}^b(k, N) = [e_{1,M+1}^b(k, N), e_{2,M+1}^b(k, N)]^t \tag{6.65}$$

where the underbar denotes a two-dimensional vector and

$$e_{1,M+1}^f(k, N) = x_k + A_{1,M+1}^t(N)X_M(k-1) \tag{6.66}$$

$$e_{2,M+1}^f(k, N) = \tilde{I}_{k-1} + A_{2,M+1}^t(N)X_M(k-1) \tag{6.67}$$

$$e_{1,M+1}^b(k, N) = x_{k-M-1} + B_{1,M+1}^t(N)X_M(k) \tag{6.68}$$

$$e_{2,M+1}^b(k, N) = \tilde{I}_{k-M-1+L} + B_{2,M+1}^t(N)X_M(k) \tag{6.69}$$

In (6.66) to (6.69) $A_{i,M}(N)$ and $B_{i,M}(N)$ for $i = 1$ and 2 represent the forward and backward prediction coefficient vectors for the DFE case and correspond to $A_M(N)$ and $B_M(N)$, respectively. Note that the permutation matrix T_{M+1} in (6.62) provides a forward prediction of x_k in (6.66) and $\tilde{I}_{k-1}$ in (6.67) from the data vector $X_M^t(k-1)$. Similarly, the permutation matrix S_{M+1} in (6.63) provides a backward prediction of $x_{k-(M+1)}$ in (6.68) and $\tilde{I}_{k-(M+1)+L}$ in (6.69) from the data vector $X_M^t(k)$.

Equation (6.64) and (6.65) can be reexpressed as

$$\underline{e}_{M+1}^f(k, N) = \left[I, \underline{A}_{M+1}^t(N) \right] T_{M+1} X_{M+1}(k) \tag{6.70}$$

$$\underline{e}_{M+1}^b(k, N) = \left[\underline{B}_{M+1}^t(N), I \right] S_{M+1} X_{M+1}(k) \tag{6.71}$$

where

$$\underline{A}_{M+1}(N) = \left[A_{1,M+1}(N), A_{2,M+1}(N) \right] \tag{6.72}$$

$$\underline{B}_{M+1}(N) = \left[B_{1,M+1}(N), B_{2,M+1}(N) \right] \tag{6.73}$$

$$I = \begin{bmatrix} 1 & 0 \\ 0 & 1 \end{bmatrix} \tag{6.74}$$

Note that $\underline{A}_{M+1}(N)$ and $\underline{B}_{M+1}(N)$ are $(2M + 1 - L) \times 2$ matrices which are selected to minimize the trace of the autocovariance matrices of $\underline{e}_{M+1}^f(k, N)$ and $\underline{e}_{M+1}^b(k, N)$ defined as

$$\sum_{i=1}^N \omega^{N-i} \underline{e}_{M+1}^{f*}(i, N) \underline{e}_{M+1}^{ft}(i, N)$$

and

$$\sum_{i=1}^N \omega^{N-i} \underline{e}_{M+1}^{b*}(i, N) \underline{e}_{M+1}^{bt}(i, N)$$

respectively. For $M = L$ in (6.66) to (6.69) the forward and backward prediction errors in the transition region can be written as

$$e_{1,L+1}^f(k, N) = x_k + A_{1,L+1}^t(N) X_L(k-1) \tag{6.75}$$

$$e_{2,L+1}^f(k, N) = \tilde{I}_{k-1} + A_{2,L+1}^t(N) X_L(k-1) \tag{6.76}$$

$$e_{1,L+1}^b(k, N) = x_{k-L-1} + B_{1,L+1}^t(N) X_L(k) \tag{6.77}$$

$$e_{2,L+1}^b(k, N) = \tilde{I}_{k-1} + B_{2,L+1}^t(N) X_L(k) \tag{6.78}$$

In order to obtain recursive solutions for (6.75) to (6.78) an analogy with previous prediction equations is required. From (4.52), (4.53), (6A.17) and (6A.18) it can be seen that:

$$e_M^f(k, N) = x_k + A_M^t(N)X_{M-1}(k - 1) \tag{6.78a}$$

$$e_M^b(k, N) = x_{k-M} + B_M^t(N)X_{M-1}(k) \tag{6.78b}$$

$$e_M(k, N) = I_k - C_M^t(N)X_M(k) \tag{6.78c}$$

Comparing (6.75) with (6.78a) and (6.77) with (6.78c), respectively, for $M = L + 1$ it can be seen that $e_{1, L+1}^f(k, N)$ and $e_{1, L+1}^b(k, N)$ have the identical optimization formulations as $e_{L+1}^f(k, N)$ and $e_{L+1}^b(k, N)$, respectively. Therefore, it can be seen that

$$e_{1, L+1}^f(k, N) = e_{L+1}^f(k, N) \tag{6.79}$$

$$e_{1, L+1}^b(k, N) = e_{L+1}^b(k, N) \tag{6.80}$$

Equations (6.76) and (6.78c) can be rewritten using ideal symbol decisions as:

$$e_{2, L+1}^f(N, N - 1) = I_{N-1} + A_{2, L+1}^t(N - 1)X_L(N - 1)$$

$$e_L(N - 1, N - 2) = I_{N-1} - C_L^t(N - 2)X_L(N - 1)$$

Comparing the above two equations it can be seen that,

$$e_{2, L+1}^f(N, N - 1) = e_L(N - 1, N - 2) \tag{6.81}$$

To obtain a recursive solution for $e_{2, L+1}^b(k, N)$, it is necessary to first express (6.69) and (6.78c) for $M = L - 1$ as:

$$e_{2, L}^b(N - 1, N - 2) = I_{N-1} + B_{2, L}^t(N - 2)X_{L-1}(N - 1)$$

$$e_{L-1}(N - 1, N - 2) = I_{N-1} - C_{L-1}^t(N - 2)X_{L-1}(N - 1)$$

The above equations then imply that

$$e_{2, L}^b(N - 1, N - 2) = e_{L-1}(N - 1, N - 2)$$

Analogous equations to (6.46) and (6.47) for $M = L + 1$ can then be written in terms of all previously defined quantities from the linear region as:

$$e_{2,L+1}^b(N, N-1) = e_{L-1}(N-1, N-2)$$

$$- \frac{k_{2,L+1}^{b*}(N-1)e_L^f(N,N-1)}{f_L(N-1)} \qquad (6.82)$$

$$k_{2,L+1}^b(N) = \omega k_{2,L+1}^b(N-1) + \alpha_L(N-1)$$

$$\times e_L^f(N,N-1)e_{L-1}^*(N-1,N-2) \qquad (6.83)$$

where $k_{2,L}^b(N)$ corresponds to the parameter $k_L(N)$ in the linear case. Then, the prediction errors, $\underline{e}_{L+1}^f(N, N-1)$ and $\underline{e}_{L+1}^b(N, N-1)$ can be written as

$$\underline{e}_{L+1}^f(N, N-1) = \left[e_{L+1}^f(N, N-1), e_L(N-1, N-2) \right]^t \qquad (6.84)$$

$$\underline{e}_{L+1}^b(N, N-1) = \left[e_{L+1}^b(N, N-1), e_{2,L+1}^b(N, N-1) \right]^t \qquad (6.85)$$

Similarly, analogous equations to the linear case for the error, order, and time updates for $M > L + 1$ are generalized to two-dimensional recursions[19]:

$$\underline{e}_M^f(N, N-1) = \underline{e}_{M-1}^f(N, N-1)$$

$$- \underline{k}_M^t(N-1)\underline{r}_{M-1}^{-1}(N-2)\underline{e}_{M-1}^b(N-1, N-2)$$

$$(6.86)$$

$$\underline{e}_M^b(N, N-1) = \underline{e}_{M-1}^b(N-1, N-2)$$

$$- \underline{k}_M^*(N-1)\underline{f}_{M-1}^{-1}(N-1)\underline{e}_{M-1}^f(N, N-1) \qquad (6.87)$$

$$\underline{k}_M(N) = \omega \underline{k}_M(N-1) + \alpha_{M-1}(N-1)$$

$$\times \underline{e}_{M-1}^{b*}(N-1, N-2)\underline{e}_{M-1}^{ft}(N, N-1) \qquad (6.88)$$

$$\underline{f}_M(N) = \omega \underline{f}_M(N-1) + \alpha_M(N-1)$$

$$\times \underline{e}_M^{f*}(N, N-1)\underline{e}_M^{ft}(N, N-1) \qquad (6.89)$$

$$\underline{r}_M(N) = \omega \underline{r}_M(N-1) + \alpha_M(N)\underline{e}_M^{b*}(N, N-1)$$

$$\times \underline{e}_M^{bt}(N, N-1) \qquad (6.90)$$

$$\hat{I}_M(N, N-1) = \hat{I}_{M-1}(N, N-1) + \underline{\mu}'_M(N-1)$$

$$\times \underline{r}_M^{-1}(N-1)\underline{e}_M^b(N, N-1) \tag{6.91}$$

$$\alpha_M(N) = \alpha_{M-1}(N) - \alpha_{M-1}^2(N)\underline{e}_{M-1}^{bt}(N, N-1)$$

$$\times \underline{r}_{M-1}^{-1}(N)\underline{e}_{M-1}^{b*}(N, N-1) \tag{6.92}$$

$$e_M(N, N-1) = I_N - \hat{I}_M(N, N-1) \tag{6.93}$$

$$\underline{\mu}_M(N) = \omega\underline{\mu}_M(N-1) + \alpha_M(N)e_{M-1}(N, N-1)$$

$$\times \underline{e}_M^{b*}(N, N-1) \tag{6.94}$$

where $\underline{k}_M$, $\underline{f}_M$, and $\underline{r}_M$ denote 2×2 matrices and $\underline{\mu}_M(N)$ is a 1×2 vector. For convenience the following definitions are provided:

$$\underline{k}_M^t(N-1)\underline{r}_{M-1}^{-1}(N-2) \equiv \begin{bmatrix} k_{11} & k_{12} \\ k_{21} & k_{22} \end{bmatrix} \tag{6.95}$$

$$\underline{k}_M^*(N-1)\underline{f}_{M-1}^{-1}(N-1) \equiv \begin{bmatrix} k'_{11} & k'_{12} \\ k'_{21} & k'_{22} \end{bmatrix} \tag{6.96}$$

and

$$\underline{\mu}_M^t(N-1)\underline{r}_M^{-1}(N-1) \equiv [\beta_{1,M} \quad \beta_{2,M}] \tag{6.96a}$$

Then the DFE lattice structure is depicted in Figure 6-11 and the one- and two-dimensional structures for the $(M-1)$-th lattice stage is depicted in Figure 6-12.

Again, if $\alpha_M(N)$ is set to one, then the least squares lattice DFE is referred to as the gradient lattice DFE. To complete the development of the lattice DFE, the initial conditions are given as follows:

$$e_0^f(N, N-1) = e_0^b(N, N-1) = x_N$$

$$f_0(N) = r_0(N) = \omega f_0(N-1) + |x_N|^2$$

$$f_0(0) = \varepsilon$$

$$r_M(0) = r_M(-1) = \varepsilon, \qquad M = 0, 1, \ldots, L$$

$$r_{-1}(N-1) = \varepsilon$$

$$\mu_M(0) = 0, \qquad M = 0, 1, \ldots, L$$

$$k_M(0) = 0, \qquad M = 0, 1, \ldots, L$$

$$e_{-1}(N, N - 1) = I_N$$

$$\alpha_{-1}(N - 1) = 1$$

$$\underline{k}_{M+1}(0) = \begin{bmatrix} 0 & 0 \\ 0 & 0 \end{bmatrix}, \qquad M = L + 1, \ldots, M_1$$

$$\underline{\mu}_M(-1) = \begin{bmatrix} 0 \\ 0 \end{bmatrix}, \qquad M = L + 1, \ldots, M_1$$

$$\underline{r}_{-1}(N - 1) = \underline{f}_M(0) = \underline{r}_M(0) = \underline{r}_M(-1)$$

$$= \begin{bmatrix} \varepsilon & 0 \\ 0 & \varepsilon \end{bmatrix}, \qquad M = L + 1, \ldots, M_1$$

Note that

$$\hat{I}_{-1} = 0$$
$$I_N = 0 \qquad \text{for } N \le 0$$

$$e_M^b(N, N - 1) = 0 \qquad \text{for } M < 0 \text{ or } N \le 0$$

$$\underline{e}_M^b(N, N - 1) = \begin{bmatrix} 0 \\ 0 \end{bmatrix} \qquad \text{for } M < 0 \text{ or } N \le 0$$

The training symbol, I_N, is used in the training mode initially to allow the equalizer to converge. Subsequently, I_N is replaced by the decision symbol, $\tilde{I}_N$, in the data (or information) mode.

In summary, the DFE lattice structure is divided into three regions. In the linear region for $M \le L$ equations (6.45) to (6.56) provide the recursion relationships. In the transition region with $M = L + 1$ equations (6.79) to (6.85) provide the needed recursions. Finally, equations (6.86) to (6.96a) are used in a two-dimensional structure for $M > L + 1$.

6.3.3.5 Kalman Method

The Kalman algorithm can also be applied to adaptive equalization.[20,21] One specific application in which this algorithm has been successfully demonstrated is the equalization of high data rate transmission over HF radio channels.[6,22,23] The algorithm is more powerful than conventional steepest-descent techniques over these channels since rapid coefficient convergence allows the equalizer to track the time-varying channel characteristics.

The derivation of the Kalman algorithm begins in a fashion similar to that previously obtained for the linear lattice structure. The Kalman equalizer tap

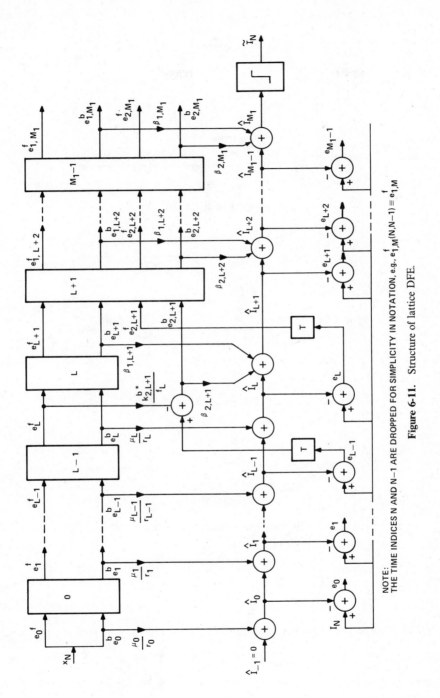

NOTE:
THE TIME INDICES N AND N−1 ARE DROPPED FOR SIMPLICITY IN NOTATION, e.g. $e_{1,M}^f(N,N−1) \equiv e_{1,M}^f$

Figure 6-11. Structure of lattice DFE.

171

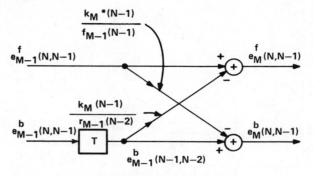

(A) ONE-DIMENSIONAL LATTICE STAGE

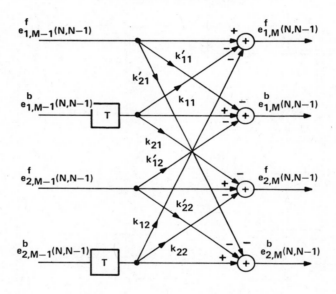

(B) TWO-DIMENSIONAL LATTICE STAGE

Figure 6-12. $(M-1)^{\text{th}}$ lattice stage.

coefficients are given by the solution of Eq. (6.30) developed in Section 6.3.3.2:

$$C_M(N) = R_{M+1}^{-1}(N)W_M(N) \tag{6.97}$$

From (6.26) and (6.27) it can be seen that the error signal $e_M(k, N)$ can be expressed as

$$e_M(k, N) = I_k - C_M^t(N)X_M(k)$$

The tap coefficients will now be computed using an error that is determined from the previous coefficient computation. Thus the error between the Nth symbol and its estimate is denoted by ε_N, and is given by

$$\varepsilon_N = e_M(N, N-1) = I_N - C_M^t(N-1)X_M(N) \tag{6.98}$$

A recursive form for the tap coefficients will be obtained by use of (4A.17):

$$R_{M+1}(N) = \omega R_{M+1}(N-1) + X_M^*(N)X_M^t(N) \tag{6.99}$$

Applying the matrix identity that follows (4A.17) to (6.99) with $A = -1$ results in

$$\omega R_{M+1}^{-1}(N) = R_{M+1}^{-1}(N-1) - \frac{R_{M+1}^{-1}(N-1)X_M^*(N)X_M^t(N)R_{M+1}^{-1}(N-1)}{\omega + X_M^t(N)R_{M+1}^{-1}(N-1)X_M^*(N)} \tag{6.100}$$

Substituting (6.100) and (6.41) in (6.97) yields

$$C_M(N) = \omega^{-1}\left[R_{M+1}^{-1}(N-1)\right.$$

$$\left. - \frac{R_{M+1}^{-1}(N-1)X_M^*(N)X_M^t(N)R_{M+1}^{-1}(N-1)}{\omega + X_M^t(N)R_{M+1}^{-1}(N-1)X_M^*(N)}\right]$$

$$\times \left[\omega W_M(N-1) + I_N X_M^*(N)\right]$$

$$= C_M(N-1) - \frac{R_{M+1}^{-1}(N-1)X_M^*(N)\omega^{-1}}{\omega + X_M^t(N)R_{M+1}^{-1}(N-1)X_M^*(N)}$$

$$\times \left[\omega X_M^t(N)C_M(N-1) - I_N(\omega + X_M^t(N)R_{M+1}^{-1}(N-1)X_M^*(N))\right.$$

$$\left. + I_N X_M^t(N)R_{M+1}^{-1}(N-1)X_M^*(N)\right]$$

$$= C_M(N-1) - \frac{R_{M+1}^{-1}(N-1)X_M^*(N)}{\omega + X_M^t(N)R_{M+1}^{-1}(N-1)X_M^*(N)}$$

$$\times \left[X_M^t(N)C_M(N-1) - I_N\right]$$

The above equation can be placed in a more familiar form by use of (6.98):

$$C_M(N) = C_M(N-1) + G_M(N)\varepsilon_N \tag{6.101}$$

where

$$G_M(N) = \frac{R_{M+1}^{-1}(N-1)X_M^*(N)}{\omega + X_M^t(N)R_{M+1}^{-1}(N-1)X_M^*(N)} \qquad (6.102)$$

The parameter $G_M(N)$ is the Kalman gain. Since the order is usually fixed, the Kalman equalizer update coefficients for time sample k are ordinarily expressed as a vector with M suppressed (see Section 6.3.3.1):

$$C_k = C_{k-1} + G_k \varepsilon_k, \qquad K = 1, 2, \dots \qquad (6.103)$$

where

$$\varepsilon_k = I_k - C_{k-1}^t X_k \qquad (6.103a)$$

The Kalman gain, G_k, is obtained from (6.102) with N replaced by k and M suppressed.

It should be noted that the form of the Kalman gain expression in Table 5-2 can be specialized to the Kalman gain formula in (6.102) by assuming $\Phi(k, k-1) = I$, $Q_k = 0$ and $R_{k\varepsilon} = \omega$ with $P_{k-1} \equiv R_{(M+1)}^{-1}(k-1)$. Note that the signal elements $x_N, \dots, x_{N-M}$ in the column vector $X_M(N)$ correspond to the elements $x_{1,k}, \dots, x_{1-M,k}$ in the first row of the received signal matrix, X_k, in Table 5-1. Therefore, the lattice formulation of the Kalman algorithm is, in fact, a special case of the more general Kalman algorithm formulation in which a prediction of the current Kalman coefficient estimate based on a dynamic model is used. Thus, in the more general case the Kalman gain is updated by use of the equation given in Table 5-2 or by means of the square root algorithm given in Section 5.4.

The form of the equations in (6.23) and (6.103) is the same. However, the Kalman algorithm performance in a time-varying channel is better than the steepest descent algorithm performance since the Kalman algorithm computes a gain that adaptively follows the time changes. This effect will be more completely described in Section 6.4.2.2.

6.4   EQUALIZER PERFORMANCE EXAMPLES

A number of specific examples are now provided using the algorithms described in Section 6.3. Performance data are given in each case. The measure of performance depends on the specific application. For example, in wireline equalization the performance is determined in terms of the reduction of amplitude and delay distortion by use of the equalizer. No specific modulation is assumed, since the equalization is intended for general-purpose use in which a variety of waveforms may be transmitted.[24,25,26] In certain multipath radio channels the performance is measured in terms of improved signal-to-noise ratio prior to demodulation. In this case only signal correlation properties are required to specify the equalizer. For other radio channel equalization examples presented, the measure of performance is the bit error rate. For these examples the signal characteristics of the digital modulation are required.

6.4.1 Linear Equalizer Performance Examples

In this section two examples are described in which a linear equalizer is utilized. The first example is based on a wireline channel where distorted amplitude and delay characteristics are compensated by use of the equalizer. In the second example a radio channel modeled by a transmission which is received over direct and reflected paths is mitigated by use of a nonrecursive equalizer that combines the energy from both paths.

6.4.1.1 Telephone Channel Example

Consider the system model shown in Figure 6-13. The cascade of the distorted telephone channel and the linear equalizer represents a new channel over which reliable digital transmission can be achieved. To accomplish this, the equalizer must be adapted to compensate for the channel distortions introduced into the received signal, x_n. This adaptation process occurs during a training period during which a probe signal, s_n, known at both the transmitter and receiver, is sent over the channel/equalizer cascade. During this training period the receiver forms the local reference signal, d_n, by computing the response of the desired channel, with the probe signal as its input.

In practice only one local reference signal is required at the receiver unless equalization is desired to yield a multiplicity of channel characteristics. Within the training period the equalizer attempts to minimize the mean square error between the local reference signal and the equalizer output. Once the training period is completed, the equalizer adaptation stops and the transmitter and receiver are switched into a data mode. Signal representations using the model in Figure 6-13 are now given.

In telephone channels the input and output signals are analog waveforms. Since the equalizer is digitally implemented, it is assumed that the analog

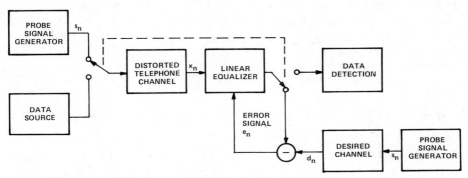

Figure 6-13. System model for general purpose wireline equalizer.

signals are converted to digital signals by sampling at or above the Nyquist rate. For example, if the baseband bandwidth of the signal is 3600 Hz, then the Nyquist rate is 7200 samples/sec. To avoid aliasing introduced by nonrectangular filter cutoffs, an 8000 sample/sec rate was selected in the analysis. In the signal descriptions provided below the sampling interval is denoted by T_s.

Probe Signal.   The probe signal consists of N_s harmonically spaced tones across the baseband signal bandwidth:

$$s_n = \sum_{k=-N_s}^{N_s} A_k e^{j(\omega_k n T_s + \phi_k)} \tag{6.104}$$

where A_k and ϕ_k are the amplitude and phase associated with the kth angular tone ω_k. The harmonic relationship among the tones is given by $\omega_k = 2\pi k f_0$ where $k = 1, \ldots, N_s$. Therefore s_n is a periodic signal with period $T_0 = 1/f_0$. Over the interval T_0, it is assumed that L samples occur so that $T_0 = LT_s$.

Equalizer Impulse Response.   The equalizer is implemented as a tapped delay line or transversal filter with $2M + 1$ tap coefficients spaced at the sampling interval and is represented by c_k for $k = -M, \ldots, M$. The equalizer impulse response, h_n, is then given by

$$h_n = \sum_{k=-M}^{M} c_k \delta(n - k) \tag{6.105}$$

Received Signal.   The channel is assumed to have tap coefficients, f_k, spaced at the sampling interval. Therefore, the received signal, x_n, can be computed from the convolution of the transmitted signal and the channel impulse response:

$$x_n = \sum_{k=-\infty}^{\infty} f_k s_{n-k} \tag{6.106}$$

An important property of linear shift invariant (LSI) systems is now used. This property states that the response of an LSI system to an exponential with frequency f is another exponential with frequency f modified by the transfer characteristic of the channel. For example, let $s_n = e^{j2\pi f n T_s}$ in (6.106):

$$x_n = e^{j2\pi f n T_s} \sum_{k=-\infty}^{\infty} f_k e^{-j2\pi f k T_s} \tag{6.107}$$

The sum in (6.107) can be recognized as the channel transfer function, $F_c(f)$:

$$F_c(f) = \sum_{k=-\infty}^{\infty} f_k e^{-j2\pi f k T_s}$$

Denoting the amplitude and phase of the channel transfer function at angular frequency ω_k by F_k and θ_k, respectively, leads to

$$F_c(\omega_k) = F_k e^{j\theta_k} \tag{6.108}$$

Applying the probe signal to the channel then results in a received signal expressed as

$$x_n = \sum_{k=-N_s}^{N_s} A_k F_k e^{j(\omega_k n T_s + \phi_k + \theta_k)} \tag{6.109}$$

Local Reference Signal.   The desired channel is chosen to have a flat amplitude characteristic in the band of interest and a phase characteristic that is linear with frequency. Since the envelope delay is the derivative of the phase characteristic with respect to frequency, the envelope delay is constant and denoted by τ. To avoid intercept distortion the phase shift at zero frequency is selected to be a multiple of π. With these assumptions the desired channel corresponds to ideal distortionless transmission in which a signal received via this channel is at worst an attenuated and delayed version of the transmitted signal. In practice the desired channel is allowed some roll-off at the band edges to allow the equalizer more rapid adaptation. The amplitude of the desired channel is denoted by D_k at angular frequency ω_k. The local reference signal can then be written as

$$d_n = \sum_{k=-N_s}^{N_s} A_k D_k e^{j(\omega_k n T_s - \omega_k \tau + \phi_k)} \tag{6.110}$$

Equalizer Error Signal.   The equalizer error signal, $\bar{e}_n$, is the difference between the local reference signal and the equalizer output:

$$\bar{e}_n = d_n - \sum_{k=-M}^{M} c_k x_{n-k} \tag{6.111}$$

The equalizer coefficients can be obtained by applying a time average version of the orthogonality principle, where the time average is taken over the probe signal period:

$$\frac{1}{L} \sum_{n=0}^{L-1} \bar{e}_n x_{n-l}^* = 0, \qquad l = -M, \ldots, M \tag{6.112}$$

Using (6.111) in (6.112) results in

$$\frac{1}{L} \sum_{n=0}^{L-1} \left(d_n - \sum_{k=-M}^{M} c_k x_{n-k} \right) x_{n-l}^* = 0, \qquad l = -M, \ldots, M$$

or

$$g'_l = \sum_{k=-M}^{M} c_k \rho'_{kl}, \qquad l = -M, \dots, M \tag{6.113}$$

where

$$\rho'_{kl} = \frac{1}{L} \sum_{n=0}^{L-1} x_{n-k} x^*_{n-l} \tag{6.114}$$

and

$$g'_l = \frac{1}{L} \sum_{n=0}^{L-1} d_n x^*_{n-l} \tag{6.115}$$

Equation (6.113) can be expressed in matrix form as

$$\boldsymbol{RC} = \boldsymbol{G} \tag{6.116}$$

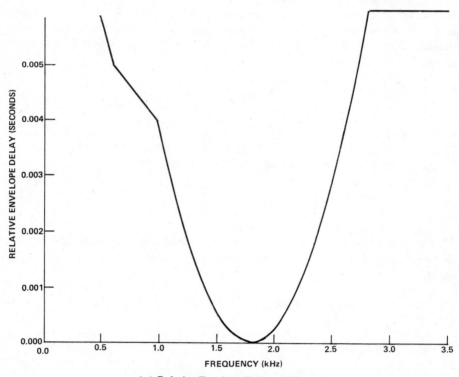

(*a*) Relative Envelope Delay of Channel

Figure 6-14. Characteristics of "Worst-Case" channel model.

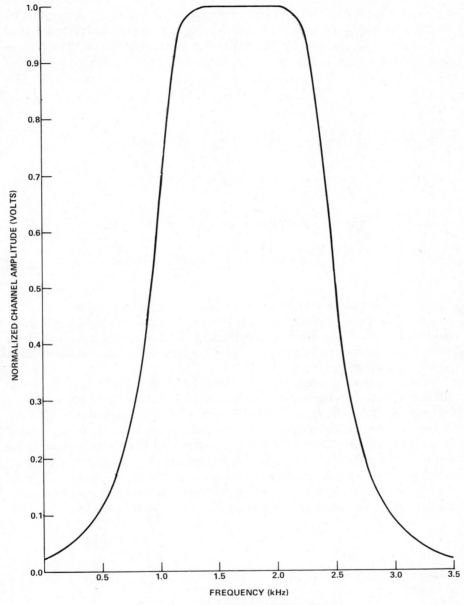

(*b*) Normalized Channel Amplitude

Figure 6-14. (*Continued*)

where R is a $(2M + 1) \times (2M + 1)$ matrix of the received autocorrelation coefficients ρ'_{kl}, G is a $2M + 1$ column vector of coefficients g'_l corresponding to the crosscorrelation between the received signal and the local reference, and C is a $2M + 1$ column vector of optimum equalizer coefficients. The correlation coefficients can be expressed in reduced form by use of the identity

$$\frac{1}{L} \sum_{n=0}^{L-1} e^{j2\pi f_0 n(i-m)T_s} = \begin{cases} 1, & i = m \\ 0, & \text{otherwise} \end{cases} \tag{6.117}$$

Substituting (6.109) into (6.114) and using (6.117) results in

$$\rho'_{kl} = \sum_{i=-N_s}^{N_s} A_i^2 F_i^2 e^{j2\pi f_0 i(l-k)T_s} \tag{6.118}$$

Similarly, substituting (6.109) and (6.110) into (6.115) yields

$$g'_l = \sum_{i=-N_s}^{N_s} A_i^2 D_i F_i e^{j[2\pi f_0 i(lT_s - \tau) - \theta_i]} \tag{6.119}$$

Using (6.118) and (6.119) in (6.116) allows the equalizer coefficients to be computed, assuming that the distorted channel response has been measured or estimated. For a large number of coefficients a receiver terminal would have excessive computation. Instead, the equalizer coefficients are iteratively computed using the steepest descent algorithm.

Typical channel characteristics are shown in Figure 6-14. Severe amplitude attenuation can be observed at the low and high ends of the band, that is, at 300 Hz and 3000 Hz, respectively. In addition, tones transmitted at different frequencies can exhibit as much as 6 ms differential delay between the band edges and midband. The equalizer is ideally designed so that the cascade of the channel and equalizer will have a flat amplitude spectrum and a constant envelope delay over the transmission bandwidth. The performance criteria are then determined by how close the channel/equalizer cascade comes to producing amplitude delay characteristics that correspond to the ideal case.

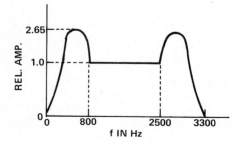

Figure 6-15. Characteristics of probe signal amplitude weighting used in equalizer simulation runs.

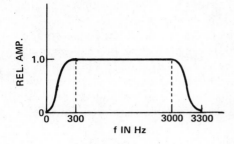

Figure 6-16. Characteristics of desired channel amplitude weighting used in equalizer simulation runs.

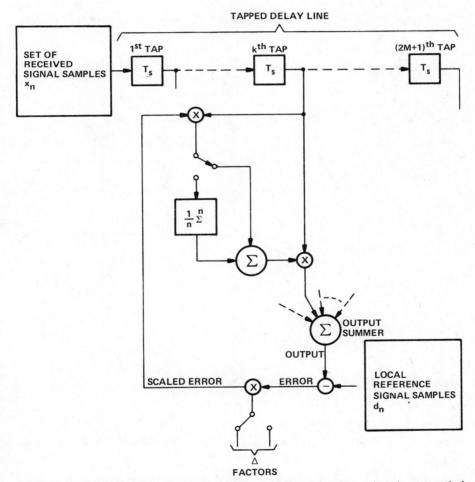

Figure 6-17. Simulation of the equalizer adjustment operation shown for one iteration at a typical tap.

The probe signal consists of 112 tones spaced at 32 Hz with the amplitude characteristic depicted in Figure 6-15. The band edges are enhanced to compensate for the attenuation in the channel and thus allow signal components at the band edges to be received with reasonable signal power. The probe waveform period is selected to be 31.25 ms and corresponds to 250 samples at 8000 samples/sec. Figure 6-16 displays the desired channel weighting with band edge roll-off introduced. In the equalization training interval only one period of the probe and local reference signal samples is stored, and these samples are used repeatedly.

The equalizer structure is shown in Figure 6-17. The coefficients are computed using Eqs. (6.23) and (6.24). In the initial interval of coefficient convergence the equalizer updates the coefficients using the product of the scaled error signal and the received signal samples. During this early stage the error will be large, and a larger scale factor is used. As equalization proceeds a better

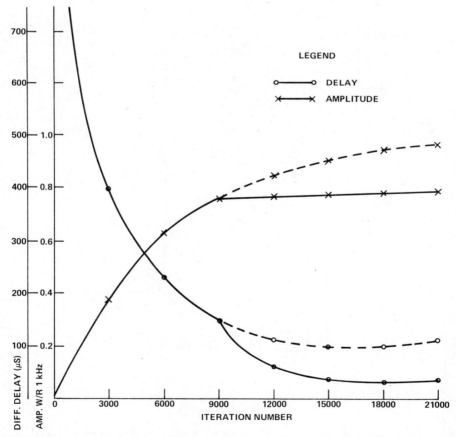

Figure 6-18. Typical results of differential delay in the band 1000–2600 Hz and band-edge amplitude convergence.

estimate of the gradient is required to reduce the error. In this subsequent stage the scale factor is reduced and averaging is performed over n iterations before updating the equalizer coefficients.

Specific performance results are obtained using computer simulation of an equalizer with 127 taps. Plots of maximum in-band amplitude and delay convergence for typical equalization simulations are shown in Figure 6-18. In the first 9000 iterations the scale factor $\Delta = 10^{-4}$, and no averaging is performed in order to produce rapid convergence. After 9000 iterations the scale factor is switched to $\Delta = 5 \times 10^{-5}$, and averaging over 8 samples is employed in order to achieve acceptable delay distortion. The dashed line shows convergence when the initial scale factor and no averaging is used throughout the simulation. Note that switching the scale factor at 9000 iterations improves the delay characteristic but degrades the amplitude characteristic.

Figure 6-19 depicts the normalized amplitude and delay of the channel/equalizer cascade. The results are obtained for a 127-tap equalizer after 21,000 iterations with the scale factors switched after 9000 iterations as described above. From the results it can be seen that the 3 dB bandwidth of the channel/equalizer cascade has increased from the 3 dB channel bandwidth of 1400 Hz to about 2800 Hz. Furthermore, the amplitude characteristic is very nearly flat from 400 Hz to 3000 Hz, and the maximum differential delay has decreased from 6 ms to about 125 μs over the band. These results have also been investigated using a 112-tap equalizer. In the latter case the differential delay and the amplitude characteristic were virtually unchanged (see Figure 6-20).

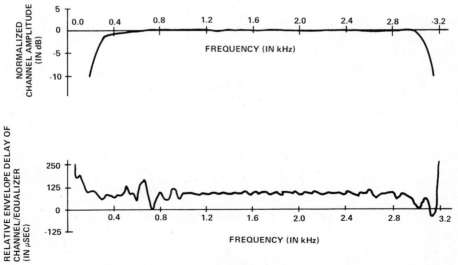

Figure 6-19.   Example of final performance using a 127-tap equalizer.

This example has demonstrated that general purpose equalization can be very effective in compensating for poor channel characteristics. This result has been obtained independent of the modulation waveform used for transmission.

6.4.1.2 Multipath Radio Channel Example

A system model for the multipath radio channel often assumes that the received signal can be represented by the sum of a "direct" and a "reflected" path.[27] If the received signal is normalized by the attenuation of the direct path signal, the complex envelope of the normalized received signal x_k can be expressed as

$$x_k = s_k + ae^{-j2\pi f_d kT_s}s_{k-D} + n_k \tag{6.120}$$

In (6.120) s_k is the complex envelope of the transmitted signal, T_s is the sampling interval, D is the differential path delay in samples between the direct and reflected paths, f_d is the differential Doppler shift between the direct and reflected paths, a is the complex valued gain of the reflected component relative to the direct component, and n_k is the complex envelope of the zero mean white Gaussian noise added within the receiver. Typical cases in which this model can be applied include the aeronautical satellite channel and transmission between aircraft in the presence of a ground reflection.

The results described below assume that the channel parameters such as gain, delay, and differential Doppler shift are known. Ordinarily, these parameters require adaptive estimation. One method used in the estimation employs ambiguity function analysis. The ambiguity function, $A_x(\lambda, f)$, of a continu-

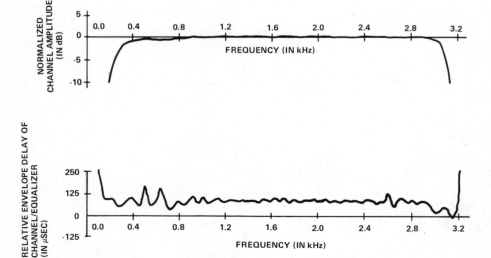

Figure 6-20.   Example of final performance using a 112-tap equalizer.

ous signal $x(t)$ is a two-parameter function in time, λ, and frequency, f, defined by

$$A_x(\lambda;f) = \frac{1}{T_I}\int_0^{T_I} x^*(t)x(t+\lambda)e^{-j2\pi ft}\,dt \qquad (6.121)$$

where T_I is the integration interval.[28] The integration interval is selected to be long enough to reduce noise arising from receiver noise and ambiguity function sidelobes, and small relative to the time variations in the channel parameters. If the ambiguity function is computed for the two-path received signal described above, where the differential delay is assumed to exceed the reciprocal of the signal bandwidth, three identifiable peaks will be observed in the plane of λ and f. A central lobe will be located at $\lambda = 0$, $f = 0$, and two secondary lobes will be located at $\lambda = D$, $f = f_d$, and $\lambda = -D$, $f = -f_d$. Therefore, estimates of the differential delay and differential Doppler shift can be obtained by determining the location in the (λ, f) plane of a secondary lobe. The relative strengths of a secondary lobe to the main lobe provide information on the relative gain and the direct path signal-to-noise ratio (SNR) parameters (see Figure 6-21).

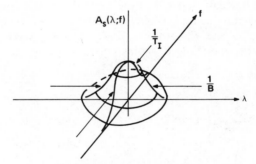

a) **Ambiguity Function of s(t)**

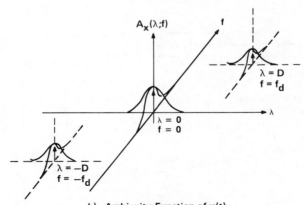

b) **Ambiguity Function of x(t)**

Figure 6-21. Ambiguity functions: (a) Ambiguity Function of $s(t)$; (b) Ambiguity Function of $x(t)$.

Two forms of equalization are now described. These forms include a recursive and a nonrecursive equalizer each of which is based on the optimization of a mean square error (MSE) criterion.

Nonrecursive Equalizer.   The nonrecursive equalizer structure shown in Figure 6-22 is implemented as a transversal filter with tap coefficients spaced at the differential channel delay. In general, the tap coefficients c_i, $i = -M, \dots, M$, are time varying, and the equalizer tap coefficient updating algorithm must be selected to track the channel time variations by using the latest channel parameter estimates. For simplicity, these time variations are neglected. The complex envelope of the equalizer output signal, $\hat{s}_k$, can be expressed as a weighted combination of delayed versions of the received signal:

$$\hat{s}_k = \sum_{i=-M}^{M} c_i x_{k-iD} \qquad (6.122)$$

The tap coefficients are selected to minimize the MSE between $\hat{s}_k$, the estimate of the transmitted signal, and s_k, the desired signal. In this formulation it is useful to identify a main tap where the desired signal energy is located. This tap will be denoted as c_p and usually corresponds to a point near or past the middle of the delay line. Equation (6.122) can now be written in a form that explicitly identifies the desired signal component, s_{k-pD}. Using (6.120) in (6.122) results in

$$\hat{s}_k = \sum_{i=-M}^{M} c_i \left[s_{k-iD} + a e^{-j2\pi f_d T_s (k-iD)} s_{k-(i+1)D} + n_{k-iD} \right]$$

$$= K_0 s_{k-pD} + J_k + N_k \qquad (6.123)$$

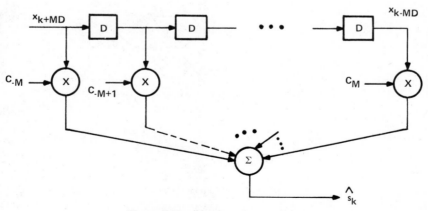

Figure 6-22.   Nonrecursive equalizer.

where

$$K_0 \equiv c_p + ae^{-j2\pi f_d T_s[k-(p-1)D]}c_{p-1}$$

$$J_k \equiv \sum_{\substack{i=-M \\ i \neq p \\ i \neq p-1}}^{M} c_i \left[s_{k-iD} + ae^{-j2\pi f_d T_s(k-iD)}s_{k-(i+1)D} \right]$$

$$+ c_{p-1}s_{k-(p-1)D} + c_p ae^{-j2\pi f_d T_s(k-pD)}s_{k-(p+1)D}$$

$$N_k \equiv \sum_{i=-M}^{M} c_i n_{k-iD}$$

Equation (6.123) expresses the equalizer output in terms of the desired signal, $K_0 s_{k-pD}$, the interference, J_k, and the noise, N_k. The power associated with each of these components obeys the relation

$$P_{\hat{s}} = P_{s_0} + P_J + P_N \tag{6.124}$$

where the equalizer output power, $P_{\hat{s}} = \frac{1}{2}\mathscr{E}[|\hat{s}_k|^2]$, the interference power, $P_J = \frac{1}{2}\mathscr{E}[|J_k|^2]$, the noise power, $P_N = \frac{1}{2}\mathscr{E}[|N_k|^2]$, and the desired signal power, $P_{s_0} = \frac{1}{2}\mathscr{E}[|K_0 s_{k-pD}|^2]$. By defining the signal power as $P_s = \frac{1}{2}\mathscr{E}[|s_k|^2]$, the desired signal power can be expressed as

$$P_{s_0} = |K_0|^2 P_s \tag{6.125}$$

The equalizer output SNR, ρ_0, can now be defined in accordance with (6.124) as

$$\rho_0 = \frac{P_{s_0}}{P_J + P_N} \tag{6.126}$$

Using (6.125) and (6.124) in (6.126), an alternate form of the output SNR is given by

$$\rho_0 = \frac{|K_0|^2 P_s}{P_{\hat{s}} - |K_0|^2 P_s} \tag{6.127}$$

The noise, n_k, is assumed to be white with zero mean and average power, σ_n^2:

$$\mathscr{E}[n_{kD}n_{lD}^*] = \begin{cases} 2\sigma_n^2 & k = l \\ 0, & k \neq l \end{cases}$$

With the above definition of noise, the output SNR can be expressed as a function of the direct path SNR, ρ_s, defined by $\rho_s = P_s/\sigma_n^2$. Computation of the output SNR requires knowledge of the equalizer coefficients.

The equalizer coefficients are determined by application of the orthogonality principle:

$$\mathscr{E}\left[e_k x^*_{k-lD}\right] = 0, \qquad -M \le l \le M \qquad (6.128)$$

where the error signal, e_k, is defined as the difference between the desired signal component and the equalizer output:

$$\varepsilon_k = s_{k-pD} - \hat{s}_k \qquad (6.129)$$

Before proceeding with the computation of the equalizer coefficients, a simple form for the output SNR will be obtained. In (6.128), it can be seen that the error is orthogonal to each delayed version of the received signal. From (6.122) it can then be seen that the error must also be orthogonal to the equalizer output signal:

$$\mathscr{E}\left[\varepsilon_k \hat{s}^*_k\right] = 0 \qquad (6.130)$$

Combining (6.129) and (6.130) results in

$$P_{\hat{s}} = \tfrac{1}{2}\mathscr{E}\left[s_{k-pD}\hat{s}^*_k\right] \qquad (6.131)$$

Further computation of the output SNR or the equalizer coefficients requires knowledge of the signal correlation properties. Here the signal correlation is assumed to be zero for delays exceeding the differential path delay:

$$\mathscr{E}\left[s_{kD}s^*_{lD}\right] = \begin{cases} 0, & k \ne l \\ 2P_s, & k = l \end{cases} \qquad (6.132)$$

The signal and noise terms are assumed to be uncorrelated. Since the noise is assumed to have zero mean, then

$$\mathscr{E}\left[s_{kD}n^*_{lD}\right] = 0 \qquad \text{for all } k \text{ and } l \qquad (6.133)$$

Combining (6.123) and (6.131) with (6.132) and (6.133) and noting that $P_{\hat{s}}$ is real results in

$$P_{\hat{s}} = \tfrac{1}{2}\mathscr{E}\left[s_{k-pD}\left(K_0^* s^*_{k-pD} + J_k^* + N_k^*\right)\right]$$

$$= K_0 P_s \qquad (6.134)$$

With (6.134) the output SNR given by (6.127) can now be written as

$$\rho_0 = \frac{K_0}{1 - K_0} \qquad (6.135)$$

where K_0 is real and is given by

$$K_0 = c_p + ae^{-j2\pi f_d T_s[k-(p-1)D]}c_{p-1}$$

The output SNR is now explicitly represented in terms of the channel parameters and the equalizer tap coefficients c_p and c_{p-1}. Prior to describing the computation of the equalizer coefficients, another form for the output SNR, expressed in terms of the minimum MSE, $P_{E_{\min}}$, will be given.

The following development for the minimum MSE parallels that provided in Section 3.2.3. The error signal power, P_E, satisfies

$$2P_E = \mathscr{E}\left[|\varepsilon_k|^2\right]$$

$$= \mathscr{E}\left[|s_{k-pD} - \hat{s}_k|^2\right]$$

$$= \mathscr{E}\left[|s_{k-pD}|^2\right] + \mathscr{E}\left[|\hat{s}_k|^2\right] - \mathscr{E}\left[s_{k-pD}\hat{s}_k^*\right] - \mathscr{E}\left[s_{k-pD}^*\hat{s}_k\right]$$

Since the orthogonality principle resulted in (6.131),

$$\mathscr{E}\left[|\hat{s}_k|^2\right] = \mathscr{E}\left[s_{k-pD}\hat{s}_k^*\right]$$

Thus, the minimum MSE can be written as

$$2P_{E_{\min}} = \mathscr{E}\left[|s_{k-pD}|^2\right] - \mathscr{E}\left[s_{k-pD}^*\hat{s}_k\right] \tag{6.136}$$

Using (6.131) and (6.134) in (6.136) results in

$$P_{E_{\min}} = (1 - K_0)P_s \tag{6.137}$$

From (6.135) the output SNR can then be written as

$$\rho_0 = \frac{K_0 P_s}{P_{E_{\min}}}$$

or

$$\rho_0 = \frac{P_s - P_{E_{\min}}}{P_{E_{\min}}} \tag{6.138}$$

Equation (6.138) indicates that the output SNR is inversely proportional to the minimum MSE.

The equalizer coefficients are now computed from (6.128). Combining (6.122), (6.128), and (6.129) results in

$$\mathscr{E}\left\{\left[s_{k-pD} - \sum_{i=-M}^{M} c_i x_{k-iD}\right]x_{k-lD}^*\right\} = 0, \qquad -M \le l \le M$$

The normal equations can then be identified as

$$g_l' = \sum_{i=-M}^{M} c_i r_{il}', \qquad -M \le l \le M \tag{6.139}$$

where

$$g_l' = \mathscr{E}\left[s_{k-pD}x_{k-lD}^*\right] \tag{6.140}$$

and

$$r'_{il} = \mathscr{E}\left[x_{k-iD}x^*_{k-lD}\right] \tag{6.141}$$

Using the signal correlation properties defined by (6.132) and (6.133) allows (6.140) and (6.141) to be expressed as

$$r'_{il} = \begin{cases} 2P_s\left(1 + |a|^2\right) + 2\sigma_n^2, & i = l \\ 2a^*P_se^{j2\pi f_dT_s(k-lD)}, & i = l + 1 \\ 2aP_se^{-j2\pi f_dT_s[k-(l-1)D]}, & i = l - 1 \\ 0, & \text{otherwise} \end{cases} \tag{6.142}$$

$$g'_l = \begin{cases} 2a^*P_se^{j2\pi f_dT_s[k-(p-1)D]}, & l = p - 1 \\ 2P_s, & l = p \\ 0, & \text{otherwise} \end{cases} \tag{6.143}$$

Using (6.142) and (6.143) in (6.139) and expressing the results in matrix notation with $(2M + 1) \times (2M + 1)$ elements leads to

$$\begin{bmatrix} 2P_s\left(1 + |a|^2\right) + 2\sigma_n^2 & 2a^*P_se^{j2\pi f_dT_s(k+MD)} & 0 & \cdots & 0 \\ 2aP_se^{-j2\pi f_dT_s(k+MD)} & 2P_s\left(1 + |a|^2\right) + 2\sigma_n^2 & 2a^*P_se^{j2\pi f_dT_s[k-(-M+1)D]} & & \vdots \\ 0 & 2aP_se^{-j2\pi f_dT_s[k-(-M+1)D]} & 2P_s\left(1 + |a|^2\right) + 2\sigma_n^2 & & \\ \vdots & & \ddots & & 0 \\ & & & & 2a^*P_se^{j2\pi f_dT_s[k-(M-1)D]} \\ 0 & \cdots & 0 & 2aP_se^{-j2\pi f_dT_s[k-(M-1)D]} & 2P_s\left(1 + |a|^2\right) + 2\sigma_n^2 \end{bmatrix}$$

$$\times \begin{bmatrix} c_{-M} \\ \\ \vdots \\ \\ c_M \end{bmatrix} = \begin{bmatrix} 0 \\ \vdots \\ 0 \\ 2a^*P_se^{j2\pi f_dT_s[k-(p-1)D]} \\ 2P_s \\ 0 \\ \vdots \\ 0 \end{bmatrix}$$

$$\tag{6.144}$$

Note that the right hand vector has the element $2P_s$ in the pth row.

The output SNR can now be evaluated from (6.135). An interim step involves the computation of the equalizer tap coefficients given by (6.144). Equations (6.144) require knowledge of the gain, differential path delay, differential Doppler shift, and direct path SNR. Procedures for estimating these parameters have been outlined previously. A simple performance calculation will be provided following a discussion of an alternate approach based on recursive equalization.

Recursive Equalizer. The recursive equalizer structure, sometimes referred to as a recursive canceller, is depicted in Figure 6-23. This structure attempts to cancel the reflected path component by reconstructing an estimate of the reflected path from the equalizer output and channel parameters. The feedback parameter, β, is a parameter selected to minimize the MSE between $\hat{s}_k$, the estimate of the transmitted signal, and the desired signal, s_k. From Figure 6-23 the recursive canceller response, $\hat{s}_k$, can be written as

$$\hat{s}_k = x_k - \beta e^{-j2\pi f_d T_s k} \hat{s}_{k-D} \tag{6.145}$$

By iterating on the above equation the response can be given explicitly in terms of the received signal:

$$\hat{s}_k = \sum_{l=0}^{\infty} \psi_l^l x_{k-lD} \tag{6.146}$$

where

$$\psi_l = -\beta e^{-j2\pi f_d T_s [k - D(l-1)/2]}$$

An expression for the mean square error, $E(\beta)$, is now derived prior to

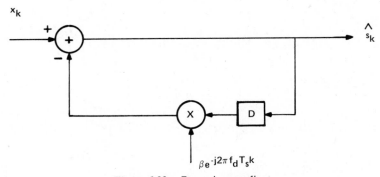

Figure 6-23. Recursive equalizer.

performing the minimization:

$$E(\beta) \equiv \mathscr{E}\left[|\varepsilon_k|^2\right] \tag{6.147}$$

where the error, ε_k, is defined by

$$\varepsilon_k = s_k - \hat{s}_k \tag{6.148}$$

Combining (6.146), (6.147), and (6.148) results in

$$E(\beta) = \mathscr{E}\left[|s_k|^2\right] + \sum_{l=0}^{\infty} \sum_{i=0}^{\infty} \psi_l^{*'} \psi_i^i \mathscr{E}\left[x_{k-iD} x_{k-lD}^*\right]$$

$$- 2R_e\left\{ \sum_{l=0}^{\infty} \psi_l^i \mathscr{E}\left[s_k^* x_{k-lD}\right]\right\} \tag{6.149}$$

where $R_e\{\ \}$ denotes the real part of the quantity in the brackets. Equation (6.149) is evaluated by using the signal correlation properties given by (6.132) and (6.133). The received signal autocorrelation has been computed previously and is given by (6.142). The crosscorrelation term can be computed using (6.120):

$$\mathscr{E}\left[s_k^* x_{k-lD}\right] = \mathscr{E}\left[s_k^*\left(s_{k-lD} + ae^{-j2\pi f_d T_s(k-lD)} s_{k-(l+1)D} + n_{k-lD}\right)\right]$$

$$= \begin{cases} 2P_s, & l = 0 \\ 0, & \text{otherwise} \end{cases} \tag{6.150}$$

Combining (6.150), (6.142), and (6.149) results in

$$E(\beta) = 2P_s - 2R_e(2P_s) + \left[2P_s(1 + |a|^2) + 2\sigma_n^2\right] \sum_{l=0}^{\infty} |\beta|^{2l}$$

$$+ 2P_s a^* \sum_{l=0}^{\infty} e^{j2\pi f_d T_s(k-lD)} \left\{ -\beta^* e^{j2\pi f_d T_s}\left[k - \left(\frac{l-1}{2}\right)D\right]\right\}^l$$

$$\times \left\{ -\beta e^{-j2\pi f_d T_s}\left[k - \frac{l}{2}D\right]\right\}^{l+1}$$

$$+ 2P_s a \sum_{i=0}^{\infty} e^{-j2\pi f_d T_s[k-iD]} \left\{ -\beta^* e^{j2\pi f_d T_s}\left[k - \frac{iD}{2}\right]\right\}^{i+1}$$

$$\times \left\{ -\beta e^{-j2\pi f_d T_s}\left[k - \left(\frac{i-1}{2}\right)D\right]\right\}^i$$

$$= -2P_s + \left[2P_s(1 + |a|^2) + 2\sigma_n^2\right] \sum_{l=0}^{\infty} |\beta|^{2l} - 2P_s a^* \beta \sum_{l=0}^{\infty} |\beta|^{2l}$$

$$- 2P_s a\beta^* \sum_{i=0}^{\infty} |\beta|^{2i} \tag{6.151}$$

Since

$$\sum_{l=0}^{\infty} |\beta|^{2l} = \frac{1}{1 - |\beta|^2}$$

then (6.151) becomes

$$E(\beta) = -2P_s + \frac{[2P_s(1 + |a|^2) + 2\sigma_n^2]}{1 - |\beta|^2} - \frac{4P_s}{1 - |\beta|^2} R_e(a*\beta) \quad (6.152)$$

It is convenient to normalize (6.152) by $2\sigma_n^2$ in order to express the results in terms of the direct path SNR:

$$\tilde{E}(\beta) = \frac{E(\beta)}{2\sigma_n^2}$$

$$= -\rho_s + \frac{\rho_s(1 + |a|^2) + 1 - 2\rho_s R_e(a*\beta)}{1 - |\beta|^2} \quad (6.153)$$

Now define the quantities a and β in magnitude and phase form according to

$$\beta = |\beta|e^{j\phi}$$

$$a = |a|e^{j\alpha}$$

Substituting the above relations in (6.153) yields

$$\tilde{E}(\beta) = -\rho_s + \frac{\rho_s(1 + |a|^2) + 1 - 2\rho_s|a||\beta|\cos(\phi - \alpha)}{1 - |\beta|^2} \quad (6.154)$$

Equation (6.152) can be minimized with respect to the phase of the feedback gain by forcing the condition $\phi = \alpha$. Using this condition the MSE can be minimized with respect to $|\beta|$ by setting the derivative to zero:

$$\frac{d\tilde{E}(\beta)}{d|\beta|} = \frac{-(1 - |\beta|^2)2\rho_s|a| + 2|\beta|[\rho_s(1 + |a|^2) + 1 - 2\rho_s|a||\beta|]}{(1 - |\beta|^2)^2}$$

$$= 0 \quad (6.155)$$

Rearranging the terms in (6.155), the optimum magnitude of the gain $|\beta|_{\text{opt}}$ satisfies

$$|\beta_{\text{opt}}|^2 - \frac{|\beta_{\text{opt}}|}{k_\beta} + 1 = 0 \quad (6.156)$$

where

$$k_\beta = \frac{|a|\rho_s}{(1 + |a|^2)\rho_s + 1}$$

Equation (6.156) has a solution given by

$$|\beta_{\text{opt}}| = \frac{1}{2k_\beta} - \sqrt{\frac{1}{4k_\beta^2} - 1} \tag{6.157}$$

To force $|\beta_{\text{opt}}|$ to be bounded by $0 \le |\beta_{\text{opt}}| \le 1$, the positive root in the solution is discarded. The parameter k_β can be estimated by the ambiguity function analysis described above.

The output SNR for the recursive canceller, ρ_0, SNR can be defined in a manner that is similar to that developed in the nonrecursive case. First the desired signal, interference, and noise terms must be identified by substituting (6.120) into (6.146):

$$\hat{s}_k = s_k + J_k + N_k \tag{6.158}$$

where

$$J_k = ae^{-j2\pi f_d T_s k} s_{k-D} + \sum_{l=1}^{\infty} \psi_l' \left[s_{k-lD} + ae^{-j2\pi f_d T_s (k-lD)} s_{k-(l+1)D} \right] \tag{6.159}$$

and

$$N_k = \sum_{l=0}^{\infty} \psi_l' n_{k-lD} \tag{6.160}$$

The interference power, P_J, can be obtained from (6.158) by use of (6.132):

$$P_J = |a|^2 P_s + P_s \sum_{l=1}^{\infty} |\beta|^{2l} + P_s |a|^2 \sum_{l=1}^{\infty} |\beta|^{2l}$$

$$- 2P_s R_e(a\beta^*) - 2P_s R_e(a\beta^*) \sum_{l=1}^{\infty} |\beta|^{2l}$$

Using the condition $\phi = \alpha$ for the optimum β in the above equation yields

$$P_J = P_s \frac{\left[|a|^2 + |\beta|^2 - 2|a||\beta| \right]}{1 - |\beta|^2} \tag{6.161}$$

The noise power, P_N, can be obtained from (6.160):

$$P_N = \sigma_n^2 \sum_{l=0}^{\infty} |\beta|^{2l}$$

$$= \frac{\sigma_n^2}{1 - |\beta|^2} \qquad (6.162)$$

From (6.161) and (6.162), the output SNR, ρ_0, can now be written as

$$\rho_0 = \frac{P_s}{P_J + P_N}$$

$$= \frac{P_s(1 - |\beta|^2)}{\sigma_n^2 + P_s[|a|^2 + |\beta|^2 - 2|a||\beta|]}$$

$$= \frac{\rho_s(1 - |\beta|^2)}{(|a|^2 - |\beta|^2)^2 \rho_s + 1} \qquad (6.163)$$

Substituting $|\beta_{opt}|$ from (6.157) into (6.163) yields the output SNR.

Performance Results.   Figure 6-24 provides a visual representation of nonrecursive equalizer performance. The transmitted signal is pulsed frequency modulation with a 0.5 duty cycle. Figure 6-24a depicts the envelope of the complex low-pass equivalent of the transmitted signal as a function of time in samples. The received signal is the sum of the transmitted signal (direct path) and a delayed attenuated version of the transmission signal (reflected path). The reflected path is attenuated to one half the strength of the direct path and is delayed by 70 samples corresponding to about 3-1/2 pulses. The ratio of the differential Doppler shift to the sampling rate is 10^{-4}. No additive noise is included here for ease of illustration. The envelopes of complex low-pass equivalents of the received and equalized signals are shown in Figures 6-24c and 6-24b, respectively. A 15-tap nonrecursive equalizer with weights obtained from knowledge of the channel parameters is used to produce the equalized signal. The results indicate that the equalized signal is an accurate reproduction of the transmitted signal with minimum error (see Figure 6-24d). Some degradation is experience with additive noise or imperfect channel estimates. This degradation is graceful and in accordance with the accuracy of the estimates.

The recursive canceller performance is now compared to the nonrecursive equalizer performance using the output SNR expressions derived above for minimum MSE. It can be shown that the performance for both the canceller and the nonrecursive equalizer are invariant with differential Doppler frequency. Curves of output SNR versus multipath gain with direct path SNR as a parameter are shown in Figure 6-25. The direct path SNR, ρ_s, can be identified from the curves by setting the multipath gain to zero which forces the output SNR to equal the direct path SNR in both the nonrecursive and recursive implementations. The recursive canceller curves are obtained by computing the output SNR, ρ_0, from (6.163) and (6.157) for assumed values of multipath gain, a, and direct path SNR, ρ_s. For comparison, the corresponding curves for a 15-tap nonrecursive equalizer with the main tap, $p = 8$, are shown for optimum tap coefficients. These curves are computed from (6.135) and (6.144). For high direct path SNR ($\sim$ 9 to 15 dB) and low multipath gains ($\sim$ 0 to 0.5) the canceller is only a fraction of a decibel poorer in performance than the nonrecursive equalizer. For low direct path SNR ($\sim$ -3 to 3 dB) and low multipath gains, the nonrecursive equalizer exhibits about 1 dB better performance. For multipath gains $\sim$ 1 or greater than 1, the nonrecursive equalizer is typically several decibels better in performance than the canceller at all direct path SNR values. The canceller then requires a separate algorithm to sense the condition that the magnitude of the gain exceeds unity in order to avoid performance degradation in this region.

The recursive and nonrecursive equalizer algorithms each has distinct advantages. The nonrecursive structure typically offers a performance advantage and does not require a separate algorithm for sensing that the magnitude of the gain exceeds unity. However, the nonrecursive structure requires a separate estimate of either the gain or direct path SNR. The recursive structure simply requires a single function of the gain and direct path SNR and is obtained directly from the ambiguity function. This channel estimation is somewhat

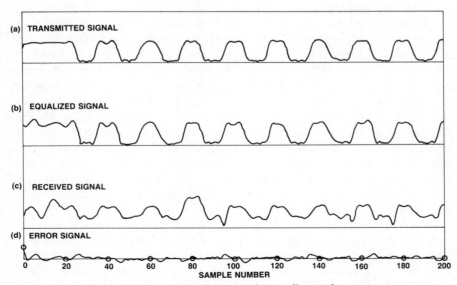

Figure 6-24. Example of nonrecursive equalizer performance.

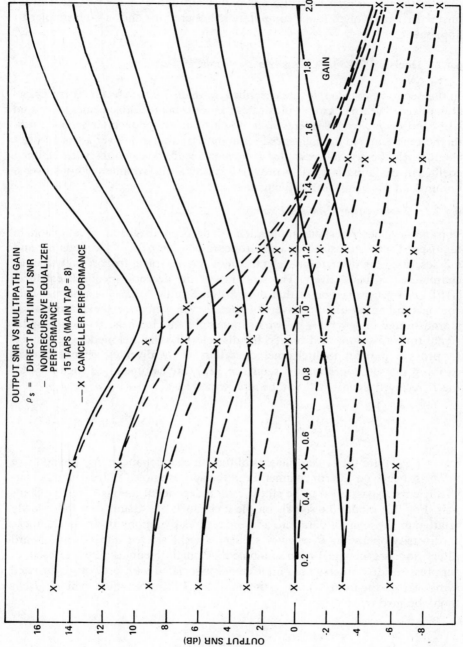

Figure 6-25. Output SNR versus multipath gain.

simpler for the recursive algorithm. Another advantage of the recursive structure is that it involves fewer computational operations and is thus simpler to implement.

6.4.2   Decision Feedback Equalizer Performance Examples

In this section two examples are described in which decision feedback equalizers are utilized. The first example illustrates the performance enhancement achieved with a decision feedback equalizer designed to mitigate intersymbol interference encountered on high-speed serial data transmission over a troposcatter channel. In the second example, a decision feedback equalizer with Kalman coefficient updating is used to provide high-data-rate communications over a nominal 3 kHz bandwidth HF channel.

6.4.2.1   Troposcatter Channel Example

This section describes the application of decision feedback equalization in high-speed serial transmission over troposcatter channels.[5,29] Sections 6.1 and 6.2 described the discrete-time channel model and typical troposcatter channel characteristics, respectively. The structure of the decision feedback equalizer (DFE) is a generalization of that provided in Section 6.3.2. The structure here combats fading by use of a diversity operation in which the same information is transmitted on several (e.g., two or four) different channels. To model digital signal transmission with diversity, the discrete-time channel model is extended to provide parallel multichannel operation. If D diversity channels, each carrying the same information sequence $\{I_k\}$, are assumed (see Figure 6-26), the D received signals, $x_k^{(i)}$, can be represented by

$$x_k^{(i)} = \sum_{m=0}^{L} f_m^{(i)} I_{k-m} + n_k^{(i)}, \qquad i = 1,\dots,D; \quad k = 0,1,\dots$$

where $\{f_m^{(i)}\}$ are the tap coefficients of the ith channel and $n_k^{(i)}$ is the additive noise sequence on the ith channel. Both the additive noise and the channel tap coefficients are assumed to be statistically independent among diversity channels. For this model the signals on these channels are assumed to fade slowly relative to the symbol duration, and independently among diversity channels. In diversity operation D received sequences $\{x_k^i\}$ are fed into D feedforward filters and are maximal-ratio combined. A single feedback filter is used to suppress residual intersymbol interference associated with previously detected symbols (see Figure 6-27). An estimate of the kth information symbol, $\hat{I}_k$, is then obtained from

$$\hat{I}_{k-d} = \sum_{i=1}^{D} \sum_{j=0}^{M_1} \alpha_j^{(i)} x_{k+j}^{(i)} - \sum_{j=1}^{M_2} \beta_j \tilde{I}_{k-d-j}, \qquad k = 0,1,\dots$$

where $\{\alpha_j^{(i)}\}$ are the $M_1 + 1$ feedforward tap gains on the ith diversity branch

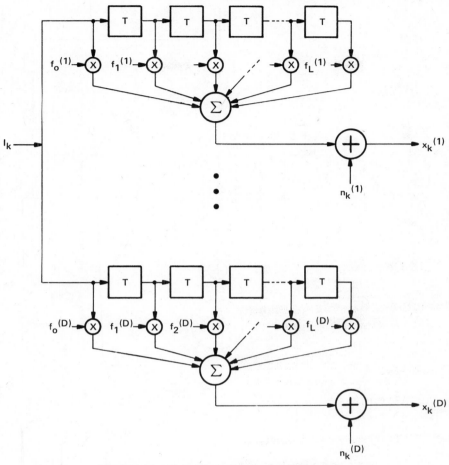

Figure 6-26. Channel model for diversity operation.

and $\{\beta_j\}$ are the M_2 feedback tap gains. Minimizing the MSE following the development in Section 6.3.2 leads to

$$\sum_{i=1}^{D}\sum_{j=0}^{M_1} \alpha_j^{(i)} B_{lj}^{(iv)} = f_{l+d}^{*(v)}, \qquad \begin{array}{l} l = 0,\ldots M_1 \\ v = 1,\ldots D \\ i = 1,\ldots D \end{array}$$

$$\sum_{i=1}^{D}\sum_{j=0}^{M_1} \alpha_j^{(i)} f_{d+l+j}^{(i)} = \beta_l, \qquad \begin{array}{l} l = 1,\ldots M_2 \\ M_2 + d \geq L + 1 \end{array}$$

$$B_{lj}^{(iv)} = \sum_{u=0}^{l+d} f_u^{*(v)} f_{u+j-l}^{(i)} + \phi_{j-l}\delta(i - v)$$

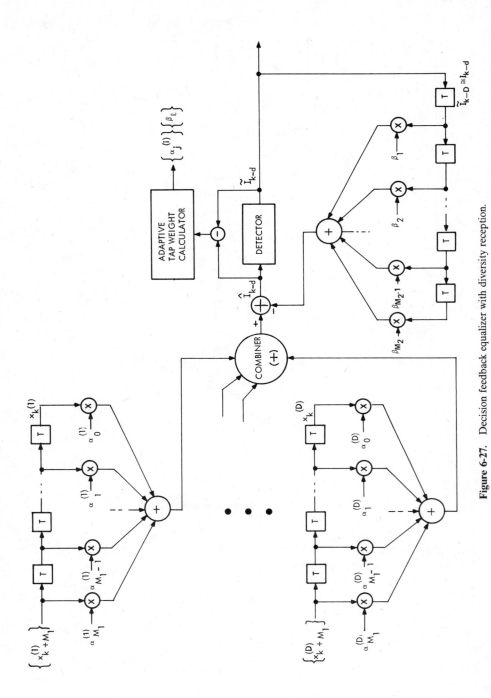

Figure 6-27. Decision feedback equalizer with diversity reception.

200

Since the channel response characteristics are unknown and time varying, the equalizer tap coefficients are adaptively computed. Several algorithms for possible use have been described in Section 6.3.3. A simple algorithm which has been successfully implemented for the troposcatter channel is the steepest descent algorithm given in Section 6.3.3.1. Use of this algorithm requires that the channel time variations be much slower than the duration of the signaling interval.

Performance Results.   DFE experimental performance operating on fading dispersive channels can be obtained by means of the Monte Carlo simulation depicted in Figure 6-28. The signal generator outputs complex low-pass equivalents of PSK modulated symbols that are obtained from an *M*-sequence generator. Uncorrelated binary or quarternary PSK signals can be generated. The channel is modeled as 1, 2, or 4 diversity branches. In simulating typical troposcatter links, the variances of the Gauss processes associated with the channel tap coefficients are selected from the delay power spectra curves shown in Figure 6-29.[5,29] Since the fading is slow relative to the symbol duration, T, a real-time simulation would require an enormous number of symbols to experience several cycles of fading. To avoid the computer costs associated with excessive numerical computation, a "snapshot" technique was used to model the fading. In this method, tap weights for a channel are selected to be samples of zero mean complex Gauss processes with variances determined by a power delay spectra curve. A block of typically 500 symbols is then passed over this channel and is referred to as a single iteration. On the next iteration a new set of tap weights is selected from the same zero mean Gauss process (i.e., with the same variances), and a new block of symbols is passed over this channel. The number of iterations is then selected to obtain an adequate statistical sample of the fading channel.

In the simulation program certain simplifications have been made for convenience. The number of taps is assumed to be the same for each diversity branch. The tap variances for each diversity branch are identical and are

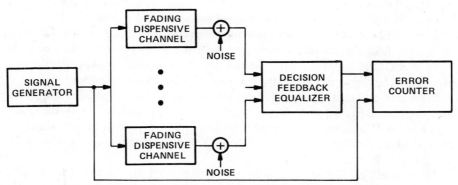

Figure 6-28.   Monte Carlo simulation of DFE in multichannel system.

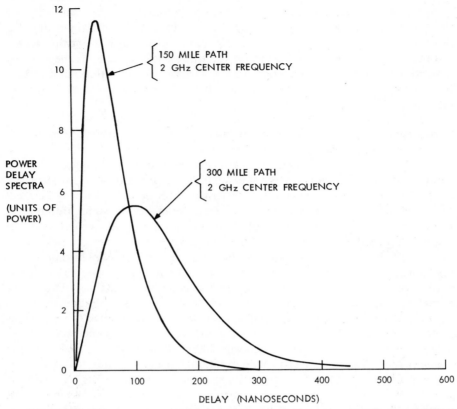

Figure 6-29. Delay power spectra for the short and long troposcatter channels (© 1979 IEEE Monsen reference 6-30).

normalized by the sum of the tap variances selected (the tap weights for each diversity branch are different). At each channel output, zero mean complex Gaussian noise with variance N_0 is added. The noise processes are assumed to be independent among the diversity branches with equal noise powers on each branch. Perfect knowledge of the channel tap weights is assumed.

It should be noted that the simulation program allows an arbitrary data rate to be specified. The variances obtained from the delay power spectra and the number of taps are the only channel parameters required to execute this program. The data rates typically selected were 2, 10 and 20 Mbps.

DFE performance results are now provided for binary PSK signaling with the notation DFE($M_1 + 1, M_2$) used to identify the number of feedforward taps ($M_1 + 1$) and the number of feedback taps (M_2) when actual decisions are fed back. Similarly, DFEI($M_1 + 1, M_2$) will be used for the case in which correct (or ideal) decisions are fed back.

Maximum likelihood sequence estimation (MLSE) using the Viterbi algorithm (VA) has also been proposed as a receiver detection technique for troposcatter applications.[29] This method is known to have better performance than the DFE technique and is therefore presented for reference purposes. VA(H) is the notation used in the performance curves where H indicates the length of the history specified in the Viterbi algorithm.

Extensive testing of the simulation program for channels with available theoretical performance was conducted as a check on the program. Figure 6-30

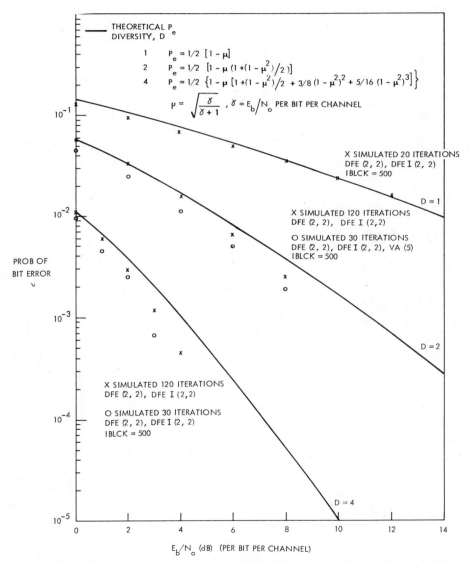

Figure 6-30. BER performance for binary coherent PSK in a nondispersive Rayleigh fading channel (© 1975 IEEE Giordano, et al., reference 6-29).

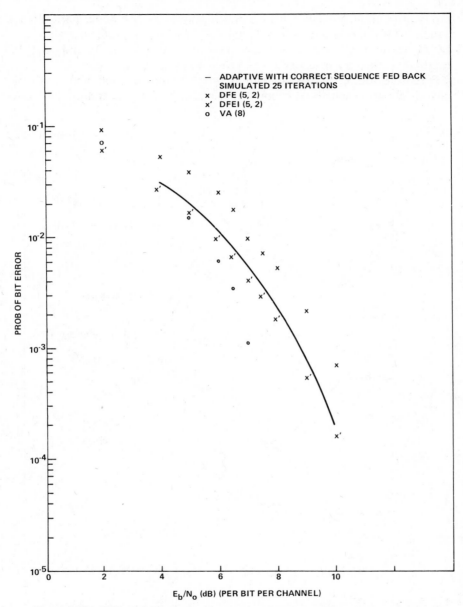

Figure 6-31. BER performance for binary coherent PSK nonfading, in a 2-tap channel $(1/\sqrt{2}, 1/\sqrt{2})$.

depicts simulated and theoretical bit error rate (BER) performance results for binary PSK signaling over a fading nondispersive channel with 1, 2, and 4 diversity branches.[30] The average bit-energy-to-noise spectral density per channel is denoted by E_b/N_0. In this case the DFE and VA yield the same performance. For 2 and 4 diversity branches, much better agreement between theory and simulation becomes apparent as the number of iterations increases

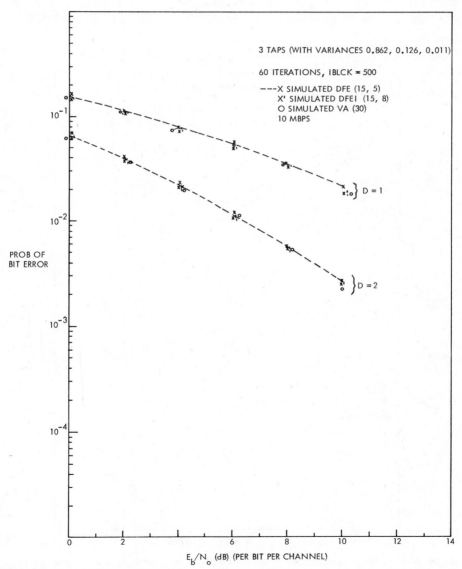

Figure 6-32. BER performance for binary coherent PSK in a Rayleigh fading channel (© 1975 IEEE Giordano, et al., reference 6-29).

from 30 to 120, due to a better statistical sampling of the fading. Figure 6-31 depicts binary PSK signaling performance for a nonfading channel with two equal strength taps. The solid line is provided for comparison and demonstrates DFE performance obtained adaptively with a steepest descent algorithm, assuming that the correct sequence is fed back.

Figures 6-32 and 6-33 depict simulated performance results for binary PSK signaling on typical long- and short-path troposcatter channels with 1 and 2 diversity branches.[29] Figure 6-32 was obtained using 3 taps with variances selected at a 10 Mbps data rate for the short-path channel. Figure 6-33 was

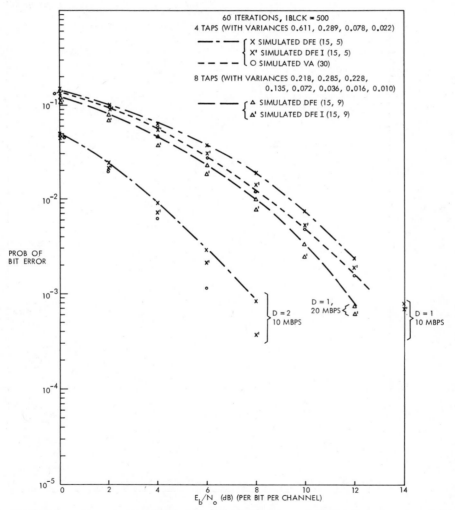

Figure 6-33. BER performance for binary coherent PSK in a Rayleigh fading channel (© 1975 Giordano, et al., reference 6-29).

obtained with variances taken from the long-path channel, where the 4-tap case corresponds to a 10 Mbps data rate and the 8-tap case is a 20 Mbps data rate. In both figures, the simulations were conducted using 60 iterations with 500 bits per iteration. The simulation was run with DFE(15, 5) and VA(30) for the 3- and 4-tap cases and with DFE(15, 9) in the 8-tap case. In Figure 6-32 the performance difference between the VA and DFE is less than 1 dB for both the 1- and 2-diversity cases. This result can be attributed to the presence of one large variance tap with little intersymbol interference. In the 4-tap cases in Figure 6-33 the performance difference is greater at larger SNR values but does not exceed about 1.5 dB. In both figures only a small degradation in performance results from feeding back actual decisions. Due to the implicit diversity available from the presence of intersymbol interference, the performance at 20 Mbps (8-tap case) is better than the performance at 10 Mbps (4-tap case).

6.4.2.2   HF Channel Example

In the troposcatter example, the DFE structure and equations for the feedforward and feedback tap coefficients presented in Section 6.3.2 were modified to permit diversity operation. In this section, a single channel is assumed and the fading is mitigated by use of a tracking algorithm that can adaptively follow the channel time variations. The DFE structure and equations for the feedforward and feedback tap coefficients presented in Section 6.3.2 apply directly here. Tracking algorithms studied for this application include variations of the steepest descent algorithm and the Kalman algorithm. Assuming knowledge of the channel coefficient at each sample instant, Eqs. (6.20) and (6.21) can be used to compute the theoretical tap coefficients. As a result of this computation, a benchmark is obtained which can be used to compare algorithm tracking performance. The adaptive algorithms are initialized by use of a training period in which the transmitted symbols are assumed to be known at the DFE receiver. When the residual distortion is small enough, the equalizer is switched into a decision-directed mode. At this point, the equalizer must self-adapt to changes in channel characteristics occurring during transmission.

Tracking Algorithms.   Table 6-1 summarizes the tracking algorithms investigated for use in the HF channel.[6,22,23] In each case, the form of the error signal at time sample k can be expressed as

$$\varepsilon_k = I_k - \tilde{C}_{k-1}^t \tilde{X}_k. \tag{6.167}$$

(See equations (6.22), (6.22a) and (6.103a).) The form of the error signal permits the tap coefficient obtained during the previous iteration to be used in the current error signal computation. For channels which vary slowly enough, the coefficient computation may be performed less often than every sample instant, thereby reducing the computational burden.

   The steepest descent algorithm was described in Section 6.3.3.1. In addition, two variations of the steepest descent algorithm are provided in the table.

TABLE 6-1.   ADAPTIVE TRACKING ALGORITHMS

Tracking Algorithm	Coefficient Updating	Ancillary Relationships
Steepest descent	$\tilde{C}_k = \tilde{C}_{k-1} - \Delta\Gamma_k$	$\Gamma_k = -\varepsilon_k \tilde{X}_k^*$
Steepest descent with VSS	$\tilde{C}_k = \tilde{C}_{k-1} - \Delta_k\Gamma_k$	$\Gamma_k = -\varepsilon_k \tilde{X}_k^*$ $\Delta_k = \dfrac{1}{\tau\mathscr{R}_k}$
Steepest descent with AGC	$\tilde{C}_k = \tilde{C}_{k-1} - \Gamma_k$	$\Gamma_k = -\varepsilon_k' Y_k^*$ $Y_k = \dfrac{X_k'}{\sqrt{\mathscr{R}_k}}$
Kalman	$\tilde{C}_k = \tilde{C}_{k-1} + G_k\varepsilon_k$	$G_k = \dfrac{[P_{k-1} + Q_k]Y_k^*}{Y_k^t[P_{k-1} + Q_k]Y_k^* + \xi_k}$ $P_k = P_{k-1} + Q_k - G_k Y_k^t[P_{k-1} + Q_k]$ $Y_k = \dfrac{X_k'}{\sqrt{\mathscr{R}_k}}$

These latter algorithms are referred to as steepest descent with variable step size (VSS) and steepest descent with automatic gain control (AGC). Each of these algorithms requires an estimate of the received signal energy at sample instant k. This estimate is denoted by $\mathscr{R}_k$, and is recursively computed as a time average according to

$$\mathscr{R}_k = \lambda X_k''^t X_k'^* + (1 - \lambda)\mathscr{R}_{k-1} \qquad (6.168)$$

where λ is a constant less than unity and $X_k''^t = (x_k, \ldots, x_{k+M_1})$. Note that $\mathscr{R}_k$ is in fact a time recursive estimate of the energy in the received signal.

In the VSS method, the scale factor, denoted by Δ_k changes at each sample instant in accordance with

$$\Delta_k = \frac{1}{\tau\mathscr{R}_k} \qquad (6.169)$$

where τ is a fixed parameter chosen in proportion to the total equalizer length. The use of a variable step size has already been introduced in the telephone example in Section 6.4.1.1, where two scale factors were introduced to improve equalizer convergence. In the steepest descent algorithm with AGC, the received signal is normalized by the square root of the estimate of the received signal energy:

$$Y_k = \frac{X_k'}{\sqrt{\mathscr{R}_k}} \qquad (6.170)$$

where the vector $Y_k^t = (y_k, \ldots, y_{k+M_1})$. This error computation is then performed using the normalized received signal:

$$\varepsilon_k' = I_k - \tilde{C}_{k-1}^t \tilde{Y}_k \tag{6.171}$$

where $\tilde{Y}_k^t = (y_k, \ldots, y_{k+M_1}, \tilde{I}_{k-1}, \ldots, \tilde{I}_{k-M_2})$. For this algorithm, the scale factor Δ is assumed to be fixed.

The Kalman algorithm was described in Section 6.3.3.5. This algorithm can be obtained from Table 5-2 by assuming $\Phi(k, k-1) = I$, $\tilde{C}(0) = 0$, $P_0 = I$, $B_k = \tilde{C}_k$, $R_{k\varepsilon} = \xi_k$, and

$$Q_k = \mathscr{E}\left\{ [\tilde{C}_k - \tilde{C}_{k-1}][\tilde{C}_k - \tilde{C}_{k-1}]^{t^*} \right\} \tag{6.172}$$

Since the receiver generally contains an AGC circuit at the front end, the received signal X_k' is replaced by Y_k, given by (6.170). Thus in the Kalman tracking and error rate performance presented in the following sections, all results assume the use of an AGC circuit.

Tracking Performance Results. Illustrative examples are now presented to compare the performance of the various tracking algorithms. In the first example, a one-tap fading channel is investigated. The channel coefficient at sample instant k, denoted by f_k, is obtained by passing white Gaussian noise through a two-pole Butterworth filter whose bandwidth is selected to correspond to the fade rate (see Figure 6-34). For HF channels, the fade rate is usually less than 1 Hz but on transauroral paths it can be 10 Hz or more. The

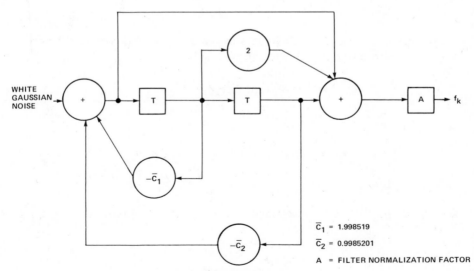

$\bar{c}_1 = 1.998519$

$\bar{c}_2 = 0.9985201$

A = FILTER NORMALIZATION FACTOR

Figure 6-34.   Two-pole Butterworth filter.

received signal can then be expressed as

$$x_k = f_k I_k + n_k$$

where n_k is zero mean additive complex white Gaussian noise with real noise power, σ_n^2. For this example, the DFE degenerates to a single feedforward adjustable tap with no feedback taps. The estimated signal, $\hat{I}_k$, is obtained by multiplying the received signal, x_k, with the equalizer tap weight, c_k, and is given by

$$\hat{I}_k = x_k c_k$$

When binary PSK is used, a symbol decision is made by quantizing the estimated signal, $\hat{I}_k$, to $+1$ or -1, corresponding to the case where $\hat{I}_k$ is positive or negative, respectively.

The theoretical equalizer tap weight can be obtained by minimizing the mean square error between the decision symbol $\tilde{I}_k$ and the estimated symbol $\hat{I}_k$. Assuming that the decision symbol $\tilde{I}_k$ is equal to the transmitted symbol, I_k, the orthogonality principle can be applied:

$$\mathcal{E}\left[(I_k - \hat{I}_k)x_j^*\right] = 0$$

Substituting the equation for the estimated signal in the above equation yields

$$\mathcal{E}\left[(I_k - x_k c_k)x_j^*\right] = 0$$

or

$$c_k = \frac{\mathcal{E}\left[I_k x_j^*\right]}{\mathcal{E}\left[x_k x_j^*\right]}$$

Computation of the ensemble averages in the equalizer tap coefficient expression using the received signal given above results in

$$\mathcal{E}\left[I_k x_j^*\right] = \mathcal{E}\left[I_k(f_j I_j + n_j)^*\right]$$

$$= f_k^*$$

$$\mathcal{E}\left[x_k x_j^*\right] = \mathcal{E}\left[(f_k I_k + n_k)(f_j I_j + n_j)^*\right].$$

$$= |f_k|^2 + 2\sigma_n^2$$

(See eq (6.9)). Note that an estimate of the received signal-to-noise ratio is given by

$$\text{SNR} = \frac{\sum_{k=0}^{L}|f_k|^2}{2\sigma^2}$$

(See eq (6.4)). Since the product of the signal bandwidth, B, and symbol duration, T, is assumed to be unity, the received signal-to-noise ratio is E_b/N_0, the energy contrast ratio. In the above equations, it is assumed that the signal and noise terms are uncorrelated and that the channel tap weight, f_k, varies slowly compared to the symbol rate (i.e., 1 Hz compared to 2400 Hz). The equalizer tap coefficient can now be expressed as

$$c_k = \frac{f_k^*}{|f_k|^2 + 2\sigma_n^2}$$

A comparison of the theoretical equalizer tap weight given above with the estimated tap weight computed by means of the VSS steepest descent algorithm is now given. Since the channel tap weight at each iteration is known, a good estimate of the short term averaged received signal energy which is used in the VSS steepest descent algorithm is

$$\mathscr{R}_k = |f_k|^2 + 2\sigma_n^2$$

Figures 6-35, 6-36, 6-37, and 6-38 illustrate the performance of the VSS steepest descent algorithm obtained from computer simulation using a single channel tap weight. This tap weight is pure real in Figures 6-35, 6-36 and 6-37, and is complex in Figure 6-38. The equalizer is trained in the first 200 iterations (symbols) by fixing the decision symbol equal to the transmitted symbol, that is, $\tilde{I}_k = I_k$. The channel fade rate is specified to be 1 Hz and the average received SNR exceeds 10 dB, so that very few errors occur. The time constant, τ, used in the step size adjustment, is indicated on the figures.

In Figure 6-35, the equalizer operates in a decision-directed mode. It can be observed that the algorithm tracks well up to about 5000 iterations. A fade then occurs and the estimate lags far behind the true tap weight, and finally converges to the negative of the actual tap weight. The negative value for the tap weight also represents a stable solution in the least mean square error criterion for decision-directed updating. Without the use of differential encoding in the data symbols or some other phase reference, this sign reversal causes long bursts of errors in the detected sequence.

In the tracking results illustrated in Figure 6-36, the same channel is used with a 100 percent training signal, that is, no information is actually transmitted. The time constants are $\tau = 50$ and $\tau = 100$ iterations. When the fade occurs in this case, the training signal helps to bring the estimated tap weight into the proper sign relationship (no sign reversal takes place here), although there is some lag. We observe that the lag can be reduced by decreasing the time constant τ at the expense of increased noise (larger variance) in the estimate of the tap coefficient.

Figure 6-37 illustrates two different strategies for adjusting the tap coefficient by using a variable step size with a time constant $\tau = 10$. In the first case

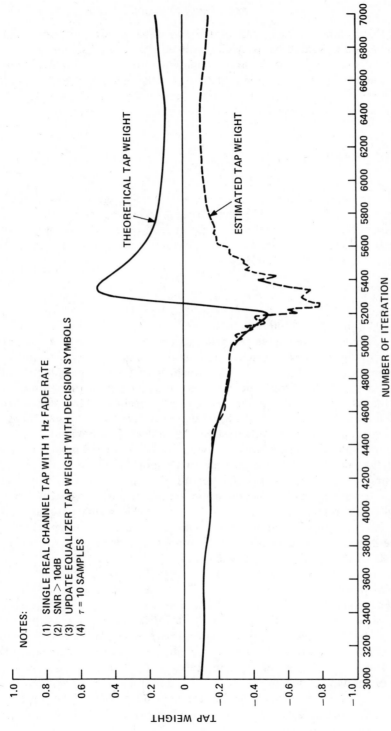

Figure 6-35. Tracking performance for VSS steepest descent algorithm (© 1980 IEEE Hsu, et al., reference 6-23).

NOTES:

(1) SINGLE REAL CHANNEL TAP WITH 1 Hz FADE RATE
(2) SNR > 10dB
(3) UPDATE EQUALIZER TAP WEIGHT WITH DECISION SYMBOLS
(4) $\tau = 10$ SAMPLES

THEORETICAL TAP WEIGHT

ESTIMATED TAP WEIGHT

TAP WEIGHT

NUMBER OF ITERATION

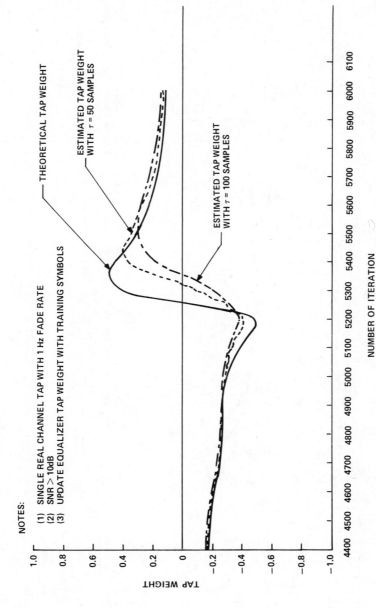

Figure 6-36. Tracking performance for VSS steepest descent algorithm (© 1980 IEEE Hsu, et al., reference 6-23).

213

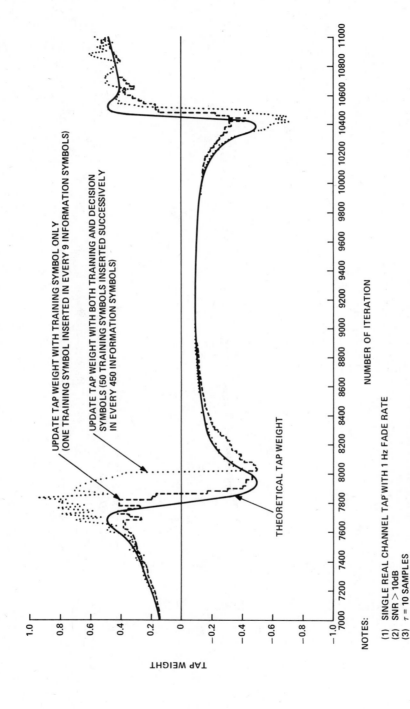

Figure 6-37. Tracking performance for VSS steepest descent tracking algorithm (© 1980 IEEE Hsu, et al., reference 6-23).

NOTES:

(1) SINGLE REAL CHANNEL TAP WITH 1 Hz FADE RATE
(2) SNR > 10dB
(3) τ = 10 SAMPLES

UPDATE TAP WEIGHT WITH TRAINING SYMBOL ONLY
(ONE TRAINING SYMBOL INSERTED IN EVERY 9 INFORMATION SYMBOLS)

UPDATE TAP WEIGHT WITH BOTH TRAINING AND DECISION
SYMBOLS (50 TRAINING SYMBOLS INSERTED SUCCESSIVELY
IN EVERY 450 INFORMATION SYMBOLS)

THEORETICAL TAP WEIGHT

NUMBER OF ITERATION

TAP WEIGHT

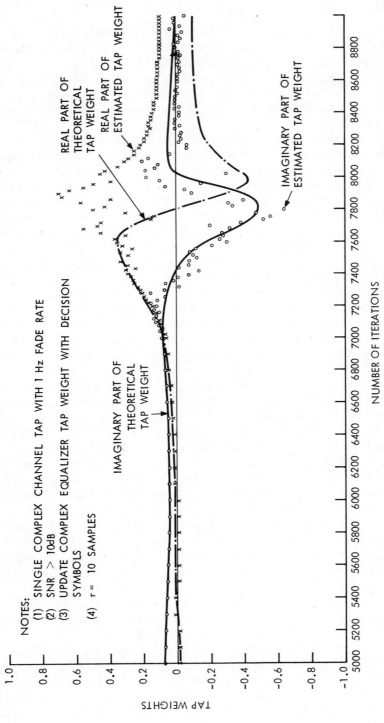

Figure 6-38. Tracking performance for VSS steepest descent tracking algorithm (© 1980 IEEE Hsu, et al., reference 6-23).

one training symbol is inserted after every nine information symbols and the tap weight is updated by the training symbols only. In the second method 50 training symbols are inserted successively after every 450 information symbols, and the tap weight is updated by both training and information symbols. The equalizer with the first training scheme tracks better than the second one in the vicinity of the deep fade region; otherwise, the second method trackes better than the first one.

Next, a channel with a single complex Gaussian tap is considered. The channel is compensated by a single complex-valued tap equalizer. In Figure 6-38, the real and imaginary parts of the estimated equalizer tap weight are compared with the theoretical tap weight. The equalizer tap weight is updated by using decision-directed operations after a training period in which 200 known information symbols are used. The data illustrates that the equalizer tracks the channel closely up to the beginning of a fade at about 7000 iterations. Beyond 7600 iterations, the equalizer fails to track for about the next 600 symbols. Finally, the equalizer begins to track the negative of the real part of the channel tap weight. These results show that the equalizer with the VSS steepest descent algorithm does not track well during periods of a deep fade. Consequently, the resulting error rate performance is poor.

In the next example, a two-tap channel model is simulated in which the tap weights spaced at the sample interval are selected to be complex with equal variance and independent Rayleigh fading amplitudes. In all cases shown subsequently, an AGC circuit was assumed, so that X_k' is replaced by Y_k. Figures 6-39 and 6-40 contrast the tracking behavior of the steepest descent algorithm using AGC with the tracking behavior of the Kalman estimator for the real part of the main equalizer tap. In the simulation the decision-feedback equalizer has four feedforward taps and one feedback tap and the equalizer is trained with all known information symbols. The theoretical equalizer tap weights are computed using Eqs. (6.20) and (6.21), with known channel tap weights and X_k' replaced by Y_k. The average SNR for the received signal is 17.5 dB, and the channel fade rate is 1 Hz. The steepest descent algorithm with AGC was empirically optimized to track as well as possible. For this case, $\Delta = 0.04$. In Figures 6-39 and 6-40, the received signal energy was estimated using (6.168) with $\lambda = 0.02$. The simulation results indicate that the performance of this algorithm during a deep fade is rather poor. On the other hand, the Kalman estimation algorithm does a good job in tracking the channel variations during a deep fade.

The parameters ξ_k and Q_k play a vital role in the Kalman algorithm; for example, $\xi_k = 0.1$, used in Figure 6-39, produces good tracking performance in rapid fading, and poor tracking performance in slow fading. On the other hand, $\xi_k = 10$, used in Figure 6-40, produces poor tracking performance in rapid fading, and good tracking performance in slow fading. Therefore, it is important to optimize ξ_k and Q_k in order to achieve good tracking performance. When this optimization is accomplished, it is clear from the results in Figures 6-39 and 6-40 that the Kalman algorithm is well suited to tracking the rapidly fading HF channel.

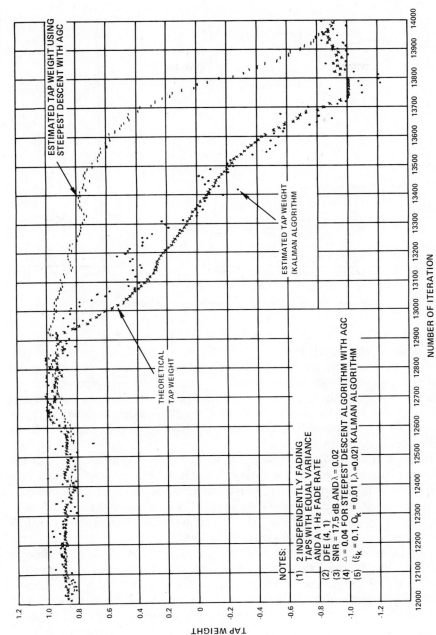

Figure 6-39. Variation of real part of main tap in period of rapid fading.

217

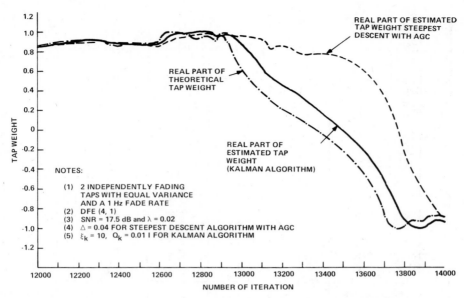

NOTES:

(1) 2 INDEPENDENTLY FADING
    TAPS WITH EQUAL VARIANCE
    AND A 1 Hz FADE RATE
(2) DFE (4, 1)
(3) SNR = 17.5 dB and λ = 0.02
(4) $\triangle$ = 0.04 FOR STEEPEST DESCENT ALGORITHM WITH AGC
(5) ξ_k = 10, Q_k = 0.01 I FOR KALMAN ALGORITHM

Figure 6-40.   Variation of real part of main tap in period of rapid fading (© 1980 IEEE Hsu, et al., reference 6-23).

Error Rate Performance Results.   The previous section described tracking algorithm performance results. In this section, error rate performance is given using either selected tracking algorithms or equalizer tap coefficients which are computed theoretically from Eqs. (6.20) and (6.21) with known channel tap coefficients. In all cases shown, an AGC circuit was assumed, so that X_k' is replaced by Y_k. The data indicates that good tracking performance results in good error rate performance.

Figures 6-41 and 6-42 depict probability of bit error, P_e, as a function of the energy contrast ratio, E_b/N_0 for binary PSK signaling. A two-tap channel with taps spaced at the sample interval with equal variance and a 1 Hz channel fade rate is simulated. The equalizer structure is DFE(4, 1) with all known information symbols used for training. The duration of the simulation is 30,000 symbols and encompasses approximately 10 fades.

Figure 6-41 illustrates DFE performance using the VSS steepest descent algorithm, the Kalman algorithm, and theoretically computed equalizer tap weights. All algorithms assume the case of a front-end AGC. The VSS steepest descent algorithm exhibits an irreducible error rate at about $P_e = 10^{-3}$. This behavior is a direct result of the poor tracking performance during deep channel fades. The Kalman estimation algorithm results in comparatively good performance with less than 2.5 dB degradation relative to that obtained with the theoretical equalizer coefficients. In this simulation, optimization of the parameters, ξ_k and Q_k, would result in improved Kalman algorithm performance.

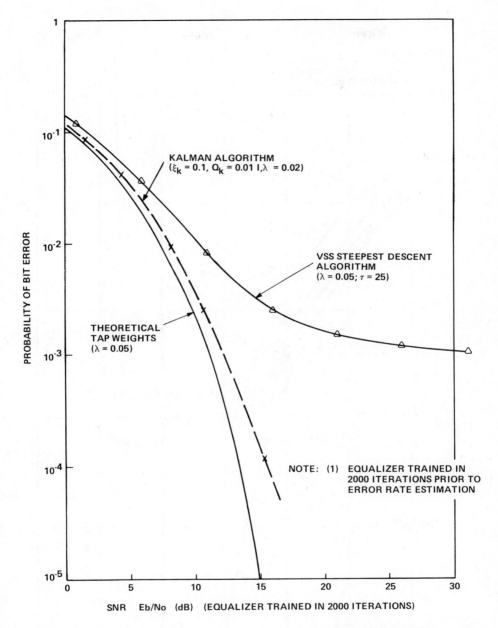

Figure 6-41. Error rate performance in two-path fading channel (© 1980 IEEE Hsu, et al., reference 6-23).

219

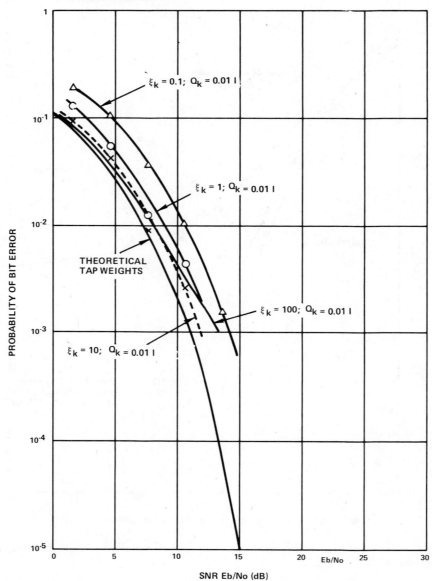

NOTES:
(1) EQUALIZER TRAINED IN 100 ITERATIONS PRIOR TO ERROR RATE ESTIMATION
(2) $\lambda = 0.02$
(3) UPDATE EQUALIZER TAPS WITH KNOWN INFORMATION SYMBOLS FEEDBACK

$\xi_k = 0.1; \; Q_k = 0.01 \; I$

$\xi_k = 1; \; Q_k = 0.01 \; I$

THEORETICAL TAP WEIGHTS

$\xi_k = 100; \; Q_k = 0.01 \; I$

$\xi_k = 10; \; Q_k = 0.01 \; I$

PROBABILITY OF BIT ERROR

SNR Eb/No (dB)

Eb/No

Figure 6-42. Error rate performance in two-path fading channel.

220

Figure 6-42 illustrates DFE performance with theoretically computed equalizer tap weights and for the Kalman algorithm with several values of the parameters ξ_k and Q_k. The results indicate that optimization of these parameters produces performance that closely matches the DFE performance obtained with the theoretical equalizer tap weights obtained from perfect knowledge of the channel. This behavior is directly attributable to the ability of the Kalman algorithm to track the channel time variations.

Other results have been obtained with larger signaling alphabets, more than two channel tap weights, and a variety of decision-directed training schemes. Figure 6-43 illustrates the probability of symbol error for one such case, in which eight-phase PSK signaling is utilized on a three-path fading channel with a DFE(6, 2) structure. Error rate results are presented for Kalman estimation with significant parameters indicated on the figure. The Kalman estimation cases include training with all known information symbols, training with all decision-directed symbols, and training with a 10 percent overhead. In the latter case, 10 known symbols are interleaved with a block of 90 information symbols that are used in a decision-directed mode. It is not surprising that some degradation is experienced in a decision-directed mode or with only partially known transmission sequences. Improved training schemes have been investigated that apply to full duplex transmission. One particular scheme that has been successfully implemented employs a technique in which a request for training sequence is initiated whenever equalizer performance is suffering. For example, the output SNR or, equivalently, the mean square error, can be monitored and used to initiate the request for training whenever a preselected threshold is exceeded.

The results presented thus far indicate that a DFE with Kalman coefficient estimation can produce excellent tracking and error rate performance on HF fading channels. Algorithms based on steepest descent techniques exhibit poor tracking, and thus degraded error rate performance. Unfortunately, the Kalman algorithm used in the results of this section requires extreme precision to achieve the illustrated performance. Simulations conducted with fixed-point arithmetic have shown that the DFE performance with conventional Kalman updating degrade rapidly as the computer word size is decreased. Fortunately, the square root form of the Kalman algorithm, presented in Section 5.4, has inherently better stabiltiiy and can better tolerate numerical inaccuracies than the conventional Kalman algorithm. The improved numerical behavior results from a reduction in the numerical ranges of the variables. An oversimplified viewpoint suggests that, for computations involving numbers between 10^{-L} and 10^{L}, where L is an integer, a reduction in the numerical range of the variables between $10^{-L/2}$ and $10^{L/2}$ allows greater accuracy to be achieved. Therefore, the square root Kalman formulation achieves accuracies that are comparable with a conventional Kalman algorithm that uses twice the numerical precision.

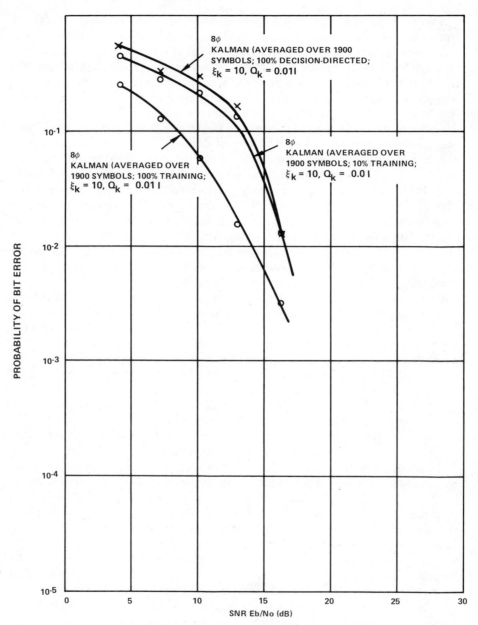

Figure 6-43. Error rate performance in three-path fading channel.

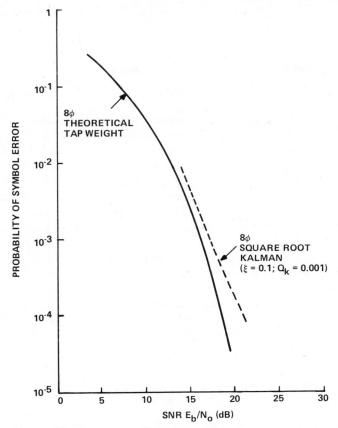

Figure 6-44. Error rate performance (© 1982 IEEE Hsu, reference 6-22).

Simulation results for the DFE with square root Kalman coefficient estimation have been obtained using 8-phase PSK. A two-tap channel with equal variances, taps spaced at the sample rate, and a channel fade rate of 1 Hz is assumed. The received signal is regulated by an AGC filter with $\lambda = 0.02$. Figure 6-44 illustrates probability of symbol error for DFE(4, 1) with $\xi_k = 0.1$ and $Q_k = 0.001$. By comparison with the theoretical results, it can be seen that good performance is achieved with the square root formulation.

6.4.3 Performance of Lattice Structures

The performance of linear and DFE lattice structures is presented in this section. For selected cases, performance results in both fading and nonfading channels are provided. The results are taken directly from Ling and Proakis.[19,31]

The two-channel models used in the performance evaluation are given in Z transform notation by

$$H_{\mathrm{NF}}(Z) = 0.408 + 0.816Z^{-1} + 0.408Z^{-2}$$

$$H_{\mathrm{F}}(Z) = a_0(t) + a_1(t)Z^{-1} + a_2(t)Z^{-2}$$

where $H_{\mathrm{NF}}(Z)$ and $H_{\mathrm{F}}(Z)$ are the channel transfer functions for nonfading and fading channels, respectively. The nonfading channel thus has 3 taps and one symbol delay separation between taps. The fading channels has time-varying tap weights $\{a_i(t)\}$ for $i = 1, 2$, and 3, with one symbol delay separation between taps. The tap weights $\{a_i(t)\}$ are generated by passing white Gaussian noise through a two-pole Butterworth filter having a 3 dB bandwidth of 0.5 Hz. In the results presented here real variables are used throughout. Typical sample functions of the three time-varying tap weights are depicted in Figure 6-45.

In the performance evaluation of the lattice equalizers, 100 samples were used as a training sequence and the succeeding 2500 samples were used to estimate performance. For comparison purposes, a DFE implemented with transversal filters and a gradient updating algorithm for the tap coefficients is also used. For the gradient transversal DFE, the first 1000 samples were used as a training sequence. Figure 6-46 illustrates typical MSE convergence of the

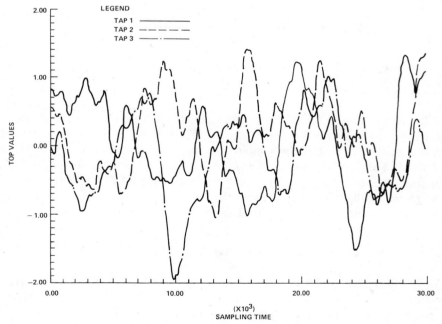

Figure 6-45. Tap values of the time-invariant channel (© 1984 IEEE Ling and Proakis, reference 6-32).

various equalizers considered using the nonfading channel. These equalizers include a linear least square (LS) lattice, a least squares DFE lattice, a gradient DFE lattice, and a gradient transversal DFE. The weighting factor ω for the least square lattice algorithms is 0.99. The step size for the gradient transversal DFE is 0.02.

It can be observed that the gradient transversal DFE requires at least 500–600 iterations to converge while the least squares DFE lattice converges in about 30–50 iterations. The gradient DFE lattice converges at about one-half the rate of the least squares lattice DFE. In Figure 6-46 it can also be seen that the final mean square error of the DFE lattice equalizers is smaller than either the linear least squares lattice equalizer or the gradient transversal DFE. Note that the final mean square error for the least square DFE lattice and the gradient DFE lattice are almost the same. Thus, it is anticipated from these

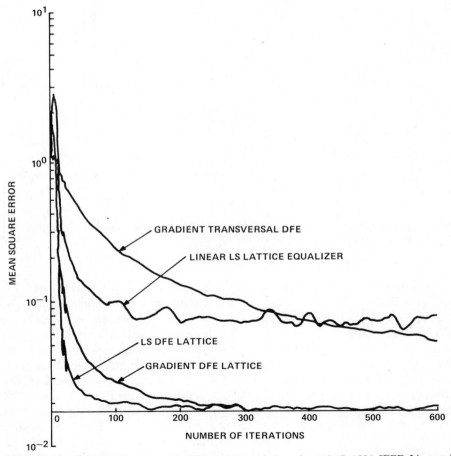

Figure 6-46. Convergence rate of equalizers for nonfading channel (© 1984 IEEE Ling and Proakis, reference 6-32).

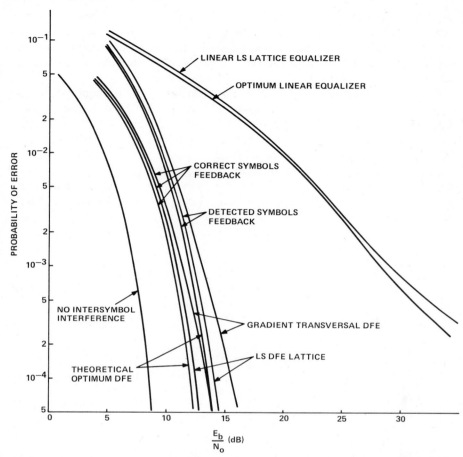

Figure 6-47.   Error rate performance of equalizers for the nonfading channel (© 1984 IEEE Ling and Proakis, reference 6-32).

results that both the least squares DFE lattice equalizer and the gradient DFE lattice will achieve rapid tracking and good performance in fading channels.

Figure 6-47 illustrates error rate performance in the nonfading channel case for the linear least squares lattice equalizer, the least squares DFE lattice equalizer, and the gradient transversal DFE. For illustrative purposes the performance for no intersymbol interference and the theoretical optimum error rate performance for linear and DFE structures are shown. Performance using either correct or detected symbols fedback is presented for the DFE cases. The optimum DFE performance is obtained by simulation using the optimum least square tap coefficients computed from a known channel. The least squares DFE lattice performance is nearly identical to the gradient DFE lattice performance, so that only one performance curve is shown for both structures. From the data it can be seen that the DFE lattice equalizers outperform the

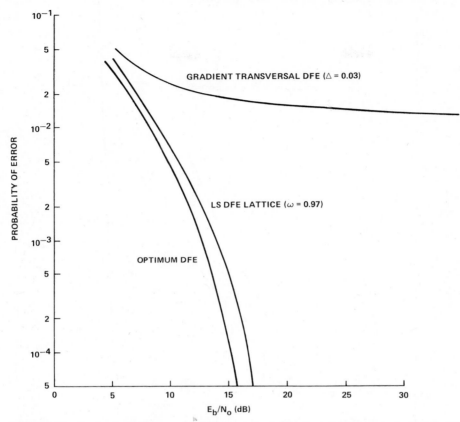

Figure 6-48. Error rate performance of DFEs for the fading channel (correct symbols fedback) (© 1984 IEEE Ling and Proakis, reference 6-32).

gradient transversal DFE and result in a performance degradation which is less than 0.5 dB from the optimum performance. Note that all DFE structures perform considerably better than the linear structures.

Figure 6-48 illustrates DFE performance in the fading channel case for the least squares DFE lattice with $\omega = 0.97$, the gradient transversal DFE with $\Delta = 0.03$, and the optimum DFE. In the simulation, all the symbols fed back are assumed to be correct. Once again the least squares DFE lattice and gradient DFE lattice are nearly identical in performance, with very little degradation from optimum performance. The gradient transversal DFE is unable to track the channel time variations and thus provides poor performance in this case.

In order to complete the discussion of lattice structures it is important to consider computational complexity. Figure 6-49, taken from Ling and Proakis,[31] provides a comparison of several algorithms. In the figure the number of multiplications per symbol required for each algorithm is plotted in terms of

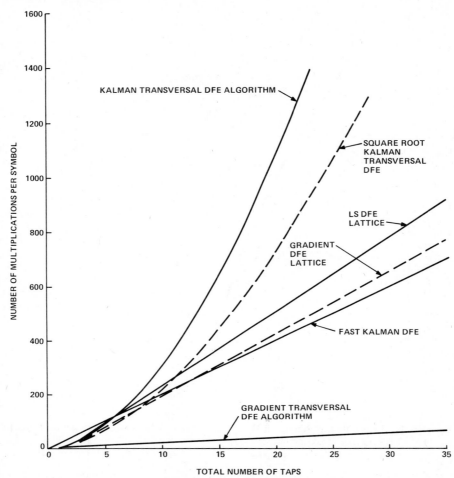

Figure 6-49. Computational complexities of adaptive DFEs (© 1984 Ling and Proakis, reference 6-32).

the total number of equalizer taps. Algorithm computational complexity is presented for the gradient transversal DFE, fast Kalman DFE,[20] gradient DFE lattice, least squares DFE lattice, square root Kalman transversal[22] DFE, and the Kalman transversal DFE.[21] Note that $M_1 + M_2 + 1$ total taps are contained in the DFE structure with $M_1 + 1$ feedforward taps and M_2 feedback taps. The computational requirements for Kalman and square root Kalman algorithms are proportional to the square of the total number of taps. The computational requirements for the remaining algorithms are linearly proportional to the number of taps. In terms of computational complexity, the gradient transversal DFE requires the fewest computations and the transversal Kalman DFE requires the most. Lattice structures are only moderately more complex than the fast Kalman algorithm.

APPENDIX: DERIVATION OF RECURSIVE EQUATIONS NEEDED IN GRADIENT LATTICE STRUCTURE FOR LINEAR EQUALIZATION

In this section several recursive equations stated without proof in Section 6.3.3.3 are derived. First, the following error recursions are proved:

$$e_M(N + 1, N + 1) = \alpha_{M+1}(N + 1)e_M(N + 1, N) \qquad (6A.1)$$

$$e_M^f(N, N) = \alpha_M(N - 1)e_M^f(N, N - 1) \qquad (6A.2)$$

$$e_M^b(N, N) = \alpha_M(N)e_M^b(N, N - 1) \qquad (6A.3)$$

where the parameter $\alpha_M(N)$ is defined by

$$\alpha_M(N) = 1 - \gamma_M(N, N) \qquad (6A.4)$$

Proof of (6A.2) requires (4A.10) and (4.52), which are repeated here for convenience:

$$\bar{A}_M(N) = \bar{A}_M(N - 1) - \frac{e_M^f(N, N)}{1 - \gamma_M(N - 1, N - 1)}\left(D_M(N - 1, N - 1)^0\right)$$

$$(4A.10)$$

$$e_M^f(k, N) = X_M^t(k)\bar{A}_M(N) \qquad (4.52)$$

Multiplying (4A.10) by $X_M^t(k)$ and using (4.52) results in

$$e_M^f(k, N) = e_M^f(k, N - 1) - \frac{X_M^t(k)e_M^f(N, N)}{1 - \gamma_M(N - 1, N - 1)}\left(D_M(N - 1, N - 1)^0\right)$$

$$(6A.5)$$

Recalling that $X_M^t(k) = (x_k \ X_{M-1}^t(k - 1))$, (6A.5) can be written as

$$e_M^f(k, N) = e_M^f(k, N - 1) - \frac{X_{M-1}^t(k - 1)D_M(N - 1, N - 1)e_M^f(N, N)}{1 - \gamma_M(N - 1, N - 1)}$$

$$(6A.6)$$

From (4A.2) it is known that

$$\gamma_M(N - 1, N - 1) = X_{M-1}^t(N - 1)D_M(N - 1, N - 1)$$

$$= X_{M-1}^t(N - 1)R_M^{-1}(N - 1)X_{M-1}^*(N - 1) \quad (4A.2)$$

Therefore, (6A.6) with k replaced by N becomes

$$e_M^f(N, N) = e_M^f(N, N - 1) - \frac{\gamma_M(N - 1, N - 1)}{1 - \gamma_M(N - 1, N - 1)} e_M^f(N, N)$$

or equivalently

$$e_M^f(N, N) = (1 - \gamma_M(N - 1, N - 1)) e_M^f(N, N - 1) \qquad (6A.7)$$

Using (6A.4) in (6A.7) yields (6A.2).

Proof of (6A.3) requires (6.40) and (4.53):

$$B_M(N) = B_M(N + 1) + \frac{R_M^{-1}(N + 1) X_{M-1}^*(N + 1)}{1 - \gamma_M(N + 1, N + 1)} e_M^b(N + 1, N + 1)$$

$$(6.40)$$

$$e_M^b(k, N) = X_M^t(k) \bar{B}_M(N) \qquad (4.53)$$

Since $\bar{B}_M(N) = \begin{pmatrix} B_M(N) \\ 1 \end{pmatrix}$, (6.40) can be written in extended form using (4A.2):

$$\bar{B}_M(N) = \bar{B}_M(N + 1) + \frac{e_M^b(N + 1, N + 1)}{1 - \gamma_M(N + 1, N + 1)} \begin{pmatrix} D_M(N + 1, N + 1) \\ 0 \end{pmatrix}$$

$$(6A.8)$$

Multiplying (6A.8) by $X_M^t(k)$ and using (4.53) and

$$X_M^t(k) = \begin{pmatrix} X_{M-1}^t(k) & x_{k-M} \end{pmatrix}$$

results in

$$e_M^b(k, N) = e_M^b(k, N + 1)$$

$$+ \frac{e_M^b(N + 1, N + 1)}{1 - \gamma_M(N + 1, N + 1)} X_{M-1}^t(k) D_M(N + 1, N + 1)$$

$$(6A.9)$$

Replacing k by $N + 1$ and using (4A.2) yields

$$e_M^b(N + 1, N) = e_M^b(N + 1, N + 1)$$

$$+ \frac{e_M^b(N + 1, N + 1)}{1 - \gamma_M(N + 1, N + 1)} \gamma_M(N + 1, N + 1)$$

or equivalently, with $N + 1$ replaced by N,

$$e_M^b(N, N) = (1 - \gamma_M(N, N))e_M^b(N, N - 1) \qquad (6A.10)$$

Using (6A.4) in (6A.10) yields (6A.3).

Proof of (6A.1) first requires a time recursive form for $C_M(N)$. The derivation uses (6.30), (4A.19), and (6.41):

$$C_M(N) = R_{M+1}^{-1}(N)W_M(N) \qquad (6.30)$$

$$R_{M+1}^{-1}(N - 1) = \left[R_{M+1}^{-1}(N) + \frac{R_{M+1}^{-1}(N)X_M^*(N)X_M^t(N)R_{M+1}^{-1}(N)}{1 - \gamma_{M+1}(N, N)}\right]_\omega$$

$$(4A.19)$$

$$W_M(N) = \omega W_M(N - 1) + I_N X_M^*(N) \qquad (6.41)$$

Using (6.30) with N replaced by $N - 1$, (4A.19), and (6.41) results in

$$C_M(N - 1) = \left[R_{M+1}^{-1}(N) + \frac{R_{M+1}^{-1}(N)X_M^*(N)X_M^t(N)R_{M+1}^{-1}(N)}{1 - \gamma_{M+1}(N, N)}\right]$$

$$\times \left[W_M(N) - I_N X_M^*(N)\right]$$

$$= C_M(N) - \frac{R_{M+1}^{-1}(N)X_M^*(N)}{1 - \gamma_{M+1}(N, N)}$$

$$\times \left[(1 - \gamma_{M+1}(N, N))I_N - X_M^t(N)C_M(N) + \gamma_{M+1}(N, N)I_N\right]$$

$$= C_M(N) - \frac{R_{M+1}^{-1}(N)X_M^*(N)}{1 - \gamma_{M+1}(N, N)}\left[I_N - X_M^t(N)C_M(N)\right]$$

$$(6A.11)$$

The error signal can be defined from (6.26) and (6.27):

$$e_M(k, N) = I_k - C_M^t(N)X_M(k) \qquad (6A.12)$$

Using (6A.12) and (4A.1) in (6A.11) yields

$$C_M(N) = C_M(N - 1) + \frac{D_{M+1}(N, N)}{1 - \gamma_{M+1}(N, N)}e_M(N, N) \qquad (6A.13)$$

Thus, the desired time recursion for $C_M(N)$ has been obtained. Multiplying (6A.13) by $X_M^t(N)$ and using (6A.12) and (4A.2) results in

$$e_M(N, N) = e_M(N, N - 1) - \frac{\gamma_{M+1}(N, N)}{1 - \gamma_{M+1}(N, N)}e_M(N, N)$$

or equivalently

$$e_M(N, N) = \left(1 - \gamma_{M+1}(N, N)\right) e_M(N, N - 1) \qquad (6A.14)$$

Using (6A.4) in (6A.14) yields (6A.1).

Now the time recursion for the parameter $\mu_M(N)$ is derived. Repeating (6.42) here and using (6A.1), (6A.3), and (6A.4) yields

$$\mu_M(N) = \omega\mu_M(N - 1) + \frac{e_{M-1}(N, N) e_M^{b*}(N, N)}{1 - \gamma_M(N, N)} \qquad (6.42)$$

$$= \omega\mu_M(N - 1) + \alpha_M(N) e_{M-1}(N, N - 1) e_M^{b*}(N, N - 1)$$

$$(6A.15)$$

An order recursion for the symbol estimate is derived by use of (6.35) where k is replaced by N and N by $N - 1$:

$$e_M(N, N - 1) = e_{M-1}(N, N - 1) - \frac{\mu_M(N - 1)}{r_M(N - 1)} e_M^b(N, N - 1)$$

$$(6A.16)$$

The symbol estimate for the kth symbol is given by (6.26):

$$\hat{I}_k = X_M^t(k) C_M(N) \qquad (6.26)$$

An alternate definition which is useful for developing a recursion relation is

$$\hat{I}_M(k, N) = X_M^t(k) C_M(N) \qquad (6A.17)$$

In this case (6A.12) can be written as

$$e_M(k, N) = I_k - \hat{I}_M(k, N) \qquad (6A.18)$$

From (6A.18) and (6A.16) a recursive order relationship for the Nth symbol estimate can be written as

$$\hat{I}_M(N, N - 1) = \hat{I}_{M-1}(N, N - 1) + \frac{\mu_M(N - 1)}{r_M(N - 1)} e_M^b(N, N - 1)$$

$$(6A.19)$$

To complete the recursion relations for the linear lattice structure, an order recursion for $\alpha_M(N)$ is needed. From (4.59) the order recursion for $\gamma_M(N, N)$

can be written as

$$\gamma_M(N, N) = \gamma_{M-1}(N, N) + \frac{|e^b_{M-1}(N, N)|^2}{r_{M-1}(N)} \tag{6A.20}$$

Using (6A.3) and (6A.4) in (6A.20) yields

$$\alpha_M(N) = \alpha_{M-1}(N) - \frac{\alpha^2_{M-1}(N)|e^b_{M-1}(N, N-1)|^2}{r_{M-1}(N)} \tag{6A.21}$$

Time recursions for the minimum forward and backward MSE, stated without proof in Section 6.3.3.3, are now derived:

$$f_M(N) = \omega f_M(N-1) + \alpha_M(N-1)|e^f_M(N, N-1)|^2 \tag{6.55}$$

$$r_M(N) = \omega r_M(N-1) + \alpha_M(N)|e^b_M(N, N-1)|^2 \tag{6.56}$$

Proof of (6.55) requires (4.32) , (4A.16) and (4A.22):

$$f_M(N) = \boldsymbol{Q}^{t*}_M(N)\boldsymbol{A}_M(N) + q_N \tag{4.32}$$

$$\boldsymbol{Q}_M(N) = \omega \boldsymbol{Q}_M(N-1) + x_N \boldsymbol{X}^*_{M-1}(N-1) \tag{4A.16}$$

$$\boldsymbol{A}_M(N) = \boldsymbol{A}_M(N-1) - \frac{e^f_M(N, N)\boldsymbol{D}_M(N-1, N-1)}{1 - \gamma_M(N-1, N-1)} \tag{4A.22}$$

Note that $q_N = \rho_N(0,0) = \sum^N_{i=1} \omega^{N-i}|x_i|^2$, so that

$$q_N = \omega q_{N-1} + |x_N|^2 \tag{6A.22}$$

Combining (4.32), (4A.16), (4A.22), (6A.22), and (6A.4) results in

$$f_M(N) = \left[\omega \boldsymbol{Q}^{t*}_M(N-1) + x^*_N \boldsymbol{X}^t_{M-1}(N-1)\right]\boldsymbol{A}_M(N) + q_N$$

$$= \omega \boldsymbol{Q}^{t*}_M(N-1)\left[\boldsymbol{A}_M(N-1) - \frac{e^f_M(N, N)}{\alpha_M(N-1)}\boldsymbol{D}_M(N-1, N-1)\right]$$

$$+ x^*_N \boldsymbol{X}^t_{M-1}(N-1)\boldsymbol{A}_M(N) + q_N$$

$$= \omega f_M(N-1) + |x_N|^2 - \frac{\omega e^f_M(N, N)}{\alpha_M(N-1)}$$

$$\times \boldsymbol{Q}^{t*}_M(N-1)\boldsymbol{D}_M(N-1, N-1)$$

$$+ x^*_N \boldsymbol{X}^t_{M-1}(N-1)\boldsymbol{A}_M(N) \tag{6A.23}$$

From (4A.21) it can be seen that

$$X_{M-1}^t(N-1)A_M(N) = e_M^f(N,N) - x_N \tag{6A.24}$$

The quantity $\omega Q_M^{t*}(N-1)D_M(N-1, N-1)$ in (6A.23) can be reexpressed by use of (4A.16), (4A.2), (4.33), (6A.24), and (6A.4):

$$\omega Q_M^{t*}(N-1)D_M(N-1, N-1)$$

$$= \left[Q_M^{t*}(N) - x_N^* X_{M-1}^t(N-1) \right] R_M^{-1}(N-1)X_{M-1}^*(N-1)$$

$$= -A_M^{t*}(N)R_M(N-1)R_M^{-1}(N-1)X_{M-1}^*(N-1)$$

$$\quad - x_N^* \gamma_M(N-1, N-1)$$

$$= -A_M^{t*}(N)X_{M-1}^*(N-1) - x_N^* \gamma_M(N-1, N-1)$$

$$= x_N^* - e_M^{f*}(N,N) - x_N^* \gamma_M(N-1, N-1)$$

$$= \alpha_M(N-1)x_N^* - e_M^{f*}(N,N) \tag{6A.25}$$

Using (6A.25) and (6A.24) in (6A.23) yields

$$f_M(N) = \omega f_M(N-1) + \frac{|e_M^f(N,N)|^2}{\alpha_M(N-1)} \tag{6A.26}$$

Combining (6A.26) and (6A.2) results in (6.55).
   Proof of (6.56) requires (4.43), (4.42), (4A.9), and (6.40):

$$r_M(N) = v_M(N) - V_M^{t*}(N)R_M^{-1}(N)V_M(N) \tag{4.43}$$

$$R_M(N)B_M(N) = -V_M(N) \tag{4.42}$$

$$V_M(N) = \omega V_M(N-1) + x_{N-M}X_{M-1}^*(N) \tag{4A.9}$$

$$B_M(N) = B_M(N-1) - R_M^{-1}(N)\frac{X_{M-1}^*(N)}{1 - \gamma_M(N,N)}e_M^b(N,N) \tag{6.40}$$

Since $v_M(N) = \rho_N(M,M) = \sum_{i=1}^N \omega^{N-i}|x_{i-M}|^2$, it can be seen that

$$v_M(N) = \omega v_M(N-1) + |x_{N-M}|^2 \tag{6A.27}$$

Combining (4.43), (4.42), and (4A.9) leads to

$$r_M(N) = \left[\omega V_M^{t*}(N-1) + x_{N-M}^* X_{M-1}^t(N) \right] B_M(N) + v_M(N)$$

(6A.28)

Using (6.40) in (6A.28) yields

$$r_M(N) = \omega V_M^{t*}(N-1) \left[B_M(N-1) - \frac{R_M^{-1}(N) X_{M-1}^*(N)}{1 - \gamma_M(N,N)} e_M^b(N,N) \right]$$

$$+ x_{N-M}^* X_{M-1}^t(N) B_M(N) + v_M(N)$$

(6A.29)

From (4A.6) it can be seen that

$$X_{M-1}^t(N) B_M(N) = e_M^b(N,N) - x_{N-M}$$

(6A.30)

Thus the quantity $\omega V_M^{t*}(N-1) R_M^{-1}(N) X_{M-1}^*(N)$ can be written using (4A.9), (4A.2), (4.42), and (6A.30) as follows:

$$\omega V_M^{t*}(N-1) R_M^{-1}(N) X_{M-1}^*(N)$$

$$= \left[V_M^{t*}(N) - x_{N-M}^* X_{M-1}^t(N) \right] R_M^{-1}(N) X_{M-1}^*(N)$$

$$= -B_M^{t*}(N) R_M(N) R_M^{-1}(N) X_{M-1}^*(N)$$

$$- x_{N-M}^* \gamma_M(N,N)$$

$$= x_{N-M}^* - e_M^{b*}(N,N) - x_{N-M}^* \gamma_M(N,N)$$

(6A.31)

Using (4.43), (4.42), (6A.27), (6A.30), (6A.31), and (6A.4) in (6A.29) leads to

$$r_M(N) = \omega r_M(N-1) + |x_{N-M}|^2 - \left[\alpha_M(N) x_{N-M}^* - e_M^{b*}(N,N) \right] \frac{e_M^b(N,N)}{\alpha_M(N)}$$

$$+ x_{N-M}^* \left[e_M^b(N,N) - x_{N-M} \right]$$

$$= \omega r_M(N-1) + \frac{|e_M^b(N,N)|^2}{\alpha_M(N)}$$

(6A.32)

Combining (6A.32) and (6A.3) results in (6.56).

REFERENCES

1. J. G. Proakis, *Digital Communications*, McGraw-Hill, New York, 1983.

2. G. D. Forney Jr., Maximum Likelihood Sequence Estimation of Digital Sequences in the Presence of Intersymbol Interference, *IEEE Trans. Information Theory IT-18*, May 1972, pp. 363–378.

3. J. G. Proakis, Advances in Equalization for Intersymbol Interference, In *Advances in Communication Systems* (A. V. Balakrisharan and A. J. Viterbi, Eds), Academic Press, San Francisco, 1975.

4. J. G. Proakis, Adaptive Filtering Techniques for Communication through Time-Dispersive Channels, NSF Grant GK-26329, June 1973, Dept. of Elect. Eng., Northeastern Univ., Boston.

5. P. Monsen, Adaptive Equalization of the Slow Fading Channel, *IEEE Trans. Communications COM-22*, Aug. 1974, pp. 1064–1075.

6. P. H. Anderson, F. M. Hsu and M. N. Sandler, A New Adaptive Modem for Long Haul HF Digital Communications at Data Rates Greater than 1 BPS/Hz, in *Proc. 1982 IEEE Military Communication Conference (MILCOM '82)*, pp. 29.2-1–29.2-7, Boston.

7. J. G. Proakis and J. H. Miller, Adaptive Receiver for Digital Signaling Through Channels with Intersymbol Interference, *IEEE Trans. Information Theory IT-15*, July 1969, pp. 484–497.

8. R. W. Lucky, Automatic Equalization for Digital Communications, *Bell System Tech. J. 44*, April 1965, pp. 547–588.

9. M. E. Austin, Decision Feedback Equalization for Digital Communication over Dispersive Channels, MIT Lincoln Laboratory, Lexington, Mass., Tech. Rep. 437, August 1967.

10. P. Monsen, Feedback Equalization for Fading Dispersive Channels, *IEEE Trans. Information Theory IT-17*, January 1971, pp. 56–64.

11. D. A. George, R. R. Bowen and J. R. Storey, An Adaptive Decision Feedback Equalizer, *IEEE Trans. Communications COM-19*, June 1971, pp. 281–293.

12. A. J. Viterbi, Error Bounds for Convolutional Codes and an Asymptotically Optimum Decoding Algorithm, *IEEE Trans. Information Theory IT-13*, April 1967, pp. 260–269.

13. G. D. Forney, The Viterbi Algorithm, *Proc. IEEE 61*, 3, 1973, pp. 268–278.

14. A. J. Viterbi and J. K. Omura, *Principles of Digital Communication and Coding*, McGraw-Hill, New York, 1979.

15. E. H. Satorius and J. D. Pack, Application of Least Square Lattice Algorithms to Adaptive Equalization, *IEEE Trans. Communications, COM-29*, Feb. 1981, pp. 136–142.

16. E. H. Satorius and S. T. Alexander, Channel Equalization Using Adaptive Lattice Algorithms, *IEEE Trans. Communications COM-27*, June 1979, pp. 899–905.

17. M. J. Shensa, A Least Square Lattice Decision Feedback Equalizer, *Int. Conf. on Communications*, Seattle, Wash., June 8–11, 1980, pp. 57.6.1–57.6.5.

18. F. Y. Ling and J. G. Proakis, Generalized Least Square Lattice Algorithm and its Application to Decision Feedback Equalization, in *Proc. ICASSP '82*, Paris, pp. 1439–1446.

19. F. Y. Ling and J. G. Proakis, Adaptive Lattice Decision Feedback Equalizers and Their Applications to Fading Dispersive Channels, in *Proc. IEEE, ICC '83*, Boston, Mass., June 1983.

20. D. D. Falconer and L. Ljung, Application of Fast Kalman Estimation to Adaptive Equalization, *IEEE Trans. Communications COM-26*, Oct. 1978, pp. 1439–1446.

21. D. Godard, Channel Equalization Using a Kalman Filter for Fast Data Transmission, *IBM J. Res. and Develop.*, May 1974, pp. 267–273.

22. F. M. Hsu, Square Root Kalman Filtering for High Speed Data Received Over Fading Dispersive HF Channels, *IEEE Trans. Information Theory IT-28*, 5, Sept. 1982.

23. F. M. Hsu, A. A. Giordano, H. dePedro and J. G. Proakis, Adaptive Equalization Techniques for High Speed Transmission on Fading Dispersive HF Channels," NTC '80, Houston, 1980.

24. R. W. Lucky and H. R. Rudin Jr., An Automatic Equalizer for General-Purpose Communication Channels, *Bell Syst. Techn. J. 46*, Nov. 1967, pp. 2179–2208.

25. H. R. Rudin Jr., A Continuously Adaptive Equalizer for General-Purpose Communication Channels," Bell Syst. Techn. J. *48*, Jul.–Aug. 1969, pp. 1865–1884.

26. W. M. Cowan and J. G. Proakis, Automatic Telephone Line Equalization, in *Proc. IEEE Int. Conf. on Communications*, Montreal, June 1971, pp. 38-12–38-17.

27. H. C. Salwen and C. B. Duncombe, Performance Evaluation of Data Modems for the Aeronautical Satellite Channel, *IEEE Trans. Communications COM-23*, 7, July 1975, pp. 695–705.

28. A. V. Oppenheim *Applications of Digital Signal Processing*, Prentice-Hall, (Ed.), Englewood Cliffs, N.J., 1978.

29. A. A. Giordano, J. G. Proakis, J. H. Lindholm, and T. A. Schonhoff, Error Rate Performance Comparison of MLSE and Decision Feedback Equalizer on Rayleigh Fading Multipath Channels, *1975 IEEE Int. Conf. on Communications*, San Francisco, June 16–18, pp. 5–1 to 5–5.

30. J. G. Proakis, On the Probability of Error for Multichannel Reception of Binary Signals, *IEEE Trans. Communications Technology COM-16*, 1, Feb. 1968, pp. 68–81.

31. F. Y. Ling and J. G. Proakis, Adaptive Lattice Decision Feedback Equalizers and Their Application to Time-Variant Multipath Channels, to be published in *IEEE Trans. Communications*.

chapter

7

POWER SPECTRAL
ESTIMATION

Applications of power spectral estimation techniques cover a wide variety of disciplines, for example, communications, digital filtering, radar/sonar, speech signal processing, geophysical exploration, radio astronomy, image processing, biomedicine, oceanography, and many others. Detailed investigations of power spectral estimation algorithms and their applications are not considered here. Instead, an overview of modern spectral analysis techniques is presented, with specific emphasis on autoregressive methods based on least squares.

7.1 HISTORICAL OVERVIEW

Harmonic analysis dates back to ancient times with studies of time.[1] More recent developments, in the eighteenth century, by Bernoulli, Euler, and Lagrange focused on studies of the wave equation and its sinusoidal solutions. In the early part of the nineteenth century, Fourier demonstrated that an arbitrary function on a finite interval possessed an infinite expansion of sines and cosines, as described in Chapter 2. Schuster introduced the notion of a periodogram based on Fourier analysis to identify sunspot number periodicities. A subsequent major discovery can be attributed to Wiener in his treatise of generalized harmonic analysis.[2] Wiener and Kinchin independently introduced the relationship between the autocorrelation function of a random process and its power spectral density.

Modern spectral estimation can be traced to the work of Blackman and Tukey.[3] These researchers applied Wiener's generalized harmonic analysis to power spectral estimation associated with sampled data sequences. Blackman and Tukey's method computes the Fourier transform of the windowed autocorrelation lags to estimate power spectrum. In 1965, Cooley and Tukey[4] published the fast Fourier transform (FFT) technique to implement the mathematically equivalent discrete Fourier Transform (DFT) more efficiently. By computing the squared magnitude of the FFT of a sampled data sequence, an estimate of the power spectrum referred to as the periodogram is obtained. Thus, power spectral estimation using the periodogram method was revolutionized by modern digital signal processing.

In addition to traditional methods, autoregressive techniques have been developed which enable an investigator to identify dominant spectral peaks that are masked in the periodogram approach. These techniques can be based on a variety of least squares algorithms, each of which offers distinct performance advantages for varying degrees of complexity. Specific algorithms include Yule-Walker, Burg,[5] unconstrained least squares[6] (lattice), and others.[7,8] The maximum entropy method[9] (MEM), formulated by Burg, is particularly attractive, since it offers high spectral resolution with short data segments and is based on the data itself rather than on autocorrelation estimates. These methods are emphasized in this chapter.

A variety of other modern methods include autoregressive moving average (ARMA), moving average (MA), Pisarenko Harmonic Decomposition (PHD), Prony's methods, Papoulis' method, and the maximum likelihood (ML) method of Capon. A good summary of these techniques and a comparative spectral estimation example for these methods is provided in reference 7. The particular method selected depends to a great extent on the intended application and the amount and quality of the available data.

In order to provide a baseline for comparison with the autoregressive methods emphasized here, some discussion of the traditional power spectral estimation technique is given.

7.2 TRADITIONAL POWER SPECTRAL ESTIMATION TECHNIQUE

Assume that a random process $x(t)$ is wide sense stationary with a continuous autocorrelation function, $R(\tau)$, defined by

$$R(\tau) = \mathscr{E}\left[x^*(t)x(t + \tau)\right] \tag{7.1}$$

The Weiner-Kinchin theorem[10] states that $P(f)$, the power spectral density of $x(t)$, is the Fourier transform (FT) of the autocorrelation function (see

Appendix Section A.3):

$$P(f) = \int_{-\infty}^{\infty} R(\tau) e^{-j2\pi f\tau}\, d\tau \qquad (7.2)$$

(see Appendix Section A.1.17.3). If the random process, $x(t)$, is also ergodic, so that ensemble averages equal time averages, then an alternate definition of the power spectrum of $x(t)$ is possible.[11] Define the random process, $X_{T_I}(f)$, associated with an ergodic process $x(t)$ by

$$X_{T_I}(f) = \int_{-T_I}^{T_I} x(t) e^{-j2\pi f t}\, dt \qquad (7.3)$$

The power spectrum is then given by

$$P(f) = \lim_{T_I \to \infty} \mathscr{E}\left\{ \frac{|X_{T_I}(f)|^2}{2T_I} \right\} \qquad (7.4)$$

or equivalently

$$P(f) = \lim_{T_I \to \infty} \mathscr{E}\left\{ \frac{1}{2T_I} \left| \int_{-T_I}^{T_I} x(t) e^{-j2\pi f t}\, dt \right|^2 \right\} \qquad (7.5)$$

For the ergodic case the autocorrelation can also be defined as a time average:

$$R(\tau) = \lim_{T_I \to \infty} \frac{1}{2T_I} \int_{-T_I}^{T_I} x^*(t) x(t + \tau)\, dt$$

A summary of the above relationships is provided in Figure 7-1. Note that ensemble averages are required to compute the autocorrelation or power spectrum of $x(t)$. No inverse relationship exists between $x(t)$ and $R(\tau)$ or between $X_{T_I}(f)$ and $P(f)$, as depicted in Figure 7-1. The Wiener-Kinchin theorem establishes that the autocorrelation and power spectrum correspond to a Fourier transform pair.

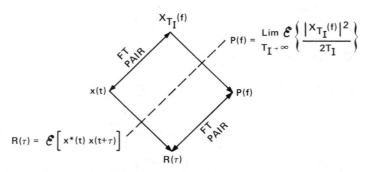

Figure 7-1.   Relationship between $x(t)$, $R(\tau)$ and $P(f)$, assuming ergodicity.

The discrete versions of the above relationships are now considered. Assume that the process $x(t)$ described above is bandlimited. If samples of a realization of this process are taken at the Nyquist rate or higher, then the original realization of the random process can be perfectly reconstructed from its samples.[11] (See A.1.18.) In the discrete case, assume that a stationary random process $\{x_k\}$ has an ensemble average autocorrelation function given by

$$r_m = \mathscr{E}\left[x_i^* x_{i+m}\right] \tag{7.6}$$

Note that ρ_m, an estimate of the autocorrelation function, r_m, can be defined using N samples from a realization of the process,[12] $\{x_k\}$:

$$\rho_m = \frac{1}{N-m} \sum_{j=0}^{N-m-1} x_j^* x_{j+m}, \qquad m = 0, 1, \ldots, L, \ldots, (N-1) \tag{7.7}$$

where L represents the number of autocorrelation lags. Usually $L \ll N$ for ρ_m to be a good estimate of r_m. Note that the negative lag estimates of the correlation function are defined by the relationship $\rho_{-m} = \rho_m^*$.

The expected value of this estimate is given by

$$\mathscr{E}[\rho_m] = \frac{1}{N-m} \sum_{j=0}^{N-m-1} \mathscr{E}\left[x_j^* x_{j+m}\right]$$

$$= \frac{1}{N-m} \sum_{j=0}^{N-m-1} r_m$$

$$= r_m, \qquad m = 0, \ldots, N-1$$

Since $\mathscr{E}[\rho_m] = r_m$, it has been shown that ρ_m is an unbiased estimate of r_m (see Appendix Section A.1.16.1). The variance of ρ_m, $V(\rho_m)$, can be computed using

$$V(\rho_m) = \mathscr{E}\left\{|\rho_m - \mathscr{E}(\rho_m)|^2\right\}$$

$$= \mathscr{E}\left\{|\rho_m|^2\right\} - |\mathscr{E}(\rho_m)|^2$$

Since the estimate, ρ_m, is unbiased, the variance can be written as

$$V(\rho_m) = \mathscr{E}\left\{|\rho_m|^2\right\} - |r_m|^2 \tag{7.8}$$

Substituting (7.7) into (7.8) results in

$$V(\rho_m) = \frac{1}{(N-m)^2} \sum_{j=0}^{N-m-1} \sum_{p=0}^{N-m-1} \mathscr{E}\left\{x_p x_j^* x_{j+m} x_{p+m}^*\right\} - |r_m|^2 \tag{7.9}$$

For a real random process an estimate of the variance can be obtained from the covariance expression on page 181 of Jenkins and Watts.[13] A simple derivation of the variance of the estimate can be obtained by assuming that $\{x_k\}$ is a zero mean Gaussian random process. The derivation requires an expansion for the fourth moment of the process. For real zero mean Gaussian random variables S_1, S_2, S_3, S_4, the expansion is[14] (see Appendix Section A.1.14)

$$\mathscr{E}(S_1 S_2 S_3 S_4) = \mathscr{E}(S_1 S_2)\mathscr{E}(S_3 S_4) + \mathscr{E}(S_1 S_3)\mathscr{E}(S_2 S_4) + \mathscr{E}(S_1 S_4)\mathscr{E}(S_2 S_3)$$

(7.10)

For complex zero mean Gaussian random variables z_1, z_2, z_3, z_4, with uncorrelated real and imaginary parts, u_l and v_l, respectively, for $l = 1, \ldots, 4$, the expansion becomes (see Appendix Section A.1.14)

$$\mathscr{E}(z_1 z_2^* z_3 z_4^*) = \mathscr{E}(z_1 z_2^*)\mathscr{E}(z_3 z_4^*) + \mathscr{E}(z_1 z_4^*)\mathscr{E}(z_2^* z_3)$$

(7.11)

where a condition that either $\mathscr{E}(u_1 u_3) = \mathscr{E}(v_1 v_3)$ or $\mathscr{E}(u_2 u_4) = \mathscr{E}(v_2 v_4)$ is required. Note that this condition forces either $\mathscr{E}(z_1 z_3)$ or $\mathscr{E}(z_2^* z_4^*)$ to be zero, resulting in only two terms in the complex version of the expansion.

Substituting (7.11) into (7.9) yields

$$V(\rho_m) = \frac{1}{(N-m)^2} \sum_{j=0}^{N-m-1} \sum_{p=0}^{N-m-1}$$

$$\times \left\{ \mathscr{E}(x_p x_j^*)\mathscr{E}(x_{j+m} x_{p+m}^*) + \mathscr{E}(x_p x_{p+m}^*)\mathscr{E}(x_j^* x_{j+m}) \right\} - |r_m|^2$$

$$= \frac{1}{(N-m)^2} \sum_{j=0}^{N-m-1} \sum_{p=0}^{N-m-1} \left\{ |r_{p-j}|^2 + |r_m|^2 \right\} - |r_m|^2$$

$$= \frac{1}{(N-m)^2} \sum_{j=0}^{N-m-1} \sum_{p=0}^{N-m-1} |r_{p-j}|^2$$

(7.12)

Performing a transformation of variables with $l = p - j$ yields

$$V(\rho_m) = \frac{1}{(N-m)^2} \sum_{l=-(N-m-1)}^{N-m-1} (N-m-|l|)|r_l|^2$$

(7.13)

For large N, (7.13) is approximately represented by

$$V(\dot{\rho}_m) \simeq \frac{1}{N} \sum_{l=-\infty}^{\infty} |r_l|^2$$

(7.14)

From (7.14), it can be seen that the variance of the estimate ρ_m approaches zero for large N, so that ρ_m is a consistent estimator (see Appendix Section A.1.16.2).

A similar result for a real zero mean Gaussian random process can be obtained using (7.10). The variance of the estimate for a pure real process $\{x_j\}$ can then be written as

$$V(\rho_m) = \frac{1}{(N-m)^2} \sum_{j=0}^{N-m-1} \sum_{p=0}^{N-m-1} \mathscr{E}(x_j x_p x_{j+m} x_{p+m}) - r_m^2$$

$$= \frac{1}{(N-m)^2} \sum_{j=0}^{N-m-1} \sum_{p=0}^{N-m-1}$$

$$\times \left\{ \mathscr{E}(x_j x_p)\mathscr{E}(x_{j+m} x_{p+m}) + \mathscr{E}(x_j x_{j+m})\mathscr{E}(x_p x_{p+m}) \right.$$

$$\left. + \mathscr{E}(x_j x_{p+m})\mathscr{E}(x_p x_{j+m}) \right\} - r_m^2$$

$$= \frac{1}{(N-m)^2} \sum_{j=0}^{N-m-1} \sum_{p=0}^{N-m-1} \left\{ r_{p-j}^2 + r_{p-j+m} r_{p-j-m} \right\} \qquad (7.15)$$

Letting $l = p - j$ in the above equation yields

$$V(\rho_m) = \frac{1}{(N-m)^2} \sum_{l=-(N-m-1)}^{N-m-1} (N - m - |l|)\left\{ r_l^2 + r_{l+m} r_{l-m} \right\}$$

For large N the variance of the estimate is approximately

$$V(\rho_m) \simeq \frac{1}{N} \sum_{l=-\infty}^{\infty} \left\{ r_l^2 + r_{l+m} r_{l-m} \right\} \qquad (7.16)$$

The result in (7.16) has been obtained by Jenkins and Watts[13] and indicates that ρ_m is a consistent estimator.

The estimate of r_m, given by (7.7), assumes that $N > m$. If the record length, N, is fixed, and the correlation lag m increases and approaches N, then the variance becomes large. In this case the estimate is poor and not useful. An alternate estimate for the correlation of a real random process x_j is denoted by $\tilde{\rho}_m$ and is defined by

$$\tilde{\rho}_m \equiv \frac{1}{N} \sum_{j=0}^{N-m-1} x_j x_{j+m}, \qquad m = 0, 1, \ldots, N - 1 \qquad (7.17)$$

Note that the negative lag estimates are determined by $\tilde{\rho}_{-m} = \tilde{\rho}_m$. For large N

the estimate given by (7.17) has a variance equal to (7.16) and is therefore a consistent estimator. From Eq. (7.17) it can be seen that this estimator is biased:

$$\mathscr{E}(\tilde{\rho}_m) = \frac{1}{N} \sum_{j=0}^{N-m-1} \mathscr{E}(x_j x_{j+m})$$

$$= \left(\frac{N-m}{N}\right) r_m, \qquad m = 0, \ldots, N-1 \qquad (7.18)$$

Since $\tilde{\rho}_{-m} = \tilde{\rho}_m$, then

$$\mathscr{E}(\tilde{\rho}_m) = \left(\frac{N-|m|}{N}\right) r_m, \qquad m = 0, \pm 1, \ldots, \pm(N-1)$$

Use of the biased estimate given above provides an estimate with a bias and variance that decreases with increasing N. As a result, the biased estimator may be preferred in certain applications. Jenkins and Watts[13] have computed the mean square error for the biased and unbiased estimators in the case of a specific first-order autoregressive process. Using this example it was demonstrated that the mean square error (MSE) for the biased estimator is less than that of the unbiased estimate. It was also shown that the MSE decreased with increasing record length in the biased estimator case and increased with increasing record length in the unbiased estimator case. Jenkins and Watts conclude that this trend is a general result, even though it was demonstrated by means of an example.

The results presented thus far indicate that for specific applications the selected estimate often results in a compromise between the allowed bias and variance even in the case where the estimators are consistent and asymptotically unbiased. This trade-off also applies to estimates of the power spectral density. In fact, autocorrelation estimates which are consistent and asymptotically unbiased may not produce good power spectral density estimates when the Fourier transform of the autocorrelation estimate is obtained. In general, the Fourier transform must include smoothing by use of a window function to achieve good power spectral density estimates.[12] Procedures for estimating the power spectral density are now described.

7.2.1 Power Spectrum Estimation Using Periodogram Methods

In Figure 7-1 the relationship between the power spectral density and autocorrelations function is defined to be a Fourier transform pair. The power spectral density, $P(f)$, associated with a discrete autocorrelation, r_m, is defined to be the discrete Fourier transform (DFT),[12] that is,

$$P(f) = \sum_{m=-\infty}^{\infty} r_m e^{-j2\pi mfT_s} \qquad (7.19)$$

where T_s is the sampling period. An estimate of the spectrum, denoted by $\hat{P}(f)$, can be obtained by taking the DFT of an estimate of the autocorrelation function:

$$\hat{P}(f) = \sum_{m=-\infty}^{\infty} \rho_m e^{-j2\pi mfT_s} \tag{7.20}$$

Let the DFT of the samples x_m be defined by $X(f)$. Then

$$X(f) = \sum_{m=-\infty}^{\infty} x_m e^{-j2\pi mfT_s}$$

For a record length N that is finite, the samples x_m for $m = 0,\ldots, N-1$ can be used to obtain a DFT. This DFT is obtained by defining a rectangular function, w_m, as

$$w_m = \begin{cases} 1, & m = 0, 1, \ldots, N-1 \\ 0, & \text{otherwise} \end{cases} \tag{7.21}$$

Then, a finite sequence of N samples, x_m^N can be represented as

$$x_m^N = \begin{cases} x_m, & m = 0, 1, \ldots, N-1 \\ 0, & \text{otherwise} \end{cases}$$

$$= x_m w_m \tag{7.22}$$

The DFT of the sequence, x_m^N, is denoted by $X_N(f)$:

$$X_N(f) = \sum_{m=0}^{N-1} x_m e^{-j2\pi mfT_s} \tag{7.23}$$

The estimate of the autocorrelation function used here is the biased estimate given by

$$\tilde{\rho}_m = \frac{1}{N} \sum_{l=0}^{N-m-1} x_l^* x_{l+m}, \qquad m = 0, 1, \ldots, N-1 \tag{7.24}$$

Rewriting this estimator using the finite data sequence results in

$$\tilde{\rho}_m = \frac{1}{N} \sum_{l=-\infty}^{\infty} x_l^{N*} x_{l+m}^N \qquad m = 0, \pm 1, \ldots$$

The estimated power spectrum can now be obtained by substituting the above equation in (7.20):

$$\hat{P}(f) = \frac{1}{N} \sum_{l=-\infty}^{\infty} \sum_{m=-\infty}^{\infty} x_l^{N*} x_{l+m}^{N} e^{-j2\pi m f T_s}$$

$$= \frac{1}{N} \sum_{l=-\infty}^{\infty} x_l^{N*} e^{j2\pi l f T_s} \sum_{m=-\infty}^{\infty} x_{l+m}^{N} e^{-j2\pi(l+m) f T_s}$$

$$= \frac{1}{N} |X_N(f)|^2 \tag{7.25}$$

The estimate $\hat{P}(f)$ given by (7.25) is referred to as the periodogram since it was first used to investigate periodicities[1] that might exist in a data sequence.

Now the statistical properties of the periodogram are determined (see Appendix Section A.1.16). The average of the estimate, $\hat{P}(f)$, can be obtained from (7.20):

$$\mathcal{E}(\hat{P}(f)) = \sum_{m=-\infty}^{\infty} \mathcal{E}(\rho_m) e^{-j2\pi m f T_s} \tag{7.26}$$

If the biased estimate given by (7.24) is used in the above equation, use of $\tilde{\rho}_{-m} = \tilde{\rho}_m^*$ results in

$$\mathcal{E}(\hat{P}(f)) = \sum_{m=-(N-1)}^{(N-1)} r_m \left(\frac{N - |m|}{N} \right) e^{-j2\pi m f T_s} \tag{7.27}$$

Define the triangular function, t_m, by

$$t_m = \begin{cases} 1 - \dfrac{|m|}{N}, & |m| \le (N-1) \\ 0, & |m| > (N-1) \end{cases} \tag{7.28}$$

The result in (7.27) can now be expressed as

$$\mathcal{E}(\hat{P}(f)) = \sum_{m=-\infty}^{\infty} r_m t_m e^{-j2\pi m f T_s} \tag{7.29}$$

Comparing (7.19) and (7.29), it can be seen that the estimate $\hat{P}(f)$ is a biased estimate of $P(f)$ as a result of the triangle function t_m. Now consider the autocorrelation function estimate defined by

$$\rho_m = \begin{cases} \dfrac{1}{N-m} \sum_{l=0}^{N-m-1} x_l^* x_{l+m}, & m = 0, 1, \dots, (N-1) \\ 0, & m > N-1 \end{cases} \tag{7.30}$$

where $\rho_{-m} = \rho_m^*$ for negative lag estimates.

Substituting (7.30) into (7.26) results in

$$\mathscr{E}(\hat{P}(f)) = \sum_{m=-\infty}^{\infty} r_m w_m e^{-j2\pi mfT_s} \tag{7.31}$$

Comparing Eqs. (7.31) and (7.19), it can be seen that the periodogram estimate is a biased estimate.

Further insight can be obtained with regard to Eqs. (7.29) and (7.31) by recalling that multiplication of two time domain sequences is equivalent to convolution in the frequency domain.[15] Thus an equivalent relationship for (7.29) is given by

$$\mathscr{E}(\hat{P}(f)) = T_s \int_{(-1/2T_s)}^{(1/2T_s)} P(\eta) T(f-\eta) \, d\eta \tag{7.32}$$

where

$$T(f) = \frac{1}{N}\left(\frac{\sin N\pi f}{\sin \pi f}\right)^2 \tag{7.33}$$

For large N the function $T(f)$ approaches an impulse function so that

$$\mathscr{E}(\hat{P}(f)) \simeq P(f) \qquad \text{for } N \text{ large}$$

Thus the estimate is asymptotically unbiased.

Unfortunately, the variance of the periodogram estimate does not become small for large N, that is, the estimate is not consistent (see Appendix Section A.1.16.2). For example, for a Gaussian process, the variance approaches the square of the power spectrum for large N,[15] that is,

$$V(\hat{P}(f)) \simeq c_0 P^2(f)$$

where c_0 is a constant that depends on the window function used in the estimate, (for example, a triangular function). Thus this estimate can produce markedly different estimates of the spectrum for different sequences extracted from the same random process even for large N. To avoid this undesirable situation, modifications of the periodogram estimate have been introduced that produce better results.

One approach to reduce the variance of the estimates utilizes averages of a number of different periodogram estimates. This method is referred to as Bartlett's procedure.[12,15] For K statistically independent identically distributed periodograms the variance of the averaged estimate would be equal to $1/K$ of the individual variances. In general, the statistical independence assumption cannot be justified so that this variance reduction is not completely achieved. However, for a large number of samples per periodogram, the correlation between adjacent segments of data is small and the variance reduction assuming statistical independence is approximately attained. An alternate procedure for reducing the variance is to introduce statistical dependence by overlapping the data segments. In this case, overlapping the data segments by a significant amount, for instance, 50 percent, results in a greater number of periodograms which can be averaged while using the same amount of data as that available in the nonoverlapped case. Studies indicate that a greater variance reduction can be achieved with the overlapped case, even though statistical dependence is introduced.[16]

An unfortunate result of averaging the periodograms is that the bias of the averaged periodogram estimate is greater than the bias of the periodogram estimate formed from all the data taken together. To demonstrate this condition, let $\hat{P}_i(f)$ for $i = 1, \ldots, K$ represent K nonoverlapping contiguous periodograms each obtained from J samples with $N = JK$ (see Figure 7-2). The averaged spectrum estimate, $\overline{P}(f)$, can then be defined as

$$\overline{P}(f) = \frac{1}{K} \sum_{i=1}^{K} \hat{P}_i(f) \tag{7.34}$$

The mean of this estimate is given by

$$\mathscr{E}(\overline{P}(f)) = \frac{1}{K} \sum_{i=1}^{K} \mathscr{E}(\hat{P}_i(f))$$

$$= \mathscr{E}(\hat{P}_i(f)) \tag{7.35}$$

Note that the average value in (7.35) can be computed from (7.32) and (7.33) where N is replaced by J. If J is used in the function in (7.33) instead of N with $J < N$, a main lobe results whose width is substantially greater, thereby producing a greater bias when (7.32) is evaluated. Thus spectral resolution is degraded as a direct result of the increased bias.

In summary, a trade-off exists between the bias (or, equivalently, spectrum resolution) and variance of the estimate. For a fixed record length, use of more periodograms in the average reduces the variance but increases the bias.

Another approach for improving spectral estimates utilizes window functions in conjunction with averaging the periodograms.[12,16] Before this approach is described, a brief discussion of window functions is required.

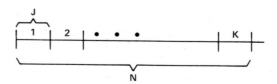

Figure 7-2.   Nonoverlapping contiguous data segments $N = JK$.

A summary of commonly used window functions is presented in References 12, 17, and 18. Two of these window functions, the triangular function, T_m (also known as the Bartlett window), and the rectangular function, w_m, occurred earlier as a direct result of using a finite data sequence. The window functions are used to taper the data (either in time or in frequency), and thereby adjust the bias and variance of the estimate. For example, the triangular window has a wider main lobe with a smaller stopband attenuation than the rectangular window. As a result, the triangular window produces a larger bias but a smaller variance of the spectral estimate than the rectangular window.

The Welch method[16] for power spectral estimation is now described. This method is a traditional power spectral estimation technique and will be used subsequently for comparison with autoregressive power spectral estimation methods. The Welch method divides a record into either overlapping or nonoverlapping segments. Each data segment is windowed before computing its periodogram. The resulting periodograms are then averaged using the Bartlett procedure. It can then be shown that the resulting estimate is asymptotically unbiased and consistent:[19]

$$E[\hat{P}(f)] \simeq P(f)$$

$$V[\hat{P}(f)] \simeq \frac{c_1}{K} P^2(f)$$

where c_1 is a constant that depends on the window function and the amount of overlap in the data. A mathematical description of this method is now provided.

Suppose segments of signal samples $\{x_i\}$ denoted by $\{x_{li}, l = 1, \ldots, K\}$ are taken to be of length J with the starting points of these segments I units apart. The parameter I allows for possible segment overlap. Specifically, the segments x_{li} are defined as follows:

$$x_{li} = x_{i+(l-1)I}, \quad i = 0, 1, \ldots, J - 1, \quad l = 1, 2, \ldots, k \quad (7.36)$$

J is generally selected to be a power of two for convenience in computing the DFT using fast Fourier transform (FFT) methods. This restriction may require that the sequence be appended with zeros to make it a power of two. The power spectral estimate, P_k, for the kth frequency component, f_k, can then be computed as

$$\hat{P}_k = \frac{J}{KU} \sum_{l=1}^{K} |G_l(k)|^2, \quad k = 0, 1, \ldots, J - 1 \quad (7.37)$$

where

$$G_l(k) = \frac{1}{J} \sum_{i=0}^{J-1} x_{li} w_i \exp\left[-j\frac{2\pi}{J} ik\right], \quad k = 0, 1, \ldots, J - 1 \quad (7.38)$$

and

$$U = \frac{1}{J} \sum_{i=0}^{J-1} w_i^2 \tag{7.39}$$

$\{w_i, i = 0, 1, \ldots, J - 1\}$ is a data window used in computing the lth periodogram $G_l(k)$. The resolution of the FFT technique (assuming no data window) is $\Delta f = f_J/J$ where f_J is the Nyquist frequency. From the discussion above it can be seen that various data windows $\{w_i\}$ can be chosen to reduce sidelobe leakage in the power spectral estimate. The selection of the window type should be a good compromise between the reduction in the sidelobes and the decreased resolution resulting from use of the window. Since the resolution decreases when a window is used, the smallest spectral separation between two spectral lines is bounded by Δf. Therefore, two spectral lines spaced at less than Δf apart appear as a single peak in the estimated spectrum.

7.3   AUTOREGRESSIVE METHOD

When the modified periodogram method, that is, the Welch method, described above, is applied to a signal with strong harmonic lines, relatively poor spectral resolution results. A marked improvement in spectral resolution can be attained by use of the autoregressive (AR) method,[5] based on the Durbin algorithm described in Chapter 3. The AR method has been applied successfully in several disciplines such as seismic exploration, geomagnetism, solar spot investigation, etc., and has led to good results.[20,21] Two AR methods are discussed in the next section, following a brief description of the approach.

7.3.1   AR Technique

Several advantages of AR techniques over traditional methods exist. In particular, for the maximum entropy method, there is no need to either know or estimate the autocorrelation function. No assumption is made about unavailable data being either zero or periodic. In fact, maximum entropy techniques assume maximum uncertainty about unknown data. In addition, AR algorithms are computationally efficient and result in high spectrum resolution.

The general approach is depicted in Figure 7-3 and involves a two-step process. First, an AR prediction algorithm is used to compute the $M + 1$ prediction error filter coefficients $\{a_i\}$ for $i = 0, \ldots, M$ (see Durbin algorithm in Chapter 3). Subsequently, $A(f)$, the DFT of the prediction coefficients, is computed:

$$A(f) = \sum_{i=0}^{M} a_i e^{-j2\pi i f T_s} \tag{7.40}$$

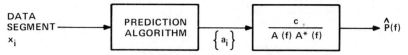

Figure 7-3. AR technique.

The power spectrum is then estimated using

$$\hat{P}(f) = \frac{c}{A(f)A^*(f)} \tag{7.41}$$

where c is a constant. Note that Eq. (7.41) contains only poles so that the model is termed an all-pole model. Improved estimates of the power spectrum can be obtained by introducing zeros in the model. These models are referred to as autoregressive moving average (ARMA) models.

A more explicit representation of the spectrum in terms of an all-pole model can be obtained by substituting (7.40) into (7.41) using $a_0 \equiv 1$:

$$\hat{P}(f) = \frac{c}{|1 + \sum_{i=1}^{M} a_i e^{-j2\pi i f T_s}|^2}$$

The above equation is a direct result of the autoregressive model which expresses the current estimate in terms of linear combination of prior samples (see Section 3.1.2):

$$\hat{x}_k = -\sum_{l=1}^{M} a_l x_{k-l}$$

A derivation of this result has been obtained by Burg in his development of the maximum entropy method (MEM).[1,9] Burg applied the concept of maximizing the entropy in estimating the spectrum and developed a procedure for determining the prediction error filter coefficients directly from only known data samples. A brief discussion of the MEM principle is now given.

Let $\{x_j, j = 0, 1, \ldots, N - 1\}$ be a sequence of time samples of a random process that is assumed to have a power spectrum $P(f)$ and autocorrelation function samples denoted by $\{r_m, m = -M, \ldots, M\}$, where $r_{-m} = r_m$ and the number of autocorrelation lags L is assumed to be equal to M. If the sequence $\{x_m\}$ of samples is taken from a stationary Gaussian process that is band-limited to $f \le f_N$, the entropy rate, H, of the process is defined by[5,22]

$$H = \frac{1}{4f_N} \int_{-f_N}^{f_N} \ln P(f) \, df \tag{7.42}$$

MEM selects the real positive function, $P(f)$, that maximizes equation (7.42) subject to the constraint equations

$$r_m = \int_{-f_N}^{f_N} P(f) \exp(j2\pi f m T_s) \, df, \qquad m = -M, \ldots, M \tag{7.43}$$

The unique solution is given by

$$\hat{P}(f) = \frac{P_{M_{\min}}}{A(f)A^*(f)} \tag{7.44}$$

where

$$A(f) = \sum_{n=0}^{M} a_n \exp[j2\pi fnT_s] \tag{7.45}$$

and (see Eq. (3.15))

$$\sum_{n=0}^{M} a_n r_{m-n} = P_{M_{\min}} \delta(m) \tag{7.46}$$

The minimum average power in the error is $P_{M_{\min}}$.

Equation (7.44) implies that maximizing the entropy results in an autoregressive model of the process. The function $A(f)$ can be identified as the Fourier transform of a set of forward prediction error filter coefficients. In this instance, the autoregressive model is given by (see Eq. (3.133))

$$e_k^f = \sum_{n=0}^{M} a_n x_{k-n}, \qquad k = 0, 1, \ldots, N - M - 1 \tag{7.47}$$

where $a_0, a_1, \ldots, a_M$, are the forward prediction error filter coefficients with $a_0 \equiv 1$ and e_k^f is the present forward prediction error at time kT_s. Similarly, $A^*(f)$ can be interpreted as the Fourier transform of a set of backward prediction error filter coefficients. A backward prediction error, e_k^b, at time kT_s can then be expressed as (see Eq. (3.134))

$$e_k^b = \sum_{n=0}^{M} a_n^* x_{k-M+n}, \qquad k = 0, 1, \ldots, N - M - 1 \tag{7.48}$$

If the process is stationary, the covariance matrix in Eq. (7.46) is Toeplitz, and a recursive relationship of the filter coefficient is given by (see Eq. (3.128))

$$a_{M,p} = a_{M-1,p} + a_{M,M} a_{M-1,M-p}^*, \qquad p = 1, 2, \ldots, M - 1 \tag{7.49}$$

where $a_{M,p}$ is the pth coefficient in the Mth iteration. Expressing Eqs. (7.47) and (7.48) in iterative form by use of Eq. (7.49) results in (see Eqs. (3.145) and (3.146))

$$e_{k,M}^f = e_{k,M-1}^f + a_{M,M} e_{k-1,M-1}^b \tag{7.50}$$

and

$$e^b_{k,M} = e^b_{k-1,M-1} + a^*_{M,M} e^f_{k,M-1} \qquad (7.51)$$

where $e^f_{k,M}$ and $e^b_{k,M}$ are the forward and backward prediction errors, respectively, at the Mth iteration. Minimizing the average sum of the forward and backward error power with respect to $a_{M,M}$ yields (see Eq. (3.155))

$$a_{M,M} = \frac{-2\sum_{k=0}^{N-M-1} e^f_{k,M-1} e^{b*}_{k-1,M-1}}{\sum_{k=0}^{N-M-1}\left[\left|e^f_{k,M-1}\right|^2 + \left|e^b_{k-1,M-1}\right|^2\right]} \qquad (7.52)$$

Note that the range for the index k results from the definitions in (7.47) and (7.48) and does not correspond to the range of k used in chapter 3. The minimum average error power P_M can then be computed approximately from Eq. (7.49) by ignoring end effects according to[23] (see 3.31)

$$P_{M_{\min}} = P_{M-1_{\min}}(1 - |a_{M,M}|^2) \qquad (7.53)$$

where the initial average error power P_0 for N samples is defined as

$$P_0 = \frac{1}{N} \sum_{k=-M}^{N-M-1} |x_k|^2 \qquad (7.54)$$

Using (7.52) and (7.53) the estimated power spectrum $P(f)$ is computed according to Eqs. (7.44) and (7.45), where $a_0 = 1$, and $a_n, n = 1, \ldots, M$, is replaced by $a_{M,p}$ ($p = 1, 2, \ldots, M$).

7.3.2 Durbin Method

The Durbin algorithm (DA) is an autoregressive scheme that is based on a linear prediction of the input time sample x_k from an estimate, $\hat{x}_k$. Minimization of the mean square error between x_k and $\hat{x}_k$ yields the Yule-Walker equations:[24]

$$\sum_{n=1}^{M} b_n r_{k-n} = r_k, \qquad k = 1, 2, \ldots, M \qquad (7.55)$$

where $\{r_k\}$, $k = 1, \ldots, M$, are the autocorrelation coefficients, and $\{b_n\}$ is the set of prediction coefficients (see Eq. (3.10)). The solution of Eq. (7.55) requires autocorrelation function estimates, ρ'_k, $k = 1, \ldots, M$, obtained by assuming the input data samples have zero mean and are zero outside the observation interval. The estimates can be computed as

$$\rho'_k = \frac{1}{N} \sum_{l=0}^{N-k-1} x_l x^*_{k-l}, \qquad k = 0, 1, \ldots, M \qquad (7.56)$$

and are used in Eq. (7.55) in place of r_k. For a stationary process, the covariance matrix $\{r_k\}$ is Toeplitz, so that the coefficients b_n are recursively computed using the Durbin algorithm (see Section 3.1.4):

$$b_{M,p} = b_{M-1,p} - b_{M,M}b^*_{M-1,M-p}, \qquad p = 1, 2, \ldots, M-1 \quad (7.57)$$

where

$$b_{M,M} = \frac{A}{B}$$

$$A = \rho'_M - \sum_{k=1}^{M-1} b_{M-1,k}\rho'_{M-k} \qquad (7.58)$$

$$B = \rho'_0 - \sum_{k=1}^{M-1} b^*_{M-1,k}\rho'_k \qquad (7.59)$$

The power spectral estimate is then obtained using Eqs. (7.44) and (7.45), with $b_n = -a_n$ and $a_0 = 1$.

7.3.3   Selection of the Order of the Prediction Filter

The autoregressive methods require careful selection of the order of the prediction filter. Too small an order results in a highly smoothed estimate, whereas too large an order causes spurious detail to appear in the estimated spectrum. Selection of the order of the filter is accomplished by means of Akaike's minimization[5] of the final prediction error, FPE(M, N) given by

$$\text{FPE}(M, N) = \frac{N + M + 1}{N - M - 1}P_{M_{\min}} \qquad (7.60)$$

The order, M, is obtained by determining the minimum of FPE(M, N). Generally, the order of the filter should be greater than the number of spectral lines, especially in the case of closely spaced spectral lines, in order to guarantee satisfactory spectral resolution.

Asymptotically, the mean and variance of power spectral estimates of an autoregressive process are given by[19]

$$E[\hat{P}(f)] = P(f)$$

$$\text{Var}[\hat{P}(f)] = \frac{2}{K}P^2(f) \qquad (7.61)$$

where K is the number of degrees of freedom. K is related to the order M of the autoregressive process and the number of samples, N, by $K = N/M$.

Therefore, the spectral estimate is asymptotically unbiased with a variance that depends on the order of the filter.

Kaveh and Cooper[19] have shown that the autoregressive estimator is asymptotically similar to the estimate obtained by means of the Welch method using a rectangular window. In summary, power spectral estimates based on either the Welch or autoregressive techniques have the following statistical properties:[19]

1. The estimates are asymptotically unbiased.
2. The estimates are asymptotically normally distributed.
3. The variance of the estimate is $(2/K)P^2(f)$ for $f \neq 0$ or $f \neq 1/2$ and is $(4/K)P^2(f)$ for $f = 0$ or $f = 1/2$ for a sampling interval normalized to unity.
4. The covariance of the estimate, $COV(P(f_1), P(f_2))$, is zero for $f_1 \neq f_2$.

Example Comparison of Spectral Estimators

In this example several spectral estimation techniques, described above, are compared using a known autoregressive process. The techniques included in the comparison are MEM, DA and the periodogram method. The example that

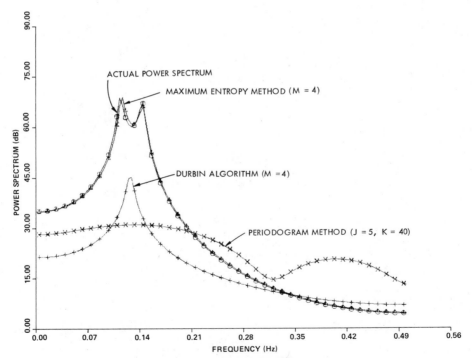

Figure 7-4. Power spectrum comparison.

is investigated is the one studied by Ulrych and Bishop[5] which is a fourth-order process:

$$v_i = x_i - 2.7607x_{i-1} + 3.8106x_{i-2} - 2.6535x_{i-3} + 0.9238x_{i-4}$$

where v_i is a random noise error signal called the innovations process. In all cases described here, the innovations process, v_i, is assumed to be a zero mean uniform pseudorandom process. The first 200 points of each realization are discarded to allow initial transients to decay.

In Figures 7-4 and 7-5 the total number of samples is $N = 200$. Thus, 200-point realizations of the autoregressive process have been used for spectral estimation. In MEM and the DA, $N = 200$ samples of data are used to obtain M filter coefficients $\{a_n, n = 0, 1, 2, \ldots, M - 1\}$. A $Q = 256$-point FFT is used to compute $A(f)$ with $a_n = 0$ for $n \geq M$. In the periodogram method, a window specified as

$$w_i = 1 - \left(\frac{i - (J - 1)/2}{(J + 1)/2} \right)^2, \qquad i = 0, 1, \ldots, J - 1$$

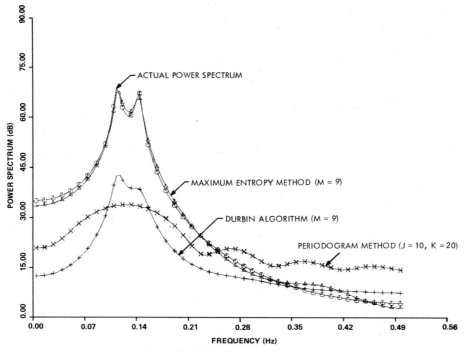

Figure 7-5. Power spectrum comparison.

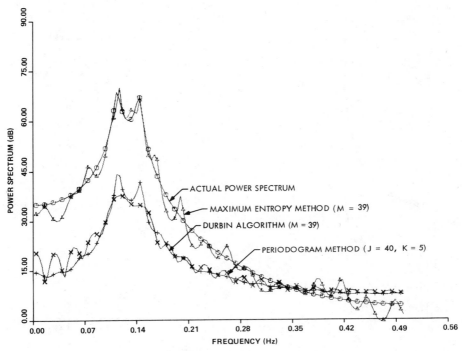

Figure 7-6. Power spectrum comparison.

is applied. Segments of x_i are taken to be of length J so that $K = 200/J$ nonoverlapping modified periodograms can be computed. A 256-point FFT is used to compute a periodogram with $x_i = 0$ for $i \geq J$. The power spectrum is then computed by averaging the $200/J$ modified periodograms. In order to make the spectra nonoverlapping, the spectra estimated by the Durbin algorithm and periodogram method are shifted down 20 dB. When the order of the filter is the same as the order of the actual process (see Figure 7-4), the MEM spectral estimate accurately fits the actual spectra, while the Durbin algorithm and periodogram method exhibit relatively poor fits. Increasing the order of the filter of the MEM increases the spurious lobes with no increase in spectral resolution (see Figures 7-5 and 7-6). In the Durbin algorithm, the best results are obtained by use of a filter of higher order than that of the autoregressive process. Once again, too high an order can result in spurious lobes. To achieve satisfactory results with either the MEM or DA, it is apparent that selection of the order of the filter by means of an acceptable criterion, (for example, Akaike), is essential. Figures 7-4, 7-5, and 7-6 indicate that the spectral resolution exhibited by the MEM is superior to either the DA or the periodogram method. The DA in Figure 7-6 displays better resolution than the smoothed periodogram obtained by averaging five segments of 40 samples each. If the periodogram is computed from a single segment of 200

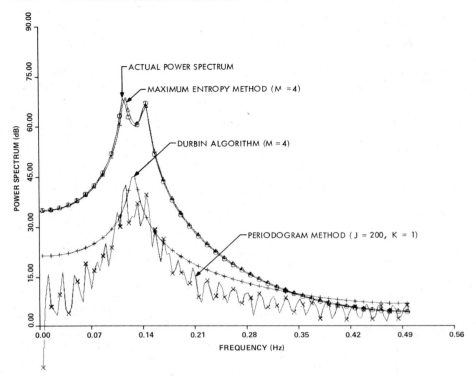

Figure 7-7. Power spectrum comparison.

samples, as in Figure 7-7, several spurious peaks occur, so that identification of the correct peaks is difficult. For the example shown, only the MEM satisfactorily resolves the peaks that occur in the actual spectrum. Consequently, the MEM is particularly valuable for spectral estimation of short records of data.

The improvement in spectral estimation performance of the MEM over the DA and the periodogram methods in this example can be explained heuristically as follows. In the MEM, no assumption about unknown data is made. In the DA, the autocorrelation function is usually estimated by Eq. (7.56), which assumes that the unknown data is zero. In the periodogram method, the data is assumed to be periodic outside the observation interval. By allowing the data to assume its maximum uncertainty (i.e., maximum entropy) outside of the observation interval, a best estimate is obtained. Note that both AR models do a better job of estimating the actual spectrum than the periodogram method. This result can be attributed to the fact that an autoregressive process can be better represented by an autoregressive model.

The performance improvement of the MEM occurs at the sacrifice of increased computational complexity. In rough terms, the MEM has about twice the computational complexity of the DA, and about an order of magnitude less than the periodogram method.

It is interesting to compare the quality of the estimate for this example. Thus, define the power spectral method for the ith data record using the MEM, DA and the periodogram estimate computed from the FFT algorithm by $\hat{P}_{i_{\text{MEM}}}(f)$, $\hat{P}_{i_{\text{DA}}}(f)$ and $\hat{P}_{i_{\text{FFT}}}(f)$, respectively. If the actual power spectrum for the ith data record is denoted by $P_i(f)$, the frequency averaged square error values in the power spectrum for the MEM, DA and FFT, denoted by ε_{MEM}, ε_{DA} and ε_{FFT}, respectively, can be represented by

$$\varepsilon_{\text{MEM}} = \int_{\text{BW}} \left[\hat{P}_{i_{\text{MEM}}}(f) - P_i(f) \right]^2 df$$

$$\varepsilon_{\text{DA}} = \int_{\text{BW}} \left[\hat{P}_{i_{\text{DA}}}(f) - P_i(f) \right]^2 df$$

$$\varepsilon_{\text{FFT}} = \int_{\text{BW}} \left[\hat{P}_{i_{\text{FFT}}}(f) - P_i(f) \right]^2 df$$

where the integrations are performed over the bandwidth, BW, of the data. In the periodogram method a Q-point transform is used to compute the FFT. In the MEM or the DA, a Q-point transform is used to compute the spectrum from the prediction coefficients (see (7.44) and (7.45)). Table 7-1 illustrates the frequency-averaged square error for Q of 256 and 512. The smallest error is obtained using the MEM method.

Now numerical estimates of the mean and variance of the sample power spectrum are computed for this example. Using K data records, the mean and variance of the sample power spectrum can be (respectively) expressed by

$$U_k(f) = \frac{1}{K} \sum_{i=1}^{K} \hat{P}_{i_k}(f), \qquad k = 1, 2, 3$$

$$V_k(f) = \frac{1}{K} \sum_{i=1}^{K} \hat{P}_{i_k}^2(f) - U_k^2(f), \qquad k = 1, 2, 3$$

The index $k = 1$, 2 and 3 corresponds to the MEM, DA, and FFT algorithms, respectively. The normalized and frequency-averaged bias and variance can be

TABLE 7-1. FREQUENCY-AVERAGED SQUARE ERRORS FOR $Q = 256$ AND $Q = 512$, IN WATTS

Q	ε_{FFT}	ε_{DA}	ε_{MEM}
256	20.79	43.10	8.46
512	20.01	42.20	6.81

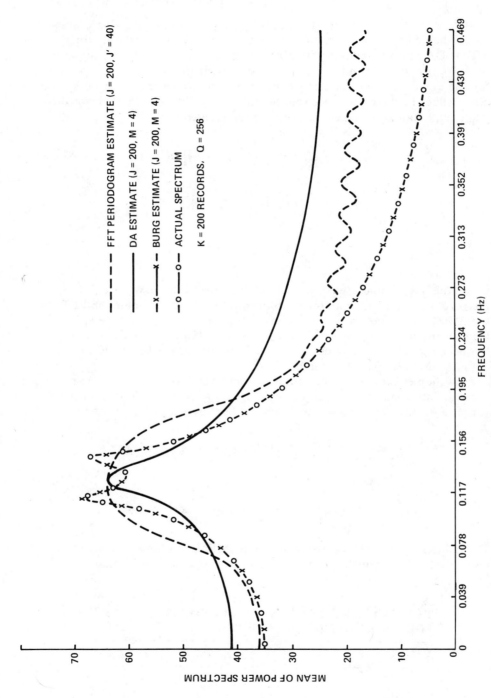

Figure 7-8. Mean of power spectral estimate.

FREQUENCY (Hz)

MEAN OF POWER SPECTRUM

FFT PERIODOGRAM ESTIMATE (J = 200, J' = 40)

DA ESTIMATE (J = 200, M = 4)

BURG ESTIMATE (J = 200, M = 4)

ACTUAL SPECTRUM

K = 200 RECORDS.   Q = 256

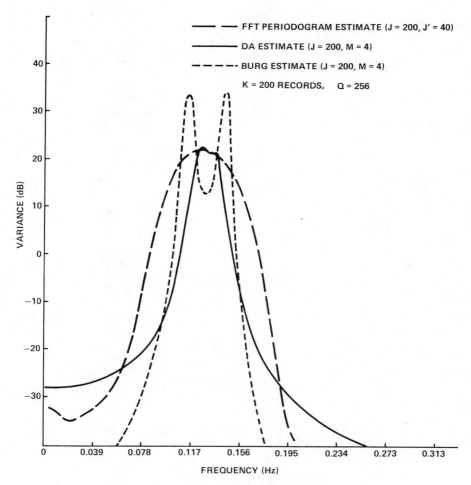

Figure 7-9. Variance of power spectral estimate.

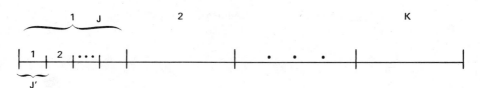

Figure 7-10. Nonoverlapping contiguous data segments.

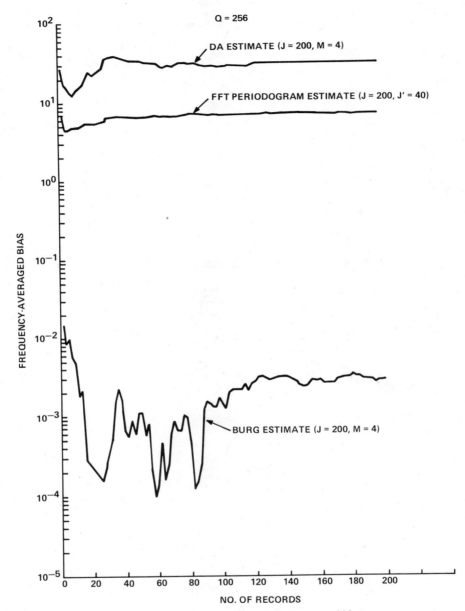

Figure 7-11. Normalized and frequency-averaged bias.

represented, respectively, by[19]

$$B_k = \frac{1}{\text{BW}} \int_{\text{BW}} \frac{U_k(f) - P(f)}{P(f)} \, df, \qquad k = 1, 2, 3$$

$$VA_k = \frac{1}{\text{BW}} \int_{\text{BW}} \frac{V_k(f)}{P^2(f)} \, df, \qquad k = 1, 2, 3$$

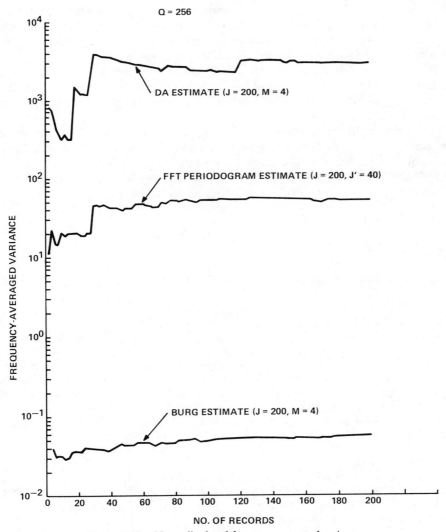

Figure 7-12. Normalized and frequency-averaged variance.

Results for the mean and variance of the power spectral estimate using $K = 200$ records are provided in Figures 7-8 and 7-9, respectivey. The relationship between principal parameters used in Figures 7-8 and 7-9 is shown in Figure 7-10. The length of each record, $J = 200$, and contiguous nonoverlapping periodograms, each with length $J' = 40$, are used. The FFT transform size, $Q = 256$, is used with $J' = 40$ and 216 zeros in the periodogram method. Then 1000 periodograms, equivalent to 5 periodograms per record for 200 records, are averaged. In the DA and MEM, $M = 4$ prediction filter coefficients (a_0 not included) and 251 zeros are used in a $Q = 256$ points FFT. Then 200 individual spectral estimates are averaged to produce the results in Figures 7-8 and 7-9. Note that the prediction filter coefficients are obtained from a segment with $J = 200$ samples. The order of the AR prediction filter is $M = 4$. In this example, the actual process and the process estimated using the MEM algorithm are virtually indistinguishable in mean, whereas the DA and FFT algorithms produce a mean that is markedly different from the actual process. Note that the variance of the MEM estimate is not necessarily smaller than the variance of the other estimates, particularly in the vicinity of the spectral peaks.

The normalized and frequency-averaged bias and variance are shown in Figures 7-11 and 7-12, respectively, and accumulatively computed as a function of the number of records. The parameters indicated on the figures are the same as those in Figures 7-8 and 7-9. Note that the frequency-averaged bias and variance are smallest for the MEM algorithm and largest for the DA.

REFERENCES

1. E. A. Robinson, A Historical Perspective of Spectrum Estimation, *Proc. IEEE 70*, 9, Sept. 1982, p. 885.

2. N. Wiener, *Extrapolation, Interpolation and Smoothing of Stationary Time Series with Engineering Applications*, MIT Press, Wiley, New York, 1949.

3. R. B. Blackman and J. W. Tukey, *The Measurement of Power Spectra from the Point of View of Communication Engineering*, Dover, New York, 1959.

4. J. W. Cooley and J. W. Tukey, An Algorithm for the Machine Calculation of Fourier Series, *Math. Comput. 19*, 1965, pp. 297–301.

5. T. J. Ulrych and T. N. Bishop, Maximum Entropy Spectral Analysis and Autoregressive Decomposition, *Rev. Geophysics 13*, 1975, pp. 183–200.

6. E. H. Satorius and J. D. Pack, Application of Least Square Lattice Algorithms to Adaptive Equalization, *IEEE Trans. Communications*, COM-29, Feb. 1981, pp. 136–142.

7. S. M. Kay and S. L. Marple Jr., Spectrum Analysis—A Modern Perspective, *Proc. IEEE*, *69*, 11, Nov. 1981, pp. 1380–1419.

8. S. L. Marple Jr., A New Autoregressive Spectrum Analysis Algorithm, *IEEE Trans. Acoust. Speech*, *Signal Process.*, ASSP-28, Aug. 1980, pp. 441–454.

9. J. P. Burg, Maximum Entropy Spectral Analysis, PhD Dissertation, Dept. Geophysics, Stanford Univ., Stanford, Calif., 1975.

10. A. B. Carlson, *Communications Systems*, McGraw-Hill, New York, 1968.

11. A. Papoulis, *Probability, Random Variables and Stochastic Processes*, McGraw-Hill, New York, 1965.

12. A. V. Oppenheim and R. W. Schafer, *Digital Signal Processing*, Prentice Hall, Englewood Cliffs, N.J., 1975.

13. G. M. Jenkins and D. G. Watts, *Spectral Analysis and Its Applications*, Holden-Day, San Francisco, 1968.

14. A. D. Whalen, *Detection of Signals in Noise*, Academic Press, New York, 1971, p. 97.

15. M. Schwartz and N. Shaw, *Signal Processing Discrete Spectral Analysis Detection and Estimation*, McGraw-Hill, New York, 1975.

16. P. D. Welch, The Use of Fast Fourier Transform for the Estimation of Power Spectra: A Method Based on Time Averaging Over Short Modified Periodograms, *IEEE Trans. Audio Electroacoustics AU-15*, June 1967, pp. 70–73.

17. L. W. Rabiner and B. Gold, *Theory and Application of Digital Signal Processing*, Prentice-Hall, Englewood Cliffs, N.J., 1975.

18. F. J. Harris, On the Use of Windows for Harmonic Analyses with the Discrete Fourier Transform, *Proc. IEEE 66*, 1, Jan. 1978, pp. 51–83.

19. M. Kaveh and G. R. Cooper, An Empirical Investigation of the Properties of the Autoregressive Spectral Estimator, *IEEE Trans. Information Theory IT-22*, 3, May 1976.

20. E. A. Robinson, *Multichannel Time Series Analysis with Digital Computer Programs*, Holden-Day, San Francisco, 1967.

21. E. A. Robinson, *Statistical Communication and Detection with Special Reference to Digital Data Processing of Radar and Seismic Signals*, Hafner, New York, 1967.

22. E. T. Jaynes, On the Rationale of Maximum Entropy Methods, *Proc. IEEE 70*, 9, Sept. 1982, pp. 939–952.

23. F. M. Hsu and A. A. Giordano, Line Tracking Using Autoregressive Spectral Estimates, *IEEE Trans. Acoustics, Speech and Signal Processing. ASSP-25*, 6, Dec. 1977, pp. 510–519.

24. G. Box and G. M. Jenkins, *Time Series Analysis, Forecasting and Control*, Holden-Day, Oakland, Calif., 1970.

chapter

$\underline{\hspace{3cm}}$

8

DIGITAL WHITENING IN SPREAD SPECTRUM (SS) COMMUNICATIONS

8.1 INTRODUCTION

This chapter describes digital whitening techniques for improving spread spectrum communications performance in the presence of narrowband jamming and interference. The techniques are based on least square algorithms described in part 1 and are applicable to pseudonoise (PN) (also referred to as direct sequence) spread spectrum modulations. First, an overview of spread spectrum communications is provided, including a description of PN spread spectrum modulation.[1,2,3,4] Subsequently, the model of a PN spread spectrum communications system with selected digital whitening algorithms is presented. Specific performance results are then provided to demonstrate improvements available by use of the least square algorithms.

8.2 OVERVIEW OF SPREAD SPECTRUM COMMUNICATIONS

Spread spectrum (SS) communications result when a signal is transmitted with a transmission bandwidth that is much wider than the information bandwidth of the signal. The band spreading is accomplished by a coded sequence which is independent of the data. This sequence can be used to produce a variety of SS modulations including pseudonoise (PN), frequency hopping (FH), time hopping (TH) or hybrids. PN modulation is used to replace each data symbol with a pseudorandom coded sequence that causes carrier phase transitions which occur more rapidly than the symbol rate. FH modulation is used to pseudorandomly shift the carrier frequency across the available bandwidth according to the coded sequence at a rate that can be faster, slower or equal to the symbol rate. TH modulation is used to produce time bursts of the signal at pseudorandom instants determined by the coded sequence. At the receiver the SS modulation is removed or despread by use of a correlation detector or matched filter that recovers the signal for subsequent data demodulation. This operation is accomplished by storing a replica of the coded sequence that is used to generate a synchronous reference of the SS modulation that is crosscorrelated with the received signal.

Ideally, the SS modulation simply introduces additional operations near the transmitter output and receiver input that are transparent to the data modulation and detection. A natural question is then, What benefit is introduced by the added complexity? The response to this question is derived from the multiplicity of SS applications including anti-jam (AJ) or anti-interference (AI) communications, low probability of intercept (LPI) communications, multiuser random access communications, high resolution ranging, multipath reduction, and many others. In AJ or AI communications, narrowband interference is reduced by spreading it across the transmitted signal bandwidth in the crosscorrelation operation. Figure 8-1 illustrates the SS concept. In Figure 8-1a the power spectral density of the narrowband interference and wideband desired signal are illustrated. Following PN correlation, the resulting spectral density, illustrated in Figure 8-1b, indicates that the interference is spread over the wide bandwidth and the signal is collapsed to the information bandwidth. Filtering the correlator output in the demodulator reduces the interference power by the ratio of the wide SS bandwidth, B_w, to the narrow information bandwidth,

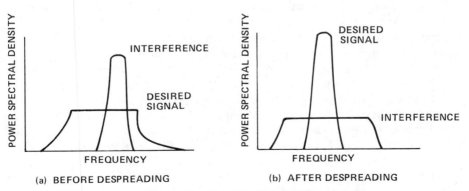

Figure 8-1. Concept of SS signalling.

B_N. This ratio, B_w/B_N, is referred to as the process gain and is a measure of the attainable improvement in interference suppression.

In low probability of intercept (LPI) communications, the transmitted signal resides below the receiver noise floor of the intercept receiver and cannot easily be detected without the use of the synchronous SS reference obtained from the coded sequence. In multiuser random access communications, users are distinguished by unique codes which are nearly orthogonal to other user codes, thereby permitting simultaneously use of the same bandwidth and transmission time. (This scheme is referred to as code division multiple access (CDMA) or spread spectrum multiple access (SSMA).) In high-resolution ranging, differences in the time of arrival of the SS signal which are on the order of the reciprocal of the SS bandwidth permit accurate location estimation of a target or platform. In the multipath channel, time delays which exceed the reciprocal of the SS bandwidth allow enhanced signal detection by either suppressing the multipath or combining the multipath components. (See Rake combining in references 4 and 5.)

In the remainder of this chapter the use of digital whitening in PN SS systems is described.[6,7,8] The improvement in performance resulting from digital whitening and the methods of implementing digital whitening algorithms are discussed.

8.3   AUGMENTATION OF PN SPREAD SPECTRUM COMMUNICATIONS WITH DIGITAL WHITENING

A heuristic description of how digital whitening improves the performance of PN spread spectrum systems is now given. The received signal consists of the sum of the SS transmitted signal, narrowband interference, and thermal receiver noise. The transmitted signal is obtained by modulating a random information signal with the spread spectrum waveform. As a result, this signal tends to have a flat spectral characteristic in a band of frequencies in the vicinity of the carrier frequency and in the short term is noncoherent over the SS signal band. The thermal receiver noise is assumed to have a flat spectral characteristic and is also noncoherent over the SS signal band. The interference, on the other hand, is generally assumed to be nonwhite over the SS signal band and is therefore coherent in the short term. This coherent component can be predicted by use of digital whitening techniques which can be implemented as a transversal filter. The filter is used to form an estimate of the received signal by utilizing linear prediction algorithms. Since the noncoherent portion of the received signal resulting from white signal components is not predictable, the estimate formed can actually be interpreted as an estimate of the interfering signal component. Once this estimate is obtained, an error signal can be computed by subtracting the estimate from the received signal. This error signal now consists of the sum of the filtered versions of the desired SS signal and thermal noise and residual interference. An improvement in performance then results from digital whitening when the narrowband inter-ference is suppressed while the correlation properties of the desired signal are degraded by a small amount. For strong narrowband interference in a poor signal-to-noise ratio condition, significant performance gains from digital whitening can be achieved.

Digital whitening is conceptually accomplished in two steps. In the first step the interfering signals are estimated and a set of coefficients are computed for use in the transversal filter. In the second step the filtering of the received signal is performed. For interference that is stationary over a specified interval, the filter coefficients can be computed and used for the entire interval. As the interference varies with time, new filter coefficients must be computed. In this way the interference is adaptively tracked and suppressed.

8.4   SS COMMUNICATION SYSTEM MODEL WITH DIGITAL WHITENING

A model of the SS communication system with and without digital whitening at the receiver is depicted in Figure 8-2, assuming binary PSK signaling. For simplicity all signals are assumed to be real in this chapter. Binary data transmission with symbol values $I_k = \pm 1$ and symbol duration T is assumed. A continuous-time representation of the transmitted signal, $s(t)$, is given by

$$s(t) = \sum_{k=-\infty}^{\infty} I_k w_k(t) \tag{8.1}$$

where

$$w_k(t) = \sum_{l=1}^{L} p_{kl} q(t - kT - lT_c)$$

In Eq. (8.1), p_{kl} for $l = 1, \ldots, L$, correspond to the PN code sequence symbols for the kth data symbol and have values ± 1. The function $q(t)$ is a pulse of duration T_c with a specified signal shaping. Each elemental waveform $p_{kl} q(t - kT - lT_c)$ is known as a chip, and the parameter L corresponds to the number of chips per data symbol, that is, $T = LT_c$. The parameter L measured in decibels is referred to as the processing gain. A discrete form of (8.1) can be obtained by sampling the transmitted signal once per chip:

$$s_i = \sum_{k=-\infty}^{\infty} I_k w_{ki}, \qquad i = 1, 2, \ldots \tag{8.2}$$

where

$$w_{ki} = \sum_{l=1}^{L} p_{kl} \delta(i - l - kL)$$

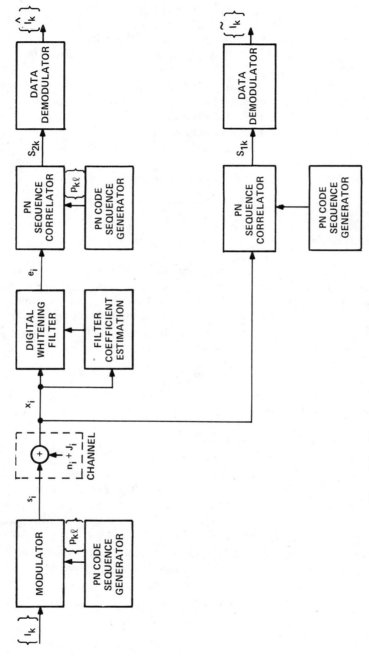

Figure 8-2. SS communication system model with and without digital whitening.

and

$$\delta(i - l - kL) = \begin{cases} 1, & i = l + kL \\ 0, & \text{otherwise} \end{cases}$$

The PN code sequence is designed to be an orthogonal sequence such that for each symbol k

$$\mathscr{E}(p_{kl}p_{kj}) = \begin{cases} 1, & \text{for } l = j \\ 0, & \text{otherwise} \end{cases}$$

Thus, if the correlation of the code symbol sequence is performed for the kth symbol in the absence of noise, the resulting output would ideally be

$$\mathscr{E}(w_{ki}w_{kj}) = \begin{cases} \sum_{l=1}^{L} p_{kl}^2, & \text{for } i = j \\ 0, & \text{otherwise} \end{cases}$$

$$= \begin{cases} L, & \text{for } i = j \\ 0, & \text{otherwise} \end{cases}$$

It is this correlation property which simultaneously extracts the desired signal from the background noise and interference, and suppresses the interference.

The received signal is obtained by adding zero mean white noise, n_i, with variance σ_n^2 and narrowband interference J_i:

$$x_i = s_i + n_i + J_i \tag{8.3}$$

The signal, noise, and interference are assumed to be mutually independent.

The interference model consists of a sum of M_J sinusoidal tones that have been sampled at the chip rate:

$$J_i = \sum_{m=1}^{M_J} A_m \cos(2\pi f_m i T_c + \phi_m) \tag{8.4}$$

where A_m, f_m, and ϕ_m are the amplitude, frequency and phase, respectively, of the mth tone. The interference model expressed by (8.4) can either be multi-tone interference or a discrete representation of a filtered narrowband Gaussian process. In the latter case, ϕ_m is assumed to be a random-phase angle distributed uniformly over the range $(-\pi, \pi)$, and the coefficient A_m can be related to the power spectrum of the interference according to

$$A_m = [2P_i(f_m)\Delta f]^{1/2}$$

where $f_m = m \, \Delta f$, $P_i(f_m)$ is the power spectrum of the interference at frequency f_m, and Δf is the frequency separation between two neighboring frequencies.

Note that the mean of the interference is zero as a result of the uniformly distributed random phase. The average power of the interference, denoted by σ_J^2, can be computed from (8.4):

$$\sigma_J^2 = \mathscr{E}\left(J_i^2 \right)$$

$$= \frac{1}{2} \sum_{m=1}^{M_J} A_m^2$$

In Figure 8-2 the received signal is processed in a conventional SS receiver to yield the output S_{1k} with corresponding symbol decisions $\tilde{I}_k$. In an SS receiver using digital whitening prior to correlation, the output S_{2k} and corresponding symbol decisions, $\hat{I}_k$ are obtained. Both receivers perform coherent detection and are assumed to be perfectly synchronized. With this model performance gains resulting from digital whitening can be quantitatively determined.

8.5   DIGITAL WHITENING ALGORITHMS

From the heuristic description provided in Section 8.3, an estimate of the interference, denoted by $\hat{J}_i$, can be obtained by predicting the coherent component in the received signal:

$$\hat{J}_i = \sum_{n=1}^{M} b_n x_{i-n} \tag{8.5}$$

where b_i for $i = 1, \ldots M$, are the linear prediction coefficients. If the estimate of the interference given by (8.5) is subtracted from the received signal in (8.3), an error signal is formed:

$$e_i = x_i - \sum_{n=1}^{M} b_n x_{i-n}$$

$$= \sum_{n=0}^{M} a_n x_{i-n} \tag{8.6}$$

where a_n for $n = 0, \ldots, M$, are the error prediction coefficients. Note that $a_n = -b_n$ for $n = 1, \ldots, M$, and $a_0 = 1$.

The prediction coefficients are computed by minimizing the mean square error associated with (8.6). Direct minimization of the MSE results in the Yule-Walker equations given by (3.10). The Yule-Walker equations can then be solved recursively by use of the Durbin algorithm or explicitly by inverting the received autocovariance matrix. In either case, the received autocorrelation function is assumed to be known, or estimates of the autocorrelation function are required. Estimation procedures for various window functions were described in Section 3.4.

An alternate formulation for computing the predictor coefficients uses forward and backward prediction to minimize the MSE. In this case, prediction coefficients are computed directly from the received data with no knowledge of the received autocorrelation function. Both the Burg and unconstrained least squares algorithms, described in Section 3.5 and Chapter 4, respectively, can be applied.

Selection of the appropriate least square estimation algorithm is based on whether or not the received signal is assumed to be stationary. In the nonstationary case, with a known or estimated autocorrelation function, Cholesky decomposition rather than the Durbin algorithm is used. In the nonstationary case with only the received data available, the unconstrained least square algorithm rather than the Burg algorithm is used.

A summary of common digital whitening algorithms is provided in Table 8-1. For completeness, algorithms are included which are not based on least squares.[9,10] Often a trade-off exists between excessive complexity and achievable performance with regard to the selection of an algorithm for a specific application.

The examples provided in this section are restricted to the stationary case, assuming least square techniques. Therefore, only the Durbin and Burg

TABLE 8-1.   DIGITAL WHITENING ALGORITHMS

Algorithm	Optimization criterion	Comments
Durbin (see Section 3.1)	Least square error	—Stationary signal —Known autocorrelation coefficients
Cholesky Decomposition (see Section 3.3)	Least square error	—Nonstationary signal —Known autocorrelation coefficients
Burg (see Section 3.5.3)	Least square error	—Stationary signal —Forward and backward prediction using only received data
Unconstrained Least Squares (see Chapter 4)	Least square error	—Nonstationary signal —Forward and backward prediction using only received data
Maximum Likelihood	Maximum likelihood	—Optimum strategy
Jamming Suppression Filter	Notch filter at jamming frequency	—Ad hoc —Sensitive to jammer frequency estimation

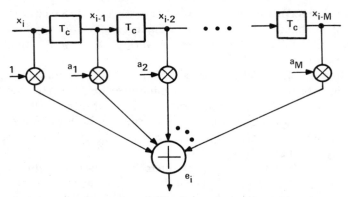

Figure 8-3.   Whitening filter.

algorithms are compared. The whitening filter in the stationary case can then be implemented as a transversal filter with taps spaced at the chip interval, T_c (see Figure 8-3). This filter is referred to as a Wiener filter in the Durbin algorithm case, and as a maximum entropy filter in the Burg case. The term "Wiener filter" refers to a filter whose coefficients are computed by solving the normal equations produced by minimizing the MSE. The term "maximum entropy filter" corresponds to a filter whose coefficients are based on an autoregressive model that results from maximizing the entropy of a process (see Chapter 7).

The concept of digital whitening is illustrated in Figure 8-4. The power spectrum of the received signal, $P_x(f)$, with a bandwidth B, illustrates the presence of narrowband jamming. The power spectrum of the whitening filter, $P_a(f)$, has notches in the vicinity of the jamming signal. As a result, the power spectrum of the error signal, $P_e(f)$, is ideally made white across the signal bandwidth by adjusting the digital whitening filter coefficients.

From a mathematical viewpoint, digital whitening requires an infinite tapped delay line. From Eq. (3.15) it can be seen that the autocorrelation

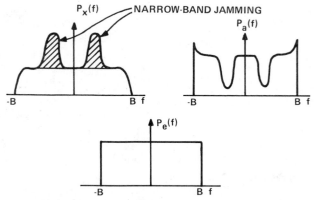

Figure 8-4.   Power spectra of $\{x\}$, $\{a\}$ and $\{e\}$.

function of the error series is $P_{M_{min}}$ for $m = 0$ and 0 for $m > 0$. Thus, the autocorrelation function is an impulse and the corresponding power spectrum is white noise which is flat across the signaling bandwidth.

8.6 DECISION VARIABLE FOR DIGITALLY WHITENED SS COMMUNICATIONS SYSTEM

The error signal given by (8.6) contains the desired spread spectrum signal. To complete the detection of the desired signal, this error signal must be correlated with the locally generated PN reference signal. If the digital whitening filter is properly designed, an improvement in SNR will occur when digital whitening is employed. This improvement in SNR will also produce an improvement in bit error rate (BER) performance.

Some additional notation is required. For the kth information symbol, the PN correlator output following digital whitening is labeled S_{2k}. The corresponding PN correlator output for no digital whitening is labeled S_{1k}. In the remaining development only the kth information symbol is considered, so that the subscript k can be dropped for notational simplicity. As a result, the decision variable S_2 at the decision instant can be written as

$$S_2 = \sum_{l=1}^{L} p_l e_l \tag{8.7}$$

The corresponding PN correlator output with no digital whitening filter can be expressed as:

$$S_1 = \sum_{l=1}^{L} p_l x_l \tag{8.8}$$

Subsequent performance results require computations of the moments of the signal outputs S_1 and S_2. The signal-to-noise ratio, SNR_i, can be defined in each case by use of the computed moments by

$$\mathrm{SNR}_i = \frac{\mathscr{E}^2(S_i)}{V(S_i)} \tag{8.9}$$

where $\mathscr{E}(S_i)$ and $V(S_i)$ correspond to the mean and variance of the output, S_i, respectively.

8.6.1 Moments of the PN Correlator Output without Digital Whitening

With no digital whitening, the PN correlator output can be written from (8.3) and (8.8) as

$$S_1 = \sum_{l=1}^{L} p_l(s_l + n_l + J_l) \tag{8.10}$$

Equation (8.10) can be expressed in a different form by recognizing that $s_l = I_k p_l$ for the kth symbol and the lth chip so that (see Eq. (8.2))

$$S_1 = I_k \sum_{l=1}^{L} p_l^2 + \sum_{l=1}^{L} p_l(n_l + J_l)$$

$$= I_k L + \sum_{l=1}^{L} p_l(n_l + J_l) \tag{8.11}$$

The moments of S_1 are now computed. Since the interference and noise have zero mean, the mean of S_1 is given by

$$\mathscr{E}(S_1) = I_k L$$

The variance of S_1 is then given by

$$V(S_1) = \mathscr{E}(S_1^2) - L^2$$

The second order moment, $\mathscr{E}(S_1^2)$, can be expanded using (8.11):

$$\mathscr{E}(S_1^2) = \mathscr{E}\left\{ \left[I_k L + \sum_{l=1}^{L} p_l(n_l + J_l) \right] \left[I_k L + \sum_{j=1}^{L} p_j(n_j + J_j) \right] \right\}$$

$$= L^2 + \sum_{l=1}^{L} \sum_{j=1}^{L} \mathscr{E}(p_l p_j) \mathscr{E}\left[(n_l + J_l)(n_j + J_j) \right]$$

$$= L^2 + \sum_{l=1}^{L} \left[\mathscr{E}(n_l^2) + \mathscr{E}(J_l^2) \right]$$

$$= L^2 + \left(\sigma_n^2 + \sigma_J^2 \right) L$$

The variance of S_1 is then

$$V(S_1) = L\left(\sigma_n^2 + \sigma_J^2 \right)$$

The SNR for the PN correlator with no digital whitening is then given by

$$\text{SNR}_1 = \frac{L}{\sigma_n^2 + \sigma_J^2}$$

$$= \frac{L}{\frac{1}{2}\sum_{m=1}^{M_J} A_m^2 + \sigma_n^2} \tag{8.12}$$

8.6.2 Moments of the PN Correlator Output with Digital Whitening

With digital whitening, the PN correlator output can be written from (8.3), (8.6) and (8.7):

$$S_2 = \sum_{l=1}^{L} p_l \left[\sum_{j=0}^{M} a_j \left(s_{l-j} + n_{l-j} + J_{l-j} \right) \right] \tag{8.13}$$

In the following development it is assumed that the processing gain, L, is much greater than the number of taps, M, in the digital whitening filter. In this case s_{l-j} can be replaced by $I_k p_{l-j}$ so that (8.13) becomes

$$S_2 = \sum_{l=1}^{L} p_l \left[\sum_{j=0}^{M} a_j \left(I_k p_{l-j} + n_{l-j} + J_{l-j} \right) \right]$$

$$= I_k \sum_{l=1}^{L} p_l^2 + I_k \sum_{j=1}^{M} a_j \sum_{l=1}^{L} p_l p_{l-j} + \sum_{l=1}^{L} \sum_{j=0}^{M} a_j p_l \left(n_{l-j} + J_{l-j} \right)$$

$$= I_k L + I_k \sum_{j=1}^{M} a_j \sum_{l=1}^{L} p_l p_{l-j} + \sum_{l=1}^{L} \sum_{j=0}^{M} a_j p_l n_{l-j} + \sum_{l=1}^{L} \sum_{j=0}^{M} a_j p_l J_{l-j}$$

$$\tag{8.14}$$

where $a_0 \equiv 1$.

In Eq. (8.14), the first term is the desired signal component and the remaining terms represent interference terms. The last two terms correspond to residual noise and interference following digital whitening. The second term represents the distortion introduced in the desired signal from the presence of the digital whitening filter and is referred to as self-noise.

The moments of S_2 are now computed. Since the mean of the interference and noise terms is zero and the PN code sequence is white, the average value of S_2 is

$$\mathscr{E}(S_2) = I_k L \tag{8.15}$$

The variance of S_2 can be computed from (8.14) assuming that the signal, noise, and interference are mutually independent and that intersymbol interference is ignored. Thus,

$$V(S_2) = \mathscr{E}(S_2^2) - L^2$$

with

$$\mathscr{E}(S_2^2) = \sum_{l=1}^{L} \sum_{j=0}^{M} \sum_{i=1}^{L} \sum_{k=0}^{M} a_j a_k$$

$$\times \left\{ \mathscr{E}[p_l p_i p_{l-j} p_{i-k}] + \mathscr{E}[p_l p_i] \mathscr{E}[n_{l-j} n_{i-k}] + \mathscr{E}[p_l p_i] \mathscr{E}[J_{l-j} J_{i-k}] \right\}$$

Using the white noise property of the PN code sequence and the relationship

$$\mathscr{E}[p_l p_i p_{l-j} p_{i-k}] = \begin{cases} 1, & j = k = 0 \\ \delta_{j-k}, & j > 0, \quad k > 0, \quad l = i \\ 0, & \text{otherwise} \end{cases}$$

allows the preceding equation to be written as

$$\mathscr{E}(S_2^2) = L^2 + L \sum_{j=1}^{M} a_j^2 + \sum_{l=1}^{L} \sum_{j=0}^{M} \sum_{k=0}^{M} \left[\mathscr{E}(n_{l-j} n_{l-k}) + \mathscr{E}(J_{l-j} J_{l-k}) \right] a_j a_k$$

(8.16)

Note that

$$\mathscr{E}(n_{l-j} n_{l-k}) = \sigma_n^2 \delta_{j-k}$$

and

$$\mathscr{E}(J_{l-j} J_{l-k}) = \frac{1}{2} \sum_{m=1}^{M_J} A_m^2 \cos[2\pi f_m T_c (j - k)]$$

Therefore, the variance of S_2 is given by

$$V(S_2) = L \sum_{j=1}^{M} a_j^2 + L\sigma_n^2 \sum_{j=0}^{M} a_j^2 + \frac{L}{2} \sum_{j=0}^{M} \sum_{k=0}^{M} a_j a_k \sum_{m=1}^{M_J} A_m^2 \cos[2\pi f_m T_c (j - k)]$$

(8.17)

The SNR for the PN correlator with digital whitening is then given by

$$\text{SNR}_2 = \frac{L}{\sum_{j=1}^{M} a_j^2 + \sigma_n^2 \sum_{j=0}^{M} a_j^2 + \sum_{j=0}^{M} \sum_{k=0}^{M} a_j a_k \left\{ \frac{1}{2} \sum_{m=1}^{M_J} A_m^2 \cos[2\pi f_m T_c (j - k)] \right\}}$$

(8.18)

Note that the last term within the braces in the denominator represents the autocorrelation of the narrowband interference.

Two measures of performance improvement resulting from digital whitening are now described. These measures include an SNR improvement factor and a BER improvement.

8.6.3 SNR Improvement Factor

The digital whitening improvement factor, W, is defined as

$$W = \frac{\text{SNR}_2}{\text{SNR}_1}$$

(8.19)

Since the means of S_1 and S_2 are equal, (8.19) can also be written as

$$W = \frac{V(S_1)}{V(S_2)}$$

In the following figures, the SNR improvement factor is plotted against SNR_1, that is, the SNR obtained with no digital whitening filter.

From Eqs. (8.12) and (8.18), the SNR improvement factor can be expressed as

$$W = \frac{\frac{1}{2}\sum_{m=1}^{M_J} A_m^2 + \sigma_n^2}{\sum_{j=1}^{M} a_j^2 + \sigma_n^2 \sum_{j=0}^{M} a_j^2 + \sum_{j=0}^{M}\sum_{k=0}^{M} a_j a_k \left\{ \frac{1}{2}\sum_{m=1}^{M_J} A_m^2 \cos\left[2\pi f_m T_c (j-k)\right] \right\}}$$

(8.20)

Note that the form exhibited by the SNR improvement factor in (8.20) is independent of the processing gain, L.

8.6.4 Bit Error Rate Computations

If the decision variable presented in (8.13) is assumed to have a Gaussian distribution with mean and variance given by (8.15) and (8.17), respectively, the BER can be computed. In this case the BER, P_b, can be computed by assuming equally likely binary PSK symbols and finding the probability that S_2 is less than zero (see Appendix Section A.1.14):

$$P_b = P[S_2 < 0]$$

$$= \int_{-\infty}^{0} \frac{1}{\sqrt{2\pi}\,\sigma_2} e^{-(S_2 - \eta_2)^2/2\sigma_2^2}\, dS_2$$

where $\eta_2 = \mathscr{E}(S_2)$ for $I_k = 1$ and $\sigma_2 = \sqrt{V(S_2)}$. The above equation is usually expressed in terms of a complementary error function:

$$P_b = \tfrac{1}{2}\text{erfc}\sqrt{\gamma_b}$$

(8.21)

where

$$\gamma_b = \frac{\eta_2^2}{2\sigma_2^2} = \frac{\text{SNR}_2}{2}$$

and

$$\text{erfc}\, Y \equiv \frac{2}{\sqrt{\pi}} \int_Y^{\infty} e^{-t^2}\, dt$$

The BER results presented are plotted as a function of signal-to-noise ratio, assuming no interference. The degradation in BER performance from the case in which no interference is present then results from components at the digitally whitened correlator output corresponding to residual noise and interference and self-noise introduced by the presence of the signal in the time-dispersive whitening filter.

8.7 DIGITAL WHITENING PERFORMANCE RESULTS

Theoretical and simulated results for the SNR improvement factor are now given for the Wiener and maximum entropy filters. In the calculations, the PN code generator is implemented with a 10-bit shift register with the octal polynomial representation 2033. Both information symbols and PN code chips assume the value ± 1. The sampling rate is one sample per chip, and the prediction error coefficients are estimated using 200 samples per estimate. All amplitude components A_m for $m = 1, \ldots, M_J$ of the interfering signal are assumed to be equal. The bandwidth of the interference is selected to be roughly one fifth of the bandwidth of the SS signal. Assuming that the SS signal bandwidth is normalized to unity, the interfering tones are uniformly spaced by $0.1/M_J$. The number of tones, M_J, is chosen to be either 10 or 100. The phases of the interfering tones are selected either to be zero or to have a random value distributed uniformly over the range $(-\pi, \pi)$. White Gaussian noise is added with a standard deviation $\sigma_n = 0.1$. The SNR is controlled by varying the amplitude A_m of the interference.

Theoretical and simulated results using the maximum entropy algorithm are compared in Figure 8-5. Simulated results based on the maximum entropy and

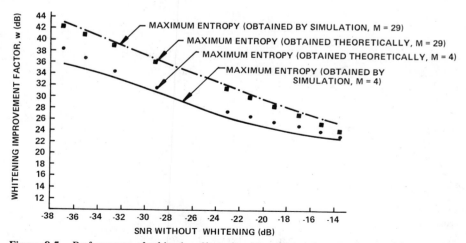

Figure 8-5. Performance of whitening filters for $M_J = 10$ evenly spaced tones ($f_m = m/100$, $m = 1, 2, \ldots, 10$), with uniform distributed random phases and 100 chips/symbol; standard deviation of the white noise, $\sigma_n = 0.1$ (© 1978 IEEE Hsu and Giordano, reference 8-6).

Wiener algorithms are compared in Figures 8-6 and 8-7. Ten chips per symbol are used in Figures 8-6 and 8-7 and 100 chips per symbol are used in Figure 8-5. The standard deviation, σ_n, of the white Gaussian noise is fixed at 0.1 in all cases. The orders of the digital whitening filters are selected to be equal to 4 or 29, and the phases of the interference are chosen to be random in Figures 8-5 and 8-6, and zero in Figure 8-7. The number of interfering tones used is 10 in Figures 8-5 and 8-7, and 100 in Figure 8-6. Computer simulation of the performance is based on Eq. (8.19) using 100 symbols. The means and variances of S_1 and S_2 are then computed by averaging 1,000 chips in the 10-chips-per-symbol case and 10,000 chips in the 100-chips-per-symbol case.

It is shown in Figure 8-5 that the performance results obtained theoretically and by simulation using the maximum entropy algorithm agree fairly well for $M = 29$. Poorer agreement is obtained for $M = 4$. These results can be explained as follows: The theoretical results obtained in Eq. (8.20) are based on an optimum whitening filter where the error signal is assumed to have a completely flat spectrum. The assumption of a flat spectrum for the error is accurate for a 29th order filter but not for a 4th order filter. Therefore, it is expected that the theoretical and simulated results will exhibit better agreement as the order of the filter is increased and becomes commensurate with the number of interfering tones.

Simulation results shown in Figures 8-6 and 8-7 indicate that there is some performance degradation at high SNR if the order of the whitening filters is not commensurate with the number of interfering tones. For example, it is shown in Figure 8-7 that a 29th order ($M = 29$) filter causes degradation over

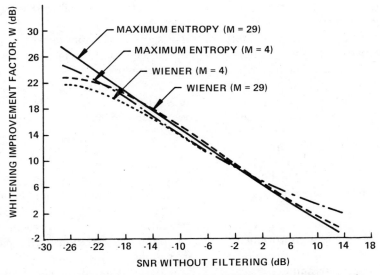

Figure 8-6. Performance of whitening filters for $M_J = 100$ evenly spaced tones ($f_m = m/1000$, $m = 1, 2, \ldots, 100$), with uniform distributed random phases and 10 chips/symbol; standard deviation of the white noise, $\sigma_n = 0.1$ (© 1978 IEEE Hsu and Giordano, reference 8-6).

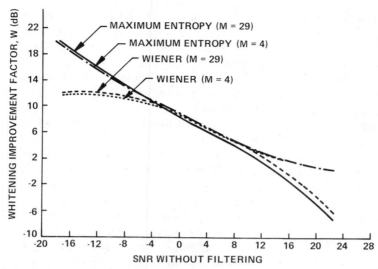

Figure 8-7. Performance of whitening filters for $M_J = 10$ evenly spaced tones ($f_m = m/100$, $m = 1, 2, \ldots, 10$), with zeros phases and 10 chips/symbol; standard deviation of the white noise, $\sigma_n = 0.1$ (© 1978 IEEE Hsu and Giordano, reference 8-6).

a fourth order filter at high SNR because there are only 10 interfering tones. If the number of tones is increased to 100, as shown in Figure 8-6, there is only slight degradation in performance for a 29th order filter. At high SNR, this degradation is enhanced because the signal is distorted to a greater extent by the filter. For a low SNR, the signal distortion is relatively small compared to the amount of interference cancellation. It is shown that the whitening filters achieve impressive performance gains at low SNR.

The maximum entropy filter displays a significant improvement in performance over the Wiener filter at low SNR. This performance improvement occurs at the sacrifice of increased signal distortion at high SNR when a 29th order filter is used. If a 4th order filter is used for whitening, the Wiener and maximum entropy filters tend to have similar performance. However, the separation of the performance of these two filters increases as the power of the interference is increased.

The heuristic description provided above described the whitening of the error signal ideally expected. For finite-duration whitening filters, this idealization is only approximately realizable. An illustration of the approximation results from examining the power spectra for the finite-duration filter case. Several examples are now provided which were obtained by computer simulation. The simulation parameters are generally the same as those described above in the SNR improvement factor computation. Variations in the parameter values from those provided earlier are supplied in the figures.

In Figures 8-8 and 8-9, the spectra of the received signal, x_k, the error signal e_k, and the PN code sequence, p_{kl} for $l = 1, \ldots, L$, are depicted. The spectra for the received and error signals are obtained by simulation and the spectrum

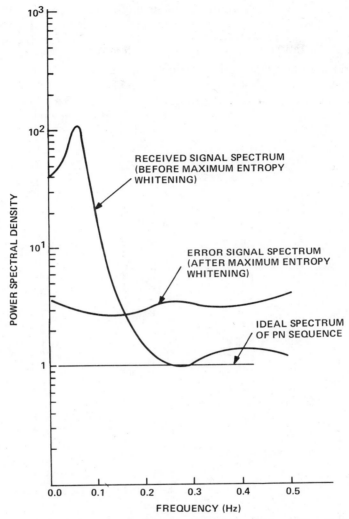

Figure 8-8. Power spectra comparison for $M_J = 100$ evenly spaced tones ($f_m = m/1000$, $m = 1, \ldots, 100$), SNR $= -11$ dB. Amplitude of the tone $A_m = 0.5$; standard deviation of the white noise $\sigma_n = 0.5$; order of the whitening filter $M = 4$. (© 1978 IEEE Hsu and Giordano, reference 8-6).

of the PN code sequence is ideal. The power spectra of the received and error signals are computed using a maximum entropy spectral estimation algorithm using four error prediction coefficients. The PN signal utilizes 10 chips per symbol. These spectra can be represented by

$$P_v(f) = \frac{P_{M_{min}} T_c}{\left| \sum_{i=0}^{4} a_i \exp\{-2\pi jifT_c\} \right|^2} \qquad \text{for } v = x_k \text{ or } e_k \qquad (8.22)$$

where a_i for $i = 0, \ldots, 4$, are the error prediction filter coefficients estimated from the signals x_k or e_k, and $P_{M_{\min}}$ is the minimum error power. The number of interfering tones in Figures 8-8 and 8-9 is chosen to be 100. The interference is concentrated over the band of frequencies from 0 to 0.1 Hz. The phases of the interfering tones are selected randomly in the interval $(-\pi, \pi)$. In Figures 8-8 and 8-9 the input SNR is assumed to be -11 dB and -2 dB, respectively. In Figure 8-8 the interference is stronger than the white Gaussian noise, and in Figure 8-9 the noise is stronger than the interference. Since only four predict-

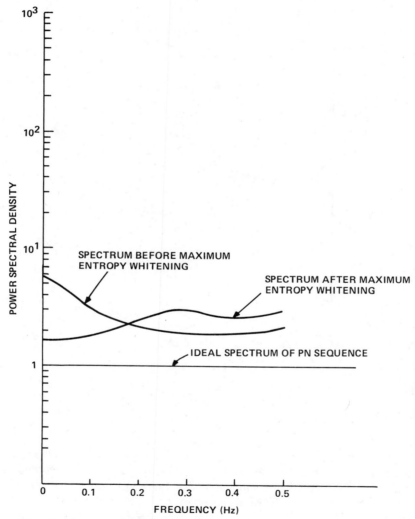

Figure 8-9. Power spectra comparison for $M_J = 100$ evenly spaced tones ($f_m = m/1000$, $m = 1, \ldots, 100$), SNR $= -2$ dB. Amplitude of the tone $A_m = 0.1$; standard deviation of the white noise $\sigma_n = 1.0$; order of the whitening filter $M = 9$. (© 1978 IEEE Hsu and Giordano, reference 8-6).

ion error filter coefficients are used for spectral estimation, only one major peak occurs in the interference band. Fine resolution of the spectrum achieved by use of a high-order filter would result in the presence of discrete interference tones in this band.

In the figures, the spectra before and after digital whitening are displayed. The spectrum of x_k, shown in Figure 8-8, is very steep in the region $0 < f < 0.1$ due to the presence of the strong interference, whereas the spectrum of e_k is relatively flat as a result of the digital whitening. In this case a significant improvement in performance can be expected. The digital whitening introduces some signal distortion, which tends to degrade the correlator output SNR. However, by suppressing the interference, the signal-to-noise ratio at the correlator output is considerably enhanced as a result of the digital whitening. In the case of Figure 8-9, the narrowband interference is relatively weak compared to the signal and white Gaussian noise. The distortion of the signal and the amount of interference cancellation tend to be similar, and no significant gain due to digital whitening can be expected.

BER performance using the assumptions given above are taken from reference 7 and reproduced here. The interference is assumed to consist of 100 tones which reside in 20 percent of the frequency band. The signal-to-interference ratio (SIR) per chip is fixed at -20 dB. A 4-tap prediction filter based on a known autocorrelation is used to suppress the interference.

Figure 8-10 illustrates the BER performance as a function of the SNR with digital whitening for various process gains. The results indicate that with such a short filter significant degradation occurs. An improvement in performance

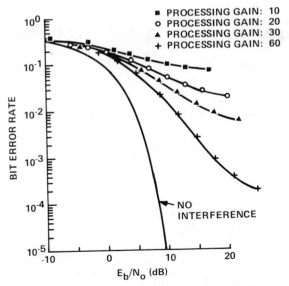

Figure 8-10. Bit error probability under the Gaussian assumption for 4-tap predictor with no matched filter. Filter coefficients computed using exact autocorrelation coefficients with -20 dB SIR, and nondispersive channel. (© 1982 IEEE Ketchum and Proakis, reference 8-7).

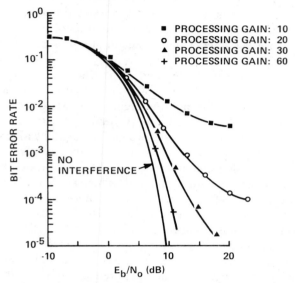

Figure 8-11. Bit error probability under the Gaussian assumption for 4-tap predictor with matched filter. Filter coefficients computed using exact autocorrelation coefficients with −20 dB SIR, and nondispersive channel. (© 1982 IEEE Ketchum and Proakis, reference 8-7).

can be achieved by adding a matched filter at the output of the digital whitening filter. These results are provided in Figure 8-11 and have been verified by simulation.

8.8   PERFORMANCE BOUND FOR SNR IMPROVEMENT FACTOR

In this section a theoretical bound on the maximum SNR improvement factor is developed. This bound was first established by Masry (reference 11) using a more general form of the development provided in Section 8.6. To develop the bound, the expression for the error signal provided by Eq. (8.6) is repeated here:

$$e_k = \sum_{l=0}^{M} a_l x_{k-l}$$

The coefficients $\{a_l\}$ are computed on the basis of minimum MSE. The resulting minimum is denoted by $P_{M_{\min}}$ and can be expressed as

$$P_{M_{\min}} = \mathcal{E}\left(e_k^2\right)$$

$$= \mathcal{E}\left\{\sum_{l=0}^{M}\sum_{i=0}^{M} a_l a_i x_{k-l} x_{k-i}\right\}$$

$$= \sum_{l=0}^{M}\sum_{i=0}^{M} a_l a_i \mathcal{E}\left\{x_{k-l} x_{k-i}\right\}$$

Substituting (8.3) into the above equation results in

$$P_{M_{min}} = \sum_{l=0}^{M} \sum_{i=0}^{M} a_l a_i \{ r_s(i-l) + r_n(i-l) + r_J(i-l) \} \qquad (8.23)$$

where $r_s(l)$, $r_n(l)$, and $r_J(l)$ are the autocorrelation of the desired signal, the noise, and the interference, respectively. Note that the signal, noise, and interference terms are assumed to be mutually independent with $r_n(0) \equiv \sigma_n^2$ and $r_J(0) \equiv \sigma_J^2$. In Section 8.6 the noise and interference autocorrelation functions assumed specific forms given, respectively, by

$$r_n(l) \equiv \sigma_n^2 \delta(l)$$

and

$$r_J(l) = \frac{1}{2} \sum_{m=1}^{M_J} A_m^2 \cos[2\pi f_m T_c l]$$

In this section the expression for the interference is maintained in a general form so that the above interference equation is not used. Further simplification of (8.23) can be performed using the orthogonality property of the PN code sequence. As a result the desired signal autocorrelation can be written as

$$r_s(i-l) = \mathscr{E}(s_{k-l} s_{k-i})$$

$$= \begin{cases} 1, & i = l \\ 0, & \text{otherwise} \end{cases} \qquad (8.24)$$

Substituting the above equation in (8.23) results in

$$P_{M_{min}} = \sum_{l=0}^{M} a_l^2 + \sum_{l=0}^{M} \sum_{i=0}^{M} a_l a_i \{ r_n(i-l) + r_J(i-l) \}$$

$$= 1 + \sum_{l=1}^{M} a_l^2 + \sum_{l=0}^{M} \sum_{i=0}^{M} a_l a_i \{ r_n(i-l) + r_J(i-l) \} \qquad (8.25)$$

The variance of the PN correlator output with digital whitening can be written with the aid of (8.16) as

$$V(S_2) = \mathscr{E}(S_2^2) - L^2$$

$$= L \sum_{l=1}^{M} a_l^2 + L \sum_{l=0}^{M} \sum_{i=0}^{M} a_l a_i \{ r_n(i-l) + r_J(i-l) \}$$

From (8.25) it can be seen that the variance of S_2 can be written as

$$V(S_2) = (P_{M_{min}} - 1)L \qquad (8.26)$$

From (8.9), (8.12), (8.15), and (8.26), the SNR improvement factor can be expressed as

$$W = \frac{\sigma_n^2 + \sigma_J^2}{P_{M_{\min}} - 1} \tag{8.27}$$

To explicitly indicate the dependence of the SNR improvement factor on M, the number of filter coefficients, W, is replaced by $W(M)$. The smallest value of the minimum MSE, denoted by P_∞^2, is obtained by using an infinte number of coefficients. Thus, a bound on the SNR improvement factor becomes:

$$W(M) \leq \frac{\sigma_n^2 + \sigma_J^2}{P_\infty - 1} \tag{8.28}$$

This bound represents the best attainable performance available by use of digital whitening implemented with realizable prediction filters based on a minimum MSE criterion. In reference 11 it is shown that this bound is exponentially tight for a wide variety of interference processes including all processes with rational power spectral densities.

The bound in (8.28) can be further evaluated using the result provided in reference 11 with the 2π normalization absorbed into the definition of the power spectral density:

$$P_{M_{\min}} = \exp\left\{\frac{1}{2\pi} \int_{-\pi}^{\pi} \ln P_x(\omega)\, d\omega\right\} \tag{8.29}$$

where $P_x(w)$ is the power spectral density of the received signal (see reference 12). From (8.3) it can be seen that the power spectral density, $P_x(\omega)$, can be expressed as

$$P_x(\omega) = P_s(\omega) + P_n(\omega) + P_J(\omega) \tag{8.30}$$

where $P_s(\omega)$, $P_n(\omega)$, and $P_J(\omega)$ are the power spectral density of the desired signal, noise, and interference, respectively. Since the autocorrelation of the signal is a unit pulse given by (8.24), the power spectral density of the signal is a constant equal to unity (see Eq. (7.19)). If the noise is assumed to be white with power σ_n^2, (8.30) becomes

$$P_x(\omega) = 1 + \sigma_n^2 + P_J(\omega) \tag{8.31}$$

The bound in (8.28) can then be written by combining (8.28), (8.29), and (8.31):

$$W(M) \leq \frac{\sigma_n^2 + \sigma_J^2}{\exp\left\{\frac{1}{2\pi} \int_{-\pi}^{\pi} \ln\left[1 + \sigma_n^2 + P_J(\omega)\right] d\omega\right\} - 1} \tag{8.32}$$

It is now assumed that the interference occupies a fraction, ρ, of the total bandwidth.[7,11] In fact, it does not matter whether the interference is contiguous or dispersed across the band. With this model for the interference, the power spectral density, $P_J(\omega)$, can be written

$$
P_J(\omega) = \begin{cases} \dfrac{\sigma_J^2}{\rho}, & -\pi\rho < \omega < \pi\rho \\ 0, & \text{otherwise} \end{cases} \tag{8.33}
$$

(see Figure 8-12.) Using the interference model given by (8.33) in (8.32) results in

$$
W(M) \le \frac{\sigma_n^2 + \sigma_J^2}{\exp\left\{(1/\pi)\int_0^{\pi\rho}\ln\left[1 + \sigma_n^2 + (\sigma_J^2/\rho)\right]\,d\omega + (1/\pi)\int_{\pi\rho}^{\pi}\ln\left[1 + \sigma_n^2\right]\,d\omega\right\} - 1}
$$

or

$$
W(M) \le \frac{\sigma_n^2 + \sigma_J^2}{\left[1 + \sigma_n^2 + \dfrac{\sigma_J^2}{\rho}\right]^{\rho}\left[1 + \sigma_n^2\right]^{1-\rho} - 1} \tag{8.34}
$$

Equation (8.34) is now expressed in terms of r, the signal-to-noise ratio per chip without digital whitening, defined by

$$
r \equiv \frac{\text{SNR}_1}{L}
$$

$$
= \frac{1}{\sigma_n^2 + \sigma_J^2} \tag{8.35}
$$

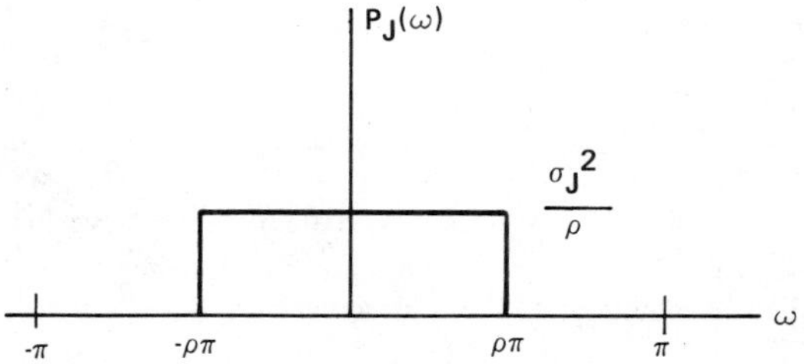

Figure 8-12. Power spectral density of the interference which occupies a fraction, ρ, of the signal bandwidth.

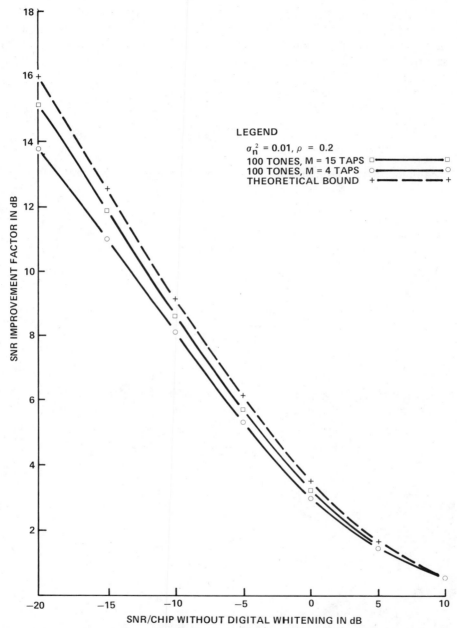

Figure 8-13. Comparison of theoretical bound and performance computed with a finite number of taps (© 1982 IEEE Ketchum and Proakis, reference 8-7).

Use of (8.35) in (8.34) results in

$$W(M) \le \cfrac{1}{r\left\{\left[1 + \sigma_n^2 + \cfrac{1}{r\rho} - \cfrac{\sigma_n^2}{\rho}\right]^\rho \left[1 + \sigma_n^2\right]^{1-\rho} - 1\right\}} \qquad (8.36)$$

Figure 8-13 illustrates the bound given by (8.36) for $\rho = 0.2$ and $\sigma_n^2 = 0.01$ plotted as a function of r. The theoretical bound is compared with the SNR improvement result shown in Figure 3 of reference 7 for a 20 percent interference band occupancy obtained from 100 sinusoidal interfering tones and linear prediction filters using 4 and 15 taps. The results demonstrate that the theoretical bound computed on the basis of an infinite number of taps gives a good approximation of the improvement regardless of the number of taps and the location of the interference band. Note that in the 15-tap case the bound differs from the computed result by approximately 1 dB for the poor SNR-per-chip region. Thus, the bound is fairly tight and represents a good estimate of the SNR improvement to be realized.

REFERENCES

1. R. C. Dixon, *Spread Spectrum Techniques*, IEEE Press, New York, 1976.

2. R. C. Dixon, *Spread Spectrum Systems*, Wiley, New York, 1976.

3. D. J. Torrieri, *Principles of Military Communications Systems*, Artech House, Dedham, Mass., 1981.

4. J. G. Proakis, *Digital Communications* (Chapter 8), McGraw-Hill, New York, 1983.

5. R. Price and P. E. Green Jr., A communication Technique for Multipath Channels, *Proc. IRE 46*, March 1956, pp. 555–570.

6. F. M. Hsu and A. A. Giordano, Digital Whitening Techniques for Improving Spread Spectrum Communication Performance in the Presence of Narrowband Jamming and Interference, *IEEE. Trans. Communications COM-26*, pp. 209–216.

7. J. W. Ketchum and J. G. Proakis, Adaptive Algorithm for Estimating and Suppressing Narrow-Band Interference in PN Spread-Spectrum Systems, *IEEE Trans. Communications COM-30*, 1982, pp. 913–927.

8. L. M. Li and L. B. Milstein, Rejection of Narrow-Band Interference in PN Spread-Spectrum Systems Using Transversal Filters, *IEEE Trans. Communications COM-30*, 1982, pp. 925–928.

9. R. A. Iltis and L. B. Milstein, Estimation Techniques for Narrowband Interference Rejection, ICC '83, Boston Mass., pp. F6.2.1–F6.2.5.

10. S. M. Sussman and E. J. Ferrari, Effects of Notch Filters on the Correlation Properties of a PN Signal, *IEEE Trans. Aerospace and Electronic Systems AES-10*, May 1974, pp. 385–390.

11. E. Masry, Closed-Form Analytical Results for the Rejection of Narrowband Interference in PN Spread Spectrum Systems, *IEEE Trans. Communications COM-32*, 1984, pp. 888–896.

12. J. Makhoul, Linear Prediction: A Tutorial Review, *Proc. IEEE*, *63*, No. 4, April 1975.

chapter

9

ADAPTIVE ARRAYS

9.1 ADAPTIVE ARRAY TECHNIQUES

In recent years, military communication systems have required improved interference suppression capability and enhanced signal-to-noise ratio (SNR). These features can be achieved by use of adaptive antenna array techniques. Several adaptive antenna array techniques have been developed in the literature using signal combining or interference concellation. Both the maximum likelihood method (MLM) and the least mean square error (MSE) criteria have been used in AM, FM, FSK, PSK and MSK modulation systems. Several adaptive algorithms have been developed for antenna steering and interference nulling techniques which offer greater interference or jamming suppression capability, faster antenna adaptability, and enhanced communication reliability in communication systems operating in bands from ELF to UHF.

An adaptive array may be defined as one that modifies its own antenna pattern, frequency response, etc., by means of internal feedback control for rejecting intentional jammers or undesired interference. Transmitting adaptive arrays have been used by Van Atta,[1] and others[2] to describe a self-phasing antenna system which reradiates a signal in the direction from which it was received. In this case the transmitting array is adaptive, because the signal reradiation occurs without any knowledge of the direction in which it is to transmit. The work on adaptive receiving arrays was done by Bryn,[3] Mermoz,[4]

Shor,[5] Applebaum,[6] Widrow et al.,[7] Capon et al.,[8] Griffiths,[9] Frost,[10] Riegler and Compton,[11] Lin,[12] and others. Bryn[3] showed that Gaussian signals in additive noise can be detected by an array with K antenna elements using the Bayes optimum detector. Mermoz,[4] Shor,[5] and Applebaum[6] proposed a similar scheme for known narrowband signals, using signal-to-noise ratio as a performance criterion. Widrow et al.[7] suggested that the variable weights of the array can be automatically adjusted by a simple least mean square (LMS) algorithm that is an application and extension of the original work on adaptive filters done by Widrow and Hoff.[13] Griffiths[9] also proposed an adaptive array technique based on an adaptive procedure using a modified gradient technique. Frost[10] used a constrained least mean squares algorithm for adjusting an array of sensors in real time to respond to a signal coming from a desired direction while discriminating against noise coming from other directions. Riegler and Compton[11] discussed an experimental adaptive array based on the feedback concept used by Widrow et al.[7] Capon et al.[8] described a large-aperture seismic array based on a multidimensional maximum-likelihood technique. Lin[12] proposed that spatial correlation coefficients be used to determine the performance of an adaptive array.

In Section 9.2, the spatial and electrical configurations of general antenna arrays are presented before signal processing and adaptive filtering of the array are introduced. Section 9.3 illustrates the concept of adaptive array signal processing techniques based on least squares algorithms. The adaptive array experimental results obtained by Riegler and Compton[11] are given in Section 9.4. Then, an example of a two-element beamforming adaptive array for side-lobe interference rejection is presented. In Section 9.5 a simple mathematical model for PSK digital signal communication with spread spectrum modulation is given. Subsequently, in Section 9.6, two adaptive antenna interference cancellers (AIC #1 and AIC #2) are developed using the mathematical model presented in Section 9.5. A baseband-antenna-array-combining technique is illustrated in Section 9.7. In Section 9.8, bit error rate performance of three adaptive array techniques is compared. Two minimum MSE adaptive algorithms are discussed in Section 9.9. The computer simulation exhibiting the speed of convergence with these two minimum MSE algorithms is shown in Section 9.10.

9.2   GENERAL CONFIGURATIONS OF ADAPTIVE ARRAYS

The spatial and electrical configurations of two general antenna arrays is presented here. Subsequently, the signal processing and adaptive filtering operations used in the array are introduced. Figure 9-1 displays an adaptive array configuration for processing N_r narrowband received signals. Since RF signal processing operations are performed, all signals are represented in continuous time. For $i = 1, 2, \ldots, N_r$ the ith antenna received signal, $x_i(t)$, and its quadrature component $x_i(t - \pi/2)$ are weighted by variable tap

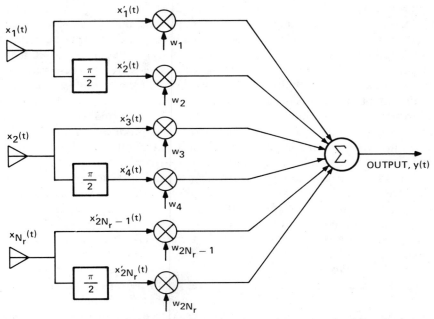

Figure 9-1. Adaptive array configuration for receiving narrowband signals.

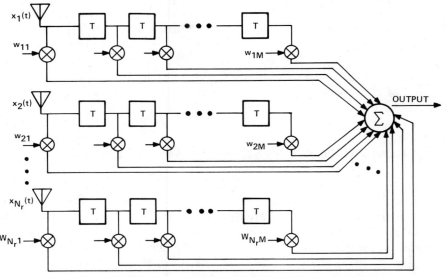

Figure 9-2. Adaptive array configuration for receiving broadband signals.

weights w_{2i-1} and w_{2i}, respectively. The two quadrature weights and $\pi/2$ time delay can provide complete amplitude and phase adjustment of the narrow-band signals received by each individual antenna element. The full array with N elements depicted in Figure 9-1 represents a general configuration for diversity combination of desired narrowband received signals and mitigation of interference. In contrast, Figure 9-2 shows an adaptive array configuration for processing N_r received signals over a wide band of frequencies. The processing requires M coefficients delayed by a symbol interval T for each received signal. Thus there are $M \times N_r$ coefficients labeled $w_{11}, \ldots, w_{N_r M}$.

Note that the in-phase and quadrature components of each antenna element shown in Figure 9-1 are replaced by a single tapped-delay line of duration MT. The tapped-delay line permits adjustment of the gain and phase as desired at a number of frequencies over the band of interest. In the following sections only the adaptive array for receiving narrowband signals shown in Figure 9-1 is considered for simplicity.

9.3 SPATIAL ADAPTIVE SIGNAL PROCESSING TECHNIQUES

Least squares adaptive array signal processing techniques for receiving narrowband signals are discussed in this section. Suppose $x'_{2i-1}(t)$ and $x'_{2i}(t)$ denote the in-phase and quadrature components of $x_i(t)$ for $i = 1, 2, \ldots, N_r$. Then, the array output, $y(t)$, is given by

$$y(t) = \sum_{i=1}^{2N_r} w_i x'_i(t) \tag{9.1}$$

where $2N_r$ is the number of weights. Let $s(t)$ be the transmitted signal; the error signal becomes

$$\varepsilon(t) = s(t) - y(t) = s(t) - \sum_{i=1}^{2N_r} w_i x'_i(t) \tag{9.2}$$

The mean square error is defined as

$$E(w) = \mathcal{E}\left[\left(s(t) - \sum_{i=1}^{2N_r} w_i x'_i(t)\right)^2\right] \tag{9.3}$$

where $\mathcal{E}$ denotes the ensemble average performed over the signal and noise statistics. To make the array adaptive, $s(t)$ is replaced by a reference signal $r(t)$, resulting in an error signal $\varepsilon'(t)$ given by

$$\varepsilon'(t) = r(t) - y(t) \tag{9.4}$$

This error signal is the input to a feedback system that controls the tap weights, w_i, $i = 1, 2, \ldots 2N_r$. The feedback circuitry is designed to adjust the tap weights to minimize the mean square error value. Alternatively stated, these signal operations have the equivalent effect of forcing the array output signal $y(t)$, to conform to the reference signal, $r(t)$, in a mean square error sense. Thus any signal not conforming to $r(t)$ appears as an error signal and the feedback mechanism adjusts the weights to suppress it in the output. As a result, an antenna pattern is produced which has a pattern null in the direction of interference and a pattern gain in the direction of the incoming desired signal. The adaptive algorithms used for adjusting the weights could be a Kalman filter, a Widrow-type algorithm, or other least squares techniques described in Part 1 of this book. The next section describes the experimental results of a two-element adaptive array obtained by Riegler and Compton[11] using the feedback configuration given by Widrow et al.[7]

9.4   ADAPTIVE ARRAY EXPERIMENTAL RESULTS OF RIEGLER AND COMPTON

Riegler and Compton[11] constructed a two-element adaptive array, depicted in Figure 9-3, for processing the received signals. Quadrature components are formed from each antenna output by inserting a 90° phase shift in one branch. For the first antenna the in-phase and quadrature components are weighted by quantities w_1 and w_2, respectively. The corresponding weights for the second antenna are w_3 and w_4, respectively. The output signal is the sum of all four weighted signals. The feedback rule is based on a steepest-descent minimization of $\overline{\varepsilon'^2}$, where — denotes a time average. Specifically, each weight w_i is adjusted according to the following rule:

$$\frac{dw_i}{dt} = -k\nabla_{w_i}\left[\overline{\varepsilon'^2(t)}\right] \tag{9.5}$$

where $\nabla_{w_i}[\overline{\varepsilon'^2(t)}]$ denotes the ith component of the gradient of $\overline{\varepsilon'^2(t)}$ with respect to the weight w_i, and k is a positive constant. Evaluating the gradient yields

$$\frac{dw_i}{dt} = 2k\overline{x_i'(t)\varepsilon'(t)}, \qquad i = 1,\ldots,4 \tag{9.6}$$

The equations in (9.6) can be expressed in integral form as

$$w_i(t) = w_i(0) + 2k\int_{t'=0}^{t} x_i'(t')\varepsilon'(t')\,dt', \qquad i = 1,\ldots,4 \tag{9.7}$$

The equations in (9.7) can be implemented as the feedback loop shown in

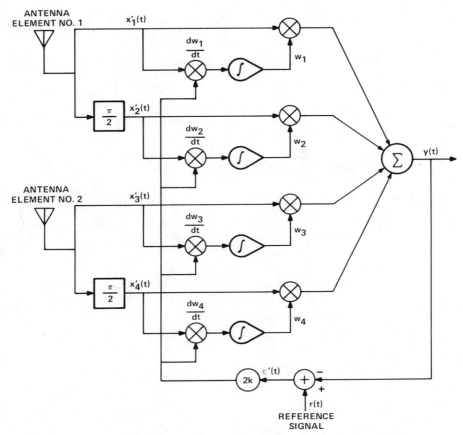

Figure 9-3. Two-element adaptive array used by Riegler and Compton.

Figure 9-3. The first experiment performed by Riegler and Compton involved a simulated test of interference rejection by the two-element array. Two continuous wave (CW) signals, representing a desired signal and an interference signal, were fed into the array. First, the desired signal in the absence of interference was injected, with equal phase on each array element. This situation corresponds to a desired signal that arrives broadside. In this case, the same signal was used for the local reference signal $r(t)$. The coefficients w_i, $i = 1, \ldots, 4$, were allowed to adapt, and the final values were used to compute the pattern labeled DESIRED SIGNAL ONLY in Figure 9-4. Next, an interfering signal was injected whose direction differed from the desired signal. The direction of the interfering signal was determined by selecting the electrical phase angle between elements to produce a signal incident 40° off broadside for a half-wavelength element spacing. This interfering signal was separated in frequency by 10 kHz from the desired signal. The weighting coefficients were again

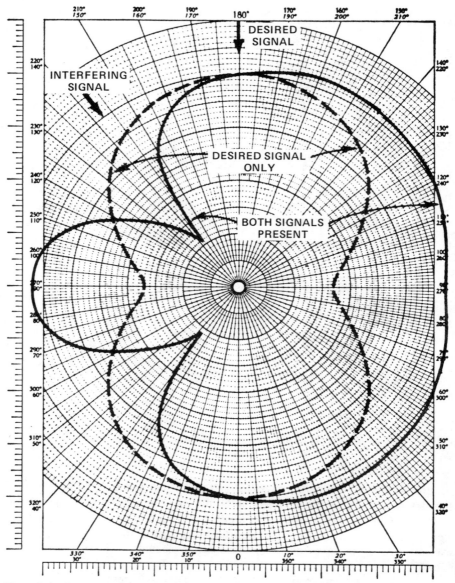

Figure 9-4. Patterns before and after adaptation (© 1973 IEEE Riegler and Compton, reference 9-11).

298

allowed to adapt and resulted in final values that were used to compute the second pattern, labeled BOTH SIGNALS PRESENT. Figure 9-4 also depicts the null formed on the interfering signal by the array.

Next, Riegler and Compton performed an interference rejection experiments in which actual antenna elements were used. The antennas used were a pair of quarter-wavelength monopoles spaced a half-wavelength apart on a rectangular ground plane. The patterns were measured at 2.1 GHz. Each element was connected directly to a mixer and downconverted to 65 MHz. Figure 9-5 shows the antenna pattern with both a desired signal and an interfering signal illuminating the antenna. Once again, it can be seen that the adaptive array forced a null in the direction of the interfering signal.

More mathematical insight for the case of a two-element array whose coefficients produce a null in the direction of the interfering signal are now provided.

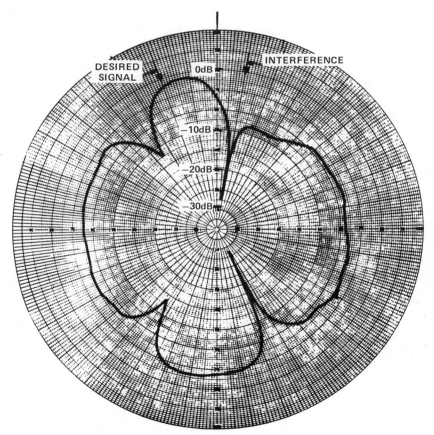

Figure 9-5. Adaptive antenna pattern; desired signal plus interference (© 1973 IEEE Riegler and Compton, reference 9-11).

Figure 9-6 shows a two-element phased array configuration for interference
rejection given by Widrow et al.[7] This figure is a modified version of Figure 9-3
constructed to illustrate an interfering signal with an angle of incidence ψ.
With respect to the midpoint between the antenna elements, the relative phase

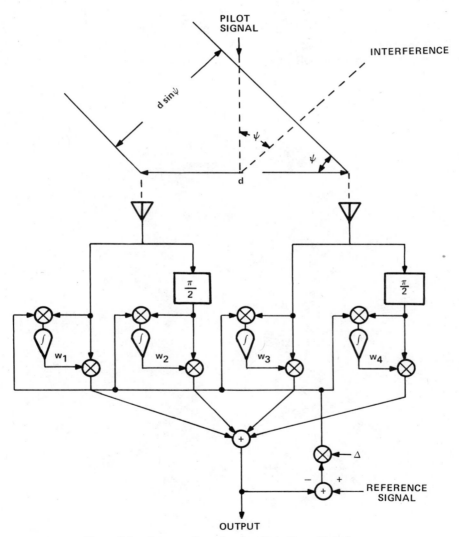

Figure 9-6. Array configuration for elimination of interference.

delays of the interference at the two antenna elements are $\pm\theta$, where

$$\theta = \frac{f_0 d\pi}{c} \sin\psi$$

$$= \frac{d\pi}{\lambda_0} \sin\psi \tag{9.8}$$

and

$$f_0 = \text{frequency of received signal}$$

$$\lambda_0 = \text{wavelength at frequency } f_0$$

$$d = \text{spacing between antenna elements}$$

$$c = \text{signal propagation velocity} = \lambda_0 f_0.$$

A simple example for illustrating the tap weight calculation is now described. Assume that a sinusoidal pilot signal of amplitude S arrives on boresite and is represented by $S\sin\omega_0 t$, where $\omega_0 = 2\pi f_0$. The interference is modeled as a sinusoid with amplitude I, that is, $I\sin\omega_0 t$, and is incident to the receiving array at an angle $\psi = \pi/6$ radians. At a point midway between the first and second elements in this representation, the signal and interference are assumed to be in phase. Assume that a very low frequency (VLF) system is under study with a carrier $f_0 = 16$ kHz and the space between the two elements $d = 30$ meters.

The array output due to the pilot signal is

$$S\left[(w_1 + w_3)\sin\omega_0 t + (w_2 + w_4)\sin(\omega_0 t - \pi/2)\right]. \tag{9.9}$$

For this output to be equal to the desired output corresponding to $S\sin\omega_0 t$, it is necessary that

$$w_1 + w_3 = 1 \tag{9.10}$$

$$w_2 + w_4 = 0 \tag{9.11}$$

Similarly, the output due to the interference is

$$I\left[w_1\sin(\omega_0 t - \theta) + w_2\sin(\omega_0 t - \pi/2 - \theta)\right.$$

$$\left. + w_3\sin(\omega_0 t + \theta) + w_4\sin(\omega_0 t - \pi/2 + \theta)\right] \tag{9.12}$$

where

$$\theta = \frac{d\pi}{\lambda_0}\sin\psi$$

$$= \frac{\pi}{1250} \tag{9.13}$$

For the response to the interference to be equal to zero, it is necessary that

$$(w_1 + w_3)\cos\theta - (w_2 - w_4)\sin\theta = 0 \tag{9.14}$$

$$(w_2 + w_4)\cos\theta + (w_1 - w_3)\sin\theta = 0 \tag{9.15}$$

The tap weights can be solved from (9.10), (9.11), (9.14), and (9.15) and are given by

$$w_1 = w_3 = 1/2$$

$$w_2 = -w_4 = \tfrac{1}{2}\cot\theta = 198.94 \tag{9.16}$$

Equation (9.16) shows that the solution is quite unstable, because $\cot\theta$ changes rapidly for small changes in θ, and therefore results in large changes in the tap weights w_2 and w_4. Rapid tap weight changes imply that the two-element VLF array cannot resolve the different arrival directions of the signal and interference. Consequently, array performance is poor. If the frequency f_0 is increased to $f_0 = 3.2$ MHz, corresponding to an HF system, and the spacing d is maintained at 30 m, then $\theta = 4\pi/25$, and the tap weights are given by

$$\left. \begin{array}{l} w_1 = w_3 = 1/2 \\ w_2 = -w_4 = 0.91 \end{array} \right\} \tag{9.17}$$

Equation (9.17) shows that the solution for the HF case is very stable, since small changes in θ produce small changes in the tap weights w_2 and w_4, thereby resulting in better array performance. Equation (9.12), describing the form of the antenna response to the interference, can be reexpressed in the general form

$$A\cos\omega_0 t + B\sin\omega_0 t$$

where A and B are given by the left-hand sides of (9.15) and (9.14), respectively. Using the tap weight values in (9.17) the antenna power pattern $\mathbb{P}$ can be defined as

$$\mathbb{P} = A^2 + B^2$$

$$= (\cos\theta - 1.82\sin\theta)^2 \tag{9.18}$$

where $\theta = 8\pi/25\sin\psi$. The antenna power pattern is plotted in Figure 9-7, where $y = \mathbb{P}\cos\psi$ and $x = \mathbb{P}\sin\psi$. The results indicate that an interfering signal with an angle of incidence $\psi = 30°$ is completely suppressed by the two-element array system. If more elements are added to the array system, more interfering signals with several angles of incidence can be suppressed. The computation of tap weights in the more general case is similar to the one illustrated in this example.

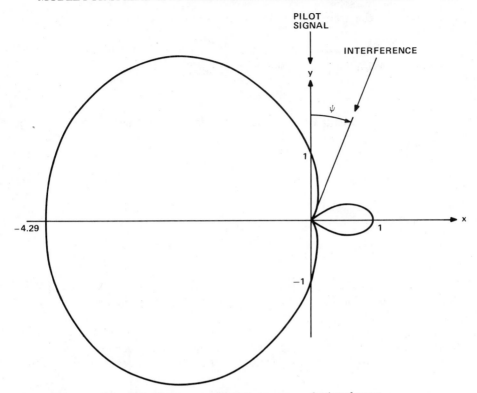

Figure 9-7.   Antenna power pattern response for interference.

In general, the angle of incidence is unknown at the receiver, and additive noise is present. In this case the tap weights should be estimated automatically by some criterion such as the MSE or the MLM. The use of the MSE criterion is illustrated in the following sections.

9.5   MATHEMATICAL MODEL FOR THE TRANSMITTED AND RECEIVED SIGNALS IN SPREAD SPECTRUM ARRAY PROCESSING APPLICATIONS

In the following sections, the adaptive array techniques are illustrated by processing signals received from two antenna elements. Both optimum and suboptimum array techniques are investigated. The signal is assumed to be binary phase-shift keying (PSK) transmission with spread spectrum PN sequence modulation. The adaptive speed, convergence stability, and error rate performance of several least mean square algorithms is evaluated by using theoretical analysis and computer simulation. The computational complexity of the algorithms is investigated to illustrate the feasibility for actual implementation. Prior to describing the array processing techniques, a mathematical description of the transmitted signal and communications channel used are provided.

In a PSK communication system with a spread spectrum PN sequence modulation, the transmitted signal can be expressed as

$$S(t) = \text{Re}\{s(t)\exp(j2\pi f_0 t)\} \tag{9.19}$$

where $\text{Re}\{x\}$ represents the real part of x. The complex envelope $s(t)$ can be represented as

$$s(t) = \sum_{k=0}^{\infty} I_k s_k(t) \tag{9.20}$$

where

$$s_k(t) = \sum_{l=kL}^{(k+1)L-1} h_l p(t - lT_c) \tag{9.21}$$

For binary phase-shift keying (PSK) transmission, $\{I_l\}$ is equal to ± 1, $\{h_l\}$ is the PN sequence for L chips per bit (or symbol), $p(t)$ is the basic transmitted pulse shape, and T_c is the chip interval.

In the following, it is assumed that the information and PN sequences are random and statistically independent. The information sequence is also assumed to be statistically independent from symbol to symbol in the same channel.

Due to signal attenuation and phase shift introduced on the communication channel, the received signals can be expressed as

$$x_l(t) = R_e\{(\alpha_l s(t) + \eta_l(t))e^{j\omega_0 t}\}$$
$$= (\alpha_{lr} s(t) + \eta_{lr}(t))\cos \omega_0 t - (\alpha_{li} s(t) + \eta_{li}(t))\sin \omega_0 t, \qquad l = 1, 2 \tag{9.22}$$

where $l = 1$ and 2 represents channels 1 and 2, respectively, and the quantities in (9.22) are defined below. The additive noise on channel l, represented by $\eta_l(t)$, for $l = 1, 2$ is jointly stationary and can be expressed in terms of its real and imaginary parts by

$$\eta_l(t) = \eta_{lr}(t) + j\eta_{li}(t) \tag{9.23}$$

In (9.23) $\eta_l(t)$ is assumed to have zero mean with uncorrelated real and imaginary parts and equal power in each part. The variance of either the real or imaginary part of $\eta_l(t)$ is given by σ_l^2. The covariance of $\eta_l(t)$ is then

$$\mathscr{E}\left[\eta_l(t_1)\eta_l^*(t_2)\right] = 2\sigma_l^2\delta(t_1 - t_2) \tag{9.23a}$$

and

$$\mathscr{E}\left[\eta_{lr}(t_1)\eta_{lr}(t_2)\right] = \mathscr{E}\left[\eta_{li}(t_1)\eta_{li}(t_2)\right] = \sigma_l^2\delta(t_1 - t_2)$$

$$\mathscr{E}\left[\eta_{lr}(t_1)\eta_{li}(t_2)\right] = 0 \tag{9.23b}$$

The cross-covariance of $\eta_1(t)$ and $\eta_2(t)$ is defined by

$$\mathscr{E}\left[\eta_1(t_1)\eta_2^*(t_2)\right] = 2\sigma_{12}^2\delta(t_1 - t_2) \tag{9.23c}$$

and

$$\mathscr{E}\left[\eta_{1r}(t_1)\eta_{2r}(t_2)\right] = \mathscr{E}\left[\eta_{1i}(t_1)\eta_{2i}(t_2)\right] = \sigma_{12}^2\delta(t_1 - t_2) \tag{9.23d}$$

$$\mathscr{E}\left[\eta_{lr}(t_1)\eta_{mi}(t_2)\right] = 0 \qquad \text{for all } l \text{ and } m \tag{9.23e}$$

where $\sigma_{12}^2 \equiv \sigma_1\sigma_2 r$.

In the following analysis a relationship between chip noise power and symbol noise power is required. Let n_l represent a noise variable obtained by integrating the input noise $\eta_l(t)$ for $l = 1, 2$ over the symbol interval:

$$n_l = \int_0^T \eta_l(t)\,dt \tag{9.23f}$$

In (9.23f) the mean of n_l is zero and the variance is defined by $\sigma_l^2 = \frac{1}{2}\mathscr{E}[|n_l|^2]$. Now let the noise variable, n_{lc}, be obtained by integrating the noise $\eta_l(t)$ for $l = 1, 2$ over the chip interval:

$$n_{lc} = \int_0^{T_c} \eta_l(t)\,dt \tag{9.23g}$$

In (9.23g) the mean of n_{lc} is zero and the variance is defined by $\sigma_{lc}^2 = \frac{1}{2}\mathscr{E}[|n_{lc}|^2]$. The chip and symbol noise variances can be related by reexpressing (9.23f) with $T = LT_c$:

$$n_l = \sum_{k=1}^{L}\int_{(k-1)T_c}^{kT_c} \eta_l(t)\,dt \tag{9.23h}$$

Since $\eta_l(t)$ is a white noise process, it can be seen that

$$\sigma_l^2 = L\sigma_{lc}^2 \tag{9.23i}$$

The differential time delay between the channels is assumed to be negligible. The factors α_1 and α_2 account for direct path attenuation and phase shift, and can be expressed as

$$\alpha_l = \alpha_{lr} + j\alpha_{li}, \qquad l = 1, 2 \tag{9.24}$$

where α_{lr} and α_{li} represent the real and imaginary parts of α_l, respectively. The noise is modeled as a combination of two zero mean Gaussian components—a correlated part and a white noise part. Using this model the noise can be written as

$$\eta_1(t) = \eta(t) + \eta_a(t)$$

$$\eta_2(t) = \eta(t) + \eta_b(t) \tag{9.25}$$

where $\eta(t)$ is the correlated part of the noise and $\eta_a(t)$, $\eta_b(t)$ and $\eta(t)$ are mutually independent. The variances of $\eta_a(t)$ and $\eta_b(t)$ are σ_a^2 and σ_b^2, respectively, with

$$\mathscr{E}\left[\eta_{ar}(t_1)\eta_{ar}(t_2)\right] = \mathscr{E}\left[\eta_{ai}(t_1)\eta_{ai}(t_2)\right] = \sigma_a^2\delta(t_1 - t_2) \tag{9.25a}$$

$$\mathscr{E}\left[\eta_{br}(t_1)\eta_{br}(t_2)\right] = \mathscr{E}\left[\eta_{bi}(t_1)\eta_{bi}(t_2)\right] = \sigma_b^2\delta(t_1 - t_2) \tag{9.25b}$$

$$\mathscr{E}\left[\eta_{ar}(t_1)\eta_{ai}(t_2)\right] = \mathscr{E}\left[\eta_{br}(t_1)\eta_{bi}(t_2)\right] = 0$$

9.6   ADAPTIVE ANTENNA INTERFERENCE CANCELLERS

An adaptive array can be formed to suppress undesired or interfering signals without a priori information about the angles of arrival of the signals.[4,5] In this section two simple adaptive array techniques are discussed which are based on interference cancellation methods at the receiver front end. If a reference signal is not available at the receiver, the signal from the second channel can be used as the reference signal. This alternate method degrades the performance significantly because the signal can be cancelled as well. Nevertheless, if the interfering signal is the dominant signal, this antenna interference canceller should work quite well.

Two adaptive antenna interference cancellers are investigated in this section. Figure 9-8 shows the configuration of the first adaptive antenna interference canceller labeled as, AIC #1 where $2k$ is the gain factor indicated in equation 9.7. The output signal $\bar{x}(t)$ is assumed to be real and can be expressed as

$$\bar{x}(t) = wx_1(t) - x_2(t) \tag{9.26}$$

where $x_l(t)$ for $l = 1, 2$ is given in (9.22) and w is a real weighting coefficient. The mean square error is defined as

$$E(w) = \mathscr{E}\left[\int_0^{T_c}\bar{x}^2(t)\,dt\right] \tag{9.27}$$

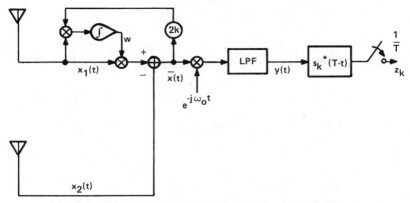

Figure 9-8. Adaptive antenna interference canceller (#1).

where $\mathscr{E}$ denotes the ensemble average and T_c is the chip time interval. The minimum mean square error solution of (9.27) is given by

$$w = \frac{\mathscr{E}\left[\int_0^{T_c} x_1(t) x_2(t)\, dt\right]}{\mathscr{E}\left[\int_0^{T_c} x_1^2(t)\, dt\right]} \tag{9.28}$$

Integrals involving double frequency terms are assumed to be zero so that equation (9.28) can be written as

$$w = \frac{\mathscr{E}\left[(\alpha_{1r}\alpha_{2r} + \alpha_{1i}\alpha_{2i})E_c + \frac{1}{2}|n_{12c}|^2\right]}{\mathscr{E}\left[|\alpha_1|^2 E_c + \frac{1}{2}|n_{1c}|^2\right]}$$

where

$$\frac{1}{2}\int_0^{T_c} s^2(t)\, dt = E_c \tag{9.29}$$

$$\int_0^{T_c}\left(\eta_{lr}^2(t) + \eta_{li}^2(t)\right) dt = |n_{lc}|^2, \qquad l = 1, 2 \tag{9.30}$$

$$\int_0^{T_c}\left(\eta_{1r}(t)\eta_{2r}(t) + \eta_{1i}(t)\eta_{2i}(t)\right) dt = |n_{12c}|^2 \tag{9.31}$$

In (9.31) the ensemble average of $|n_{12c}|^2$ is defined by $\sigma_{12c}^2 = \frac{1}{2}\mathscr{E}[|n_{12c}|^2]$. Using a similar representation to that given by (9.23h) it can be seen that

$$\sigma_{12}^2 = L\,\sigma_{12c}^2 \tag{9.32}$$

Consequently, it can be shown that

$$w = \frac{(\alpha_{1r}\alpha_{2r} + \alpha_{1i}\alpha_{2i})E_c + r\sigma_{1c}\sigma_{2c}}{|\alpha_1|^2 E_c + \sigma_{1c}^2} \tag{9.33}$$

where the normalized correlation coefficient, r, is given by

$$r = \frac{\sigma_{12}^2}{\sigma_1 \sigma_2} = \frac{\sigma_{12c}^2}{\sigma_{1c}\sigma_{2c}} \tag{9.34}$$

By combining (9.26), and (9.22), the output signal can be expressed as

$$\bar{x}(t) = ((w\alpha_{1r} - \alpha_{2r})s(t) + w\eta_{1r}(t) - \eta_{2r}(t))\cos\omega_0 t$$

$$- ((w\alpha_{1i} - \alpha_{2i})s(t) + w\eta_{1i}(t) - \eta_{2i}(t))\sin\omega_0 t \tag{9.35}$$

The low-pass equivalent signal, after mixing and low-pass filtering using a filter gain of 2, becomes

$$y(t) = (w\alpha_1 - \alpha_2)s(t) + w\eta_1(t) - \eta_2(t) \tag{9.36}$$

Note that $y(t)$ is processed with a filter that is matched to the transmitted signal pulse and the PN sequence. The output of the receiver filter is periodically sampled at the symbol rate, $1/T$. The corresponding discrete time signal output at the symbol level, denoted by z_k, can be represented as

$$z_k = \int_0^T y(t)s_k^*(t)\,dt \tag{9.36a}$$

or

$$z_k = (w\alpha_1 - \alpha_2)I_k 2E + wn_1(k) - n_2(k) \tag{9.37}$$

where

$$n_1(k) = \int_0^T \eta_1(t)s_k^*(t)\,dt \tag{9.38}$$

$$n_2(k) = \int_0^T \eta_2(t)s_k^*(t)\,dt \tag{9.39}$$

$$E = \tfrac{1}{2}\int_0^T |s_k(t)|^2\,dt$$

Note that z_k can also be expressed as

$$z_k = |w\alpha_1 - \alpha_2|e^{j\phi}I_k 2E + wn_1(k) - n_2(k) \tag{9.40}$$

where

$$\phi = \tan^{-1}\frac{w\alpha_{1i} - \alpha_{2i}}{w\alpha_{1r} - \alpha_{2r}} \tag{9.40a}$$

The signal z_k is phase compensated by multiplying it by $e^{-j\phi}$. Thus, the decision output variable z_k' becomes

$$z_k' = \text{Re}\left(z_k e^{-j\phi}\right)$$

$$= |w\alpha_1 - \alpha_2|I_k 2E + \text{Re}\left(wn_1(k)e^{-j\phi} - n_2(k)e^{-j\phi}\right) \qquad (9.41)$$

The variance computation of z_k' uses the properties given by equations (9.23b), (9.23c) and (9.23d). In the computation the noise variances are unchanged by the presence of an arbitrary phase ϕ. Using (9.23i), (9.32) and (9.34) the mean and variance of z_k', given $I_k = 1$, are then

$$m_D = 2E|w\alpha_1 - \alpha_2| \qquad (9.42)$$

$$\sigma_D^2 = 2EL\left[w^2\sigma_{1c}^2 + \sigma_{2c}^2 - 2w\sigma_{1c}\sigma_{2c}r\right] \qquad (9.43)$$

Therefore, the error probability is given by (see Appendix Section A.1.14)

$$P_e(\text{AIC \#1}) = P\left[z_k' < 0\right]$$

or

$$P_e(\text{AIC \#1}) = \tfrac{1}{2}\text{erfc}\sqrt{\left(m_D^2/2\sigma_D^2\right)}$$

$$= \tfrac{1}{2}\text{erfc}\sqrt{\left(\gamma_1|w - \beta_1|^2/\left(w^2 + \beta_{2c}^2 - 2wr\beta_{2c}\right)\right)} \qquad (9.44)$$

where w is given in (9.22) and γ_1, β_1, and β_{2c} are given by

$$\gamma_1 = \frac{E}{L\sigma_{1c}^2}|\alpha_1|^2, \qquad \beta_1 = \frac{\alpha_2}{\alpha_1}, \qquad \beta_{2c} = \frac{\sigma_{2c}}{\sigma_{1c}} \qquad (9.45)$$

Equations (9.44) and (9.45) were developed by defining the MSE in terms of the chip interval implying that the coefficient w is updated at the chip rate. (See (9.27)). If the MSE is defined in terms of the symbol interval, equation (9.44) is once again obtained with

$$\gamma_1 = \frac{E}{\sigma_1^2}|\alpha_1|^2, \qquad \beta_1 = \frac{\alpha_2}{\alpha_1} \quad \text{and} \quad \beta_2 = \frac{\sigma_2}{\sigma_1} = \beta_{2c}$$

In this case the coefficient w is updated at the symbol rate. Since an equivalence between chip and symbol performance has been established, the remainder of the analysis in this chapter uses quantities expressed in terms of the symbol interval.

Figure 9-9 shows the configuration of the second antenna interference canceller labeled as AIC #2. The technique applies to narrowband interfering

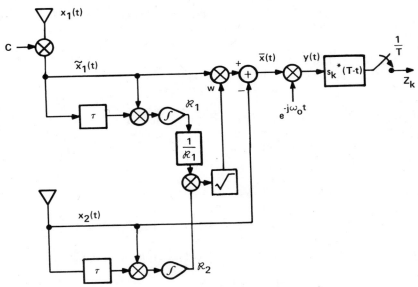

Figure 9-9. Antenna interference canceller (#2).

signals only. The signals from the first and second antennas $\tilde{x}_1(t)$ and $x_2(t)$ are delayed by τ units, where $\tilde{x}_1(t)$ is determined by weighting $x_1(t)$ with a scale factor c, that is, $\tilde{x}_1(t) = cx_1(t)$. The delayed signals $\tilde{x}_1(t - \tau)$ and $x_2(t - \tau)$ are then correlated with the original signals $\tilde{x}_1(t)$ and $x_2(t)$, respectively. Choosing τ as the chip time interval T_c, a minimum correlation output is produced. Since the interfering signals are narrowband, the autocorrelation functions $\mathscr{R}_1$ and $\mathscr{R}_2$, shown in Figure 9-9, are dependent on the interfering signals only. The signal inputs $\tilde{x}_1(t)$ and $x_2(t)$ are then given by

$$\tilde{x}_1(t) = R_e\{c(\alpha_1 s(t) + \eta_1(t))e^{j\omega_0 t}\}$$

$$= c(\alpha_{1r}s(t) + \eta_{1r}(t))\cos\omega_0 t - c(\alpha_{1i}s(t) + \eta_{1i}(t))\sin\omega_0 t \quad (9.46)$$

$$x_2(t) = R_e\{(\alpha_2 s(t) + \eta_2(t))e^{j\omega_0 t}\}$$

$$= ((\alpha_{2r}s(t) + \eta_{2r}(t))\cos\omega_0 t - (\alpha_{2i}s(t) + \eta_{2i}(t))\sin\omega_0 t \quad (9.47)$$

The autocorrelation outputs $\mathscr{R}_1$ and $\mathscr{R}_2$ are computed by assuming that the double frequency terms average to zero and that the factor $\dfrac{\cos\omega_0\tau}{2}$ is suppressed by the integration gain. The PN signal properties for $\tau \geq T_c$ result in

$$\int_0^T s(t)s(t - \tau)\,dt = 0$$

It is also assumed that integrals involving signal and noise products are zero. For example,

$$\int_0^T s(t)\eta_r(t-\tau)\,dt = 0$$

The autocorrelation outputs $\mathscr{R}_1$ and $\mathscr{R}_2$ can then be expressed as

$$\mathscr{R}_1 = c^2 R_1(\tau) \tag{9.48}$$

$$\mathscr{R}_2 = R_2(\tau) \tag{9.49}$$

where for $l = 1, 2$,

$$R_l(\tau) = \int_0^T \left[\eta_{rl}(t)\eta_{rl}(t-\tau) + \eta_{il}(t)\eta_{il}(t-\tau)\right]dt \tag{9.50}$$

Assuming that $R_1(\tau) = R_2(\tau)$ the weighting factor w is selected to be

$$w = \sqrt{\mathscr{R}_2/\mathscr{R}_1} = \frac{1}{c} \tag{9.51}$$

Hence, the output $\bar{x}(t)$ can be expressed as

$$\begin{aligned}
\bar{x}(t) &= w\tilde{x}_1(t) - x_2(t) \\
&= \left[(\alpha_{1r} - \alpha_{2r})s(t) + \eta_{1r}(t) - \eta_{2r}(t)\right]\cos\omega_0 t \\
&\quad - \left[(\alpha_{1i} - \alpha_{2i})s(t) + \eta_{1i}(t) - \eta_{2i}(t)\right]\sin\omega_0 t \tag{9.52}
\end{aligned}$$

Note that the interfering signals are completely eliminated in $\bar{x}(t)$ (see (9.25)). Neglecting double frequency terms and using a low pass filter with a gain of 2 the low-pass equivalent signal $y(t)$ becomes

$$y(t) = (\alpha_1 - \alpha_2)s(t) + \eta_a(t) - \eta_b(t) \tag{9.53}$$

where $\eta_a(t)$ and $\eta_b(t)$ are white Gaussian noise processes. A similar detection procedure to that described for AIC #1 can be used again. Using the definition of ϕ in (9.40a) with $w = 1$ the sampled symbol output z_k' can be written as

$$z_k' = |\alpha_1 - \alpha_2|I_k 2E + R_e\left[n_a(k)e^{-j\phi} - n_b(k)e^{-j\phi}\right] \tag{9.54}$$

where

$$n_a(k) = \int_0^T \eta_a(t)s_k^*(t)\,dt \tag{9.55}$$

$$n_b(k) = \int_0^T \eta_b(t)s_k^*(t)\,dt. \tag{9.56}$$

Using (9.25a) and (9.25b) the mean and variance of the decision variable given $I_k = 1$ are

$$m_D = 2E|\alpha_1 - \alpha_2| \tag{9.57}$$

$$\sigma_D^2 = 2E\left(\sigma_a^2 + \sigma_b^2\right) \tag{9.58}$$

Hence, the error probability is given by (see Appendix Section A.1.14)

$$P_e(\text{AIC} \#2) = \tfrac{1}{2}\text{erfc}\left(\frac{m_D}{\sqrt{2}\,\sigma_D}\right)$$

$$= \tfrac{1}{2}\text{erfc}\sqrt{\frac{E|\alpha_1 - \alpha_2|^2}{\sigma_a^2 + \sigma_b^2}} \tag{9.59}$$

To evaluate the error probability in (9.59) the noise sequences $\{n_l(k)\}$ for $l = 1, 2$ at the correlation outputs are defined as follows:

$$n_1(k) = n(k) + n_a(k) \tag{9.60}$$

$$n_2(k) = n(k) + n_b(k) \tag{9.60a}$$

where

$$n(k) = \int_0^T \eta(t) s_k^*(t)\, dt \tag{9.60b}$$

The moments in (9.59) can be expressed in terms of the variance of the noise sequences $\{n_1(k)\}$ and $\{n_2(k)\}$ and the correlation coefficient $\bar{r}$ where

$$\bar{r} = \frac{\mathscr{E}\left(n_1(k) n_2^*(k)\right)}{\sqrt{\mathscr{E}\left[|n_1(k)|^2\right]\mathscr{E}\left[|n_2(k)|^2\right]}}$$

The noise sequences are zero mean complex Gauss processes with variances

$$\mathscr{E}\left[|n_i(k)|^2\right] = 4E\sigma_i^2, \qquad i = 1, 2$$

Note that $n(k)$ corresponds to the correlated part of the noise sequence and is completely eliminated by this canceller. The variance of $n(k)$ is $4E\sigma_n^2$ and can be related to variances of $n_1(k)$ and $n_2(k)$ by

$$\sigma_1^2 = \sigma_n^2 + \sigma_a^2$$

$$\sigma_2^2 = \sigma_n^2 + \sigma_b^2$$

The correlation coefficient can then be written as

$$\bar{r} = \frac{\sigma_n^2}{\sigma_1 \sigma_2}$$

Equation (9.59) can now be reexpressed as (see Appendix Section A.1.14)

$$P_e(\text{AIC } \#2) = \tfrac{1}{2}\text{erfc}\sqrt{\frac{\gamma_2|1 - \beta_1|^2}{\beta^2 + |\beta_1|^2 - 2\bar{r}\beta|\beta_1|}} \tag{9.61}$$

where

$$\gamma_2 = \frac{E|\alpha_2|^2}{\sigma_2^2}, \qquad \beta_1 = \frac{\alpha_2}{\alpha_1}, \qquad \beta = \frac{|\beta_1|}{\beta_2}, \qquad \text{and} \qquad \beta_2 = \frac{\sigma_2}{\sigma_1}$$

The performance of AIC #1 and AIC #2 is given in Section 9.8.

9.7 BASEBAND ANTENNA ARRAY COMBINING TECHNIQUES

The RF/IF received signals given in (9.22) may be heterodyned and low-pass filtered to baseband, and combined at the receiver. The received equivalent low-pass signal from the first and second channel can be expressed as

$$x_i(t) = \alpha_i s(t) + \eta_i(t), \qquad i = 1, 2 \tag{9.62}$$

where $i = 1$ and $i = 2$ represent the signals at the first and second antennas, respectively, and $\eta_i(t)$ is the additive receiver noise. One possible characterization for the noise processes is to assume that they are statistically independent between channels and are each zero mean white and Gaussian. On the other hand, the noise processes may be highly correlated between channels. For example, interference from local thunderstorms or jammers can introduce this correlation. In this instance the noise processes are assumed to be zero mean, correlated and Gaussian.

The received signals $x_1(t)$ and $x_2(t)$ are processed with filters that are matched to the transmitted signal pulse and the PN sequence. The outputs of these filters are periodically sampled at the symbol rate $1/T$ where $T = LT_c$. The resulting discrete time signal, denoted by $y_i(k)$, can be represented as

$$y_i(k) = \int_0^T x_i(t)s_k^*(t)\, dt$$

$$= \int_0^T \left[\alpha_i\left(\sum_{m=0}^{\infty} I_m s_m(t)\right) + \eta_i(t)\right] s_k^*(t)\, dt$$

$$= 2E\alpha_i I_k + n_i(k), \qquad i = 1, 2 \tag{9.63}$$

where

$$\frac{1}{2} \int_0^T s_m(t) s_k^*(t) \, dt = \begin{cases} E, & m = k \\ 0, & m \neq k \end{cases} \tag{9.64}$$

and

$$\int_0^T \eta_i(t) s_k^*(t) \, dt = n_i(k) \tag{9.65}$$

The complex valued factors α_1 and α_2 account for direct path attenuation and phase shift (see (9.24)). Note that for a binary PSK modulation and perfect symbol synchronization, α_1 and α_2 are real quantities. The noise sequences $\{n_1(k)\}$ and $\{n_2(k)\}$ are zero mean complex Gaussian processes, with variances

$$\mathscr{E}\left[|n_i(k)|^2\right] = 4E\sigma_i^2, \qquad i = 1, 2 \tag{9.66}$$

9.7.1 Minimum Mean Square Error (MSE) Diversity Antenna Combining Technique with Independent Gaussian Noise

Using the minimum MSE criterion with independent Gaussian noise processes $\{n_1(k)\}$ and $\{n_2(k)\}$, an estimate of I_k is formed by combining the received signal samples $y_1(k)$ and $y_2(k)$ so that the mean square error is a minimum. Figure 9-10 depicts the baseband diversity combining technique. The estimate of the information symbol can be expressed as

$$D = \text{Re}\{w_1 y_1(k) + w_2 y_2(k)\} \tag{9.67}$$

where w_1 and w_2 are the weights to be determined. For the estimate specified by (9.67), given that $I_k = 1$ is transmitted, the weights w_1 and w_2 are chosen to minimize the MSE defined as

$$E(w) = \mathscr{E}\left[|2E - (w_1 y_1(k) + w_2 y_2(k))|^2\right]$$

$$= \mathscr{E}\left[|2E - 2E\alpha_1 w_1 - 2E\alpha_2 w_2 - w_1 n_1(k) - w_2 n_2(k)|^2\right] \tag{9.68}$$

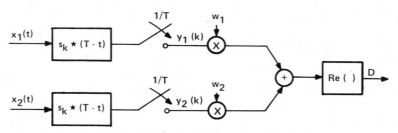

Figure 9-10. Baseband diversity combining technique.

In the MSE expression $2E$ is the matched filter output when there is no channel and represents a desired reference. Minimization of $E(w)$ with respect to w_1 and w_2 leads to

$$w_1 = \frac{E\alpha_1^* \sigma_2^2}{E|\alpha_1|^2\sigma_2^2 + E|\alpha_2|^2\sigma_1^2 + \sigma_1^2\sigma_2^2} \qquad (9.69)$$

$$w_2 = \frac{E\alpha_2^* \sigma_1^2}{E|\alpha_1|^2\sigma_2^2 + E|\alpha_2|^2\sigma_1^2 + \sigma_1^2\sigma_2^2} \qquad (9.70)$$

Substituting (9.69) and (9.70) into (9.67), the estimate becomes

$$D = \frac{1}{Den}\left[2E^2\left(|\alpha_1|^2\sigma_2^2 + |\alpha_2|^2\sigma_1^2\right)\right.$$

$$\left. + E\sigma_2^2|\alpha_1||n_1(k)|\cos(\psi_1 - \theta_1) + E\sigma_1^2|\alpha_2||n_2(k)|\cos(\psi_2 - \theta_2)\right] \qquad (9.71)$$

where

$$Den = E|\alpha_1|^2\sigma_2^2 + E|\alpha_2|^2\sigma_1^2 + \sigma_1^2\sigma_2^2 \qquad (9.72)$$

$$\alpha_1 = |\alpha_1|e^{j\theta_1}$$

$$\alpha_2 = |\alpha_2|e^{j\theta_2}$$

$$n_1(k) = |n_1(k)|e^{j\psi_1}$$

$$n_2(k) = |n_2(k)|e^{j\psi_2}.$$

Hence, the mean and variance are given by

$$m_D = \mathscr{E}[D]$$

$$= \frac{2E^2}{Den}\left[|\alpha_1|^2\sigma_2^2 + |\alpha_2|^2\sigma_1^2\right] \qquad (9.73)$$

$$\sigma_D^2 = \mathscr{E}\left[|D - m_D|^2\right]$$

$$= \frac{2E^3\sigma_1^2\sigma_2^2}{Den^2}\left[|\alpha_1|^2\sigma_2^2 + |\alpha_2|^2\sigma_1^2\right] \qquad (9.74)$$

The error probability for the minimum MSE combining technique can be expressed as (see Appendix Section A.1.14)

$$P_e(\text{MSE}) = \tfrac{1}{2}\text{erfc}\sqrt{\frac{m_D^2}{2\sigma_D^2}}$$

$$= \tfrac{1}{2}\text{erfc}\sqrt{\gamma_1 + \gamma_2}$$

$$= \tfrac{1}{2}\text{erfc}\sqrt{\gamma_1(1 + \beta^2)} \tag{9.75}$$

where γ_1, γ_2 and r are given as

$$\gamma_1 = \frac{E|\alpha_1|^2}{\sigma_1^2}$$

$$\gamma_2 = \frac{E|\alpha_2|^2}{\sigma_2^2}$$

$$\beta_1 = \frac{|\alpha_2|}{|\alpha_1|}$$

$$\beta_2 = \frac{\sigma_2}{\sigma_1}$$

$$\beta = \frac{\beta_1}{\beta_2} \tag{9.76}$$

9.7.2   Minimum Mean Square Error (MSE) Diversity Antenna Combining Technique with Correlated Gaussian Noise

Using the minimum MSE criterion with correlated Gaussian noise processes $\{n_1(k)\}$ and $\{n_2(k)\}$, the signals can be combined in similar fashion to the method illustrated in Section 9.7.1. In order to simplify the analysis, it is assumed that each channel has a separate phase-locked loop so that each phase shift θ_1 and θ_2 introduced by the channel has been estimated and compensated for. The combining technique is shown in Figure 9-11. The received signal samples $y_1(k)$ and $y_2(k)$ are now real quantities. The analysis can be extended to baseband complex equivalents, that is, without relative phase-shift compensation (see the appendix at the end of this chapter). The estimate of the information symbol I_k is

$$D = \text{Re}\{w_1 y_1(k) + w_2 y_2(k)\} \tag{9.77}$$

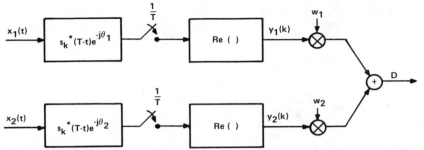

Figure 9-11. Antenna diversity combining technique with compensation for individual phase shifts.

where

$$y_1(k) = 2E|\alpha_1|I_k + n_{1r}$$

$$y_2(k) = 2E|\alpha_2|I_k + n_{2r} \qquad (9.78)$$

and $n_{lr} = \mathrm{Re}(n_l(k))$ for $l = 1, 2$, and w_1 and w_2 are the weights to be determined. Assuming $I_k = 1$, the weights w_1 and w_2 are chosen to minimize the MSE defined as

$$E(w) = \mathscr{E}\left[\left(2E - (w_1 y_1(k) + w_2 y_2(k))\right)^2 \right]$$

$$= \mathscr{E}\left[\left(2E - 2Ew_1|\alpha_1| - 2Ew_2|\alpha_2| - w_1 n_{1r} - w_2 n_{2r}\right)^2 \right] \qquad (9.79)$$

Minimization of the MSE with respect to w_1 and w_2 leads to

$$w_1 = \frac{2E\left(\sigma_2'^2|\alpha_1| - r'\sigma_1'\sigma_2'|\alpha_2|\right)}{Den} \qquad (9.80)$$

$$w_2 = \frac{2E\left(\sigma_1'^2|\alpha_2| - r'\sigma_1'\sigma_2'|\alpha_1|\right)}{Den} \qquad (9.81)$$

where

$$Den = 2E\sigma_1'^2|\alpha_2|^2 + 2E\sigma_2'^2|\alpha_1|^2 + \sigma_1'^2\sigma_2'^2 - 4Er'\sigma_1'\sigma_2'|\alpha_1||\alpha_2| - r'^2\sigma_1'^2\sigma_2'^2 \qquad (9.82)$$

and σ_1', σ_2' and r' are given as

$$2E\sigma_1'^2 = \mathscr{E}\{n_{1r}^2\}, \qquad 2E\sigma_2'^2 = \mathscr{E}\{n_{2r}^2\}, \qquad r' = \frac{\mathscr{E}\{n_{1r}n_{2r}\}}{2E\sigma_1'\sigma_2'} \qquad (9.83)$$

Substituting (9.80) and (9.81) into (9.78), it can be shown that the mean and variance of D, given $I_k = 1$, are

$$
m_D = \mathscr{E}(D)
$$

$$
= \frac{4E^2}{\text{Den}} \left[\sigma_1'^2 |\alpha_2|^2 + \sigma_2'^2 |\alpha_1|^2 - 2r'\sigma_1'\sigma_2'|\alpha_1||\alpha_2| \right] \tag{9.84}
$$

$$
\sigma_D^2 = \text{Var}(D)
$$

$$
= \frac{8E^3}{\text{Den}^2} \left[\sigma_1'^2\sigma_2'^2 (1 - r'^2)\left(\sigma_2'^2 |\alpha_1|^2 + \sigma_1'^2 |\alpha_2|^2 - 2r'\sigma_1'\sigma_2'|\alpha_1||\alpha_2| \right) \right]
$$

$$
\tag{9.85}
$$

Hence, the error probability for the minimum MSE diversity combining technique is given by (see Appendix Section A.1.14)

$$
P_e(\text{MSE}) = \tfrac{1}{2}\text{erfc}\left[\sqrt{\gamma_1' \frac{\beta_1^2 + \beta_2'^2 - 2r'\beta_1\beta_2'}{(1 - r'^2)\beta_2'^2}} \right]
$$

$$
= \tfrac{1}{2}\text{erfc}\left[\sqrt{\gamma_1' \frac{\beta'^2 - 2r'\beta' + 1}{(1 - r'^2)}} \right] \tag{9.86}
$$

where γ_1', β_1, β_2' and β are given as

$$
\beta_1 = \frac{|\alpha_2|}{|\alpha_1|}, \qquad \beta_2' = \frac{\sigma_2'}{\sigma_1'}, \qquad \beta' = \frac{\beta_1}{\beta_2'}, \qquad \gamma_1' = \frac{E}{\sigma_1'^2}|\alpha_1|^2 \tag{9.87}
$$

The above analysis is now extended to include the case where the received signal samples $y_1(k)$ and $y_2(k)$ are complex quantities. In this analysis the phase shift on each channel is not compensated by a phase tracking loop but is instead estimated by the complex tap weights w_1 and w_2. It can then be shown that the tap gains w_1 and w_2 are given by (see appendix at the end of this chapter)

$$
w_1 = c\left(\beta_2^2 \alpha_1^* - \beta_2 \bar{r}\alpha_2^* \right) \tag{9.88}
$$

$$
w_2 = c\left(\alpha_2^* - \beta_2 \bar{r}\alpha_1^* \right) \tag{9.89}
$$

where

$$c = \frac{E\sigma_1^2}{Den} \tag{9.90}$$

$$Den = E\sigma_2^2|\alpha_1|^2 + E\sigma_1^2|\alpha_2|^2 + \sigma_1^2\sigma_2^2(1 - \bar{r}^2) - 2E\bar{r}\sigma_1\sigma_2\,\text{Re}(\alpha_1\alpha_2^*) \tag{9.91}$$

$$\beta_2 = \frac{\sigma_2}{\sigma_1} \tag{9.92}$$

$$\mathscr{E}(n_1(k)n_2^*(k)) = 4E\bar{r}\sigma_1\sigma_2 \tag{9.93}$$

The error probability for the minimum MSE diversity combining technique is then expressed as (see Appendix Section A.1.14)

$$P_e(\text{MSE}) = \tfrac{1}{2}\text{erfc}\left[\sqrt{\gamma_1 \frac{\beta_1^2 + \beta_2^2 - 2\bar{r}\beta_2\delta}{(1 - \bar{r}^2)\beta_2^2}}\,\right] \tag{9.94}$$

where $\delta = \text{Re}(\alpha_1\alpha_2^*)/|\alpha_1|^2$ and $\gamma_1 = E|\alpha_1|^2/\sigma_1^2$. Equation (9.94) is identical to (9.86) if α_1 and α_2 are real quantities.

9.8 ERROR RATE PERFORMANCE

The performance of three adaptive array techniques for independent Gaussian noise and correlated Gaussian Noise is given in this section. Since the minimum MSE technique is optimum and equivalent to the MLM for Gaussian noise, it can be used as a benchmark for comparison purposes.

The results for the bit error rate performance are presented in Figures 9-12 to 9-17. The probability of a bit error as a function of signal-to-noise ratio (E_b/N_0), corresponding to γ_1 for channel #1 (or, γ_2 for channel #2), is illustrated for several values of β, β_1, β_2, α_1, and α_2. In Figure 9-12, the performance loss obtained from 9.94 for correlation coefficients $\bar{r} = 0.5$ and $\bar{r} = 1.0$ are 1.75 dB and 3.0 dB, respectively, relative to $\bar{r} = 0.0$ for the minimum MSE technique. The performance loss in the case of correlated noise is due to the assumed channel condition $\beta = 1$. For example, $\bar{r} = 1.0$ implies that the received signals $y_1(k)$ and $y_2(k)$ in both channels are identical and that no gain due to diversity combining is possible. However, this channel condition is not expected to occur often. In Figure 9-13, the signals in both channels have different signals strengths. Hence, impressive performance gains for the minimum MSE technique can be achieved by diversity combining as a

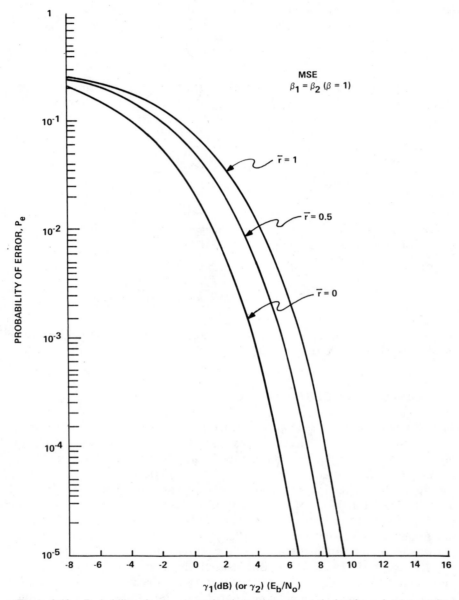

Figure 9-12. Probability of error versus signal-to-noise power ratio for channel #1 (or #2).

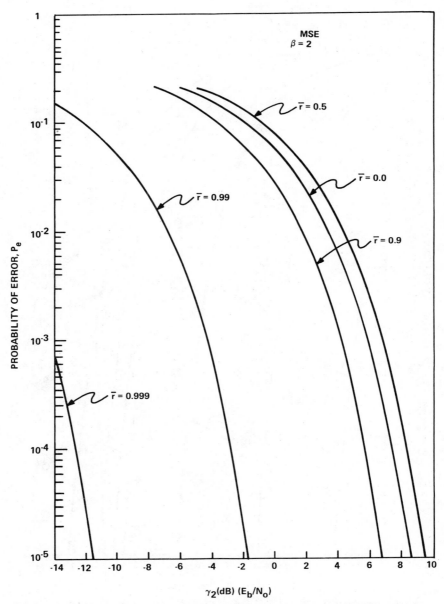

Figure 9-13. Probability of error versus signal-to-noise power ratio for channel #2.

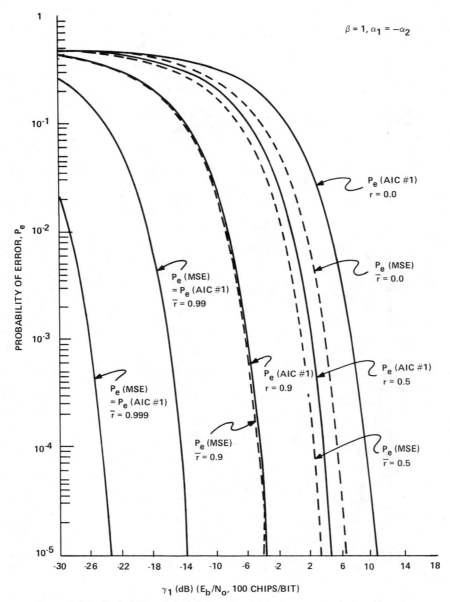

Figure 9-14. Probability of error versus signal-to-noise power ratio for channel #1.

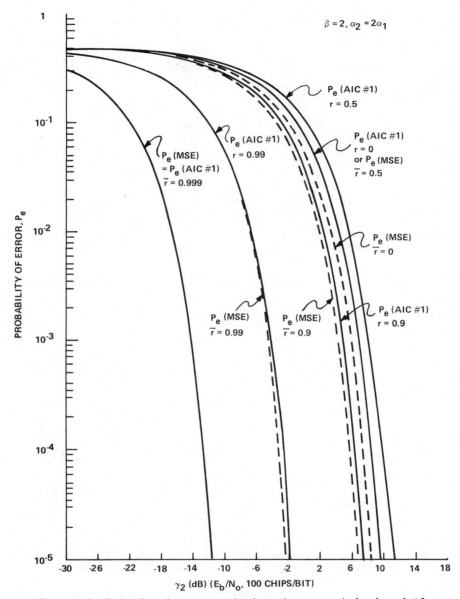

Figure 9-15. Probability of error versus signal-to-noise power ratio for channel #2.

323

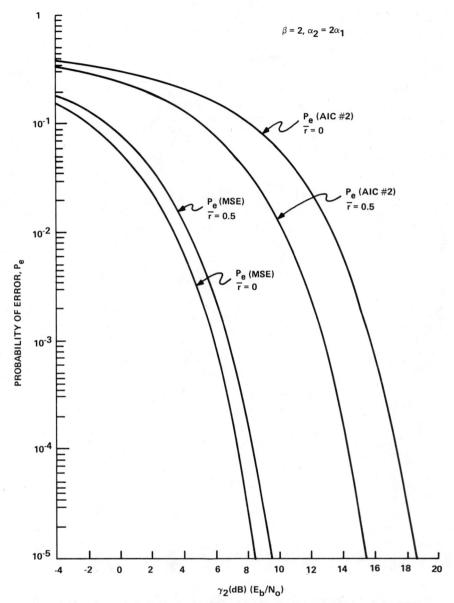

Figure 9-16. Probability of error versus signal-to-noise power ratio for channel #2.

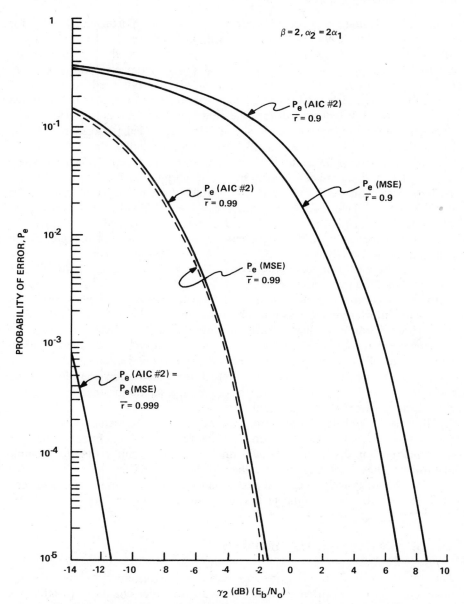

Figure 9-17. Probability of error versus signal-to-noise power ratio for channel #2.

result of correlated noise processes. In Figure 9-13 the performance gains for $\bar{r} = 0.0, 0.9, 0.99, 0.999$ are 1.0 dB, 2.7 dB, 11.2 dB and 21 dB, respectively, relative to the performance obtained for $\bar{r} = 0.5$. Note that $\bar{r} = 0.5$ corresponds to the worst-case performance and results from choosing $\beta = 2$. In Figures 9-14 and 9-15 the performance of the AIC #1 obtained from 9.44 and minimum MSE technique for the 100-chip-per-bit case is depicted. It is noted that the AIC #1 tends to cancel the interference when the interference is stronger. Otherwise, it cancels the signal. Thus degraded performance occurs for SNR values in excess of 20 dB. It is anticipated that the AIC #1 will perform well for $r \geq 0.9$ and a SNR(E_b/N_0) ≤ 20 dB, except in the case where $\alpha_1 = \alpha_2$ (α_1 and α_2 are assumed to be real in all figures). In Figures 9-16 and 9-17 the performance for the AIC #2 obtained from 9.61 and that for the minimum MSE technique are exhibited. The AIC #2 will cancel the narrow-band interfering signal but may also cancel the desired signal. In Figures 9-16 and 9-17 the signals from both channels are partly cancelled so that significant degradation occurs between AIC #2 and the MSE technique for $\bar{r} \leq 0.9$. However, the performance of the AIC #2 for $\bar{r} > 0.9$ is close to that of the optimum method (minimum MSE).

In this section the performance of the three adaptive antenna array techniques for interference rejection has been presented. The AIC methods can be used as a sidelobe canceller or an interference canceller. In the latter case the interference canceller may cause the signal and interference to be simultaneously cancelled. The minimum MSE adaptive array technique is optimum for a Gaussian process but requires greater implementation complexity than the AIC methods. All of the diversity combining techniques can be used for enhancing the SNR or for anti-jam purposes.

The tap weights of the array can be updated by a steepest descent or a square root Kalman filter algorithm. The square-root Kalman filter can be used for tracking rapidly fading signals and fast phase shifts due to channel instability. These methods are described in the next section.

Adaptive array techniques and performance for more than two elements can be developed in a way similar to that given in previous sections.

9.9 MINIMUM MSE ALGORITHMS

The minimum MSE criterion provides optimum performance for the Gaussian noise channel and is often close to optimum for other channels. In addition, this criterion results in algorithms which can be implemented with acceptable complexity. Specific algorithms which implement the minimum MSE criterion are now considered. Two algorithms, the steepest descent algorithm and the square root Kalman filter algorithm, are discussed in this section.

The steepest descent algorithm is based on a very simple iterative procedure in which one begins by choosing an initial vector $(0, 0)$ (or $(1, 0)$) for (w_1, w_2). The symbol-by-symbol change in the jth tap gain is proportional to the size of the jth gradient component. Thus, succeeding values of the coefficients $w_j = w_j(k)$ for $j = 1, 2$ at the kth iteration are updated according to the relation

$$w_j(k) = w_j(k-1) + \Delta\varepsilon(k)y_j^*(k), \qquad j = 1, 2 \qquad (9.95)$$

where $\varepsilon(k) = \tilde{I}_k - \hat{I}_k$ is the error signal at the kth iteration (see Figure 9-18), $\{y_j(k)\}$ are the discrete time equivalent received signal samples that make up the estimate $\hat{I}_k$, Δ is a positive number chosen small enough to ensure covergence of the iterative procedure, and $\hat{I}_k$ is given by

$$\hat{I}(k) = w_1(k)y_1(k) + w_2(k)y_2(k) \qquad (9.96)$$

The implementation of the steepest-descent algorithm is given in Figure 9-18.

Often the simple steepest descent algorithm does not provide tracking capabilities required to follow rapid signal fading or fast phase shift changes. A Kalman filter, on the other hand, provides a considerably faster tracking rate resulting in successful tracking of a time-varying channel (see Section 6.4.2.2). Unfortunately, the Kalman algorithm has been found to be sensitive to computer roundoff errors, with a numerical accuracy that sometimes degrades

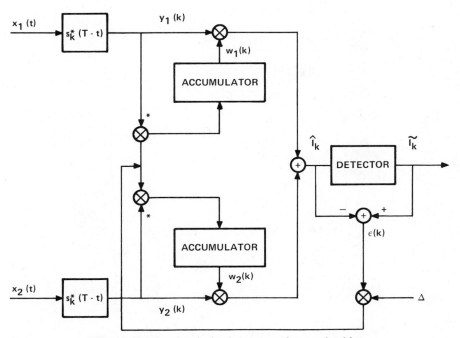

Figure 9-18. Baseband adaptive steepest descent algorithm.

performance to the point where the results are not meaningful. In other applications, researchers have shown that the square root formulation of the Kalman algorithm has inherently better stability and numerical accuracy than the conventional Kalman algorithm.[14] The Kalman U-D covariance factorization filter proposed for updating the weights is depicted in Figure 9-19. Note that a switch is provided in the figure and is used to accomplish array training with a known information sequence. The computation for succeeding values of the coefficients $w_j = w_j(k)$ for $j = 1, 2$ at the kth iteration from the received signal samples $\{y_j(k)\}$ are obtained according to the relation

$$w_j(k) = w_j(k-1) + G_j, \qquad j = 1, 2 \tag{9.97}$$

where $\{G_j\}$ are computed according to the equations provided in Section 5.4.3, for $N = 2$. These equations can then be written in the notation provided in this section as

$$\varepsilon(k) = \tilde{I}_k - w_1(k)y_1(k) - w_2(k)y_2(k)$$

$$f_1(k-1) = y_1^*(k)$$

$$f_2(k-1) = u_{1,2}(k-1)y_1^*(k) + y_2^*(k)$$

$$g_{1,2} = d_1(k-1)f_1(k-1)$$

$$g_{2,3} = d_2(k-1)f_2(k-1)$$

$$\alpha_1' = \xi + g_{1,2}f_1^*(k-1)$$

$$\alpha_2' = \alpha_1' + g_{2,3}f_2^*(k-1)$$

$$d_1(k) = \frac{d_1(k-1)\xi}{\alpha_1'} \tag{9.98}$$

$$\lambda_2 = \frac{-f_2(k-1)}{\alpha_1'}$$

$$d_2(k) = d_2(k-1)\alpha_1'/\alpha_2'$$

$$u_{1,2}(k) = u_{1,2}(k-1) + g_{1,2}^*\lambda_2$$

$$g_{1,3} = g_{1,2} + g_{2,3}u_{1,2}^*(k-1)$$

$$\varepsilon' = \varepsilon(k)/\alpha_2'$$

$$G_j = g_{j,j+1}\varepsilon', \qquad j = 1, 2$$

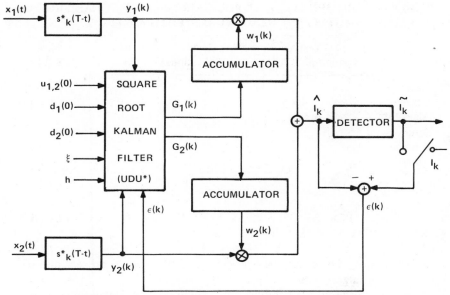

Figure 9-19. Baseband adaptive square root Kalman (U-D) filter.

The quantities $f_1(k-1)$, $f_2(k-1)$, $g_{1,2}$, $g_{2,3}$, λ_2, $u_{1,2}(k)$ and ε' are complex intermediate variables, and d_1, d_2, α'_1, and α'_2, are real intermediate variables. The performance is quite sensitive to the parameter, ξ. Consequently, this parameter should be adjusted in order to obtain optimum performance. The initial values of $u_{1,2}(k)$, $d_1(k)$ and $d_2(k)$ in the square root algorithm are given as

$$d_1(0) = d_2(0) = 1.0, \qquad u_{1,2}(0) = (0.0, 0.0) \tag{9.99}$$

where the notation (a, b) denotes the real part labeled a and the imaginary part labeled b of a complex quantity. The algorithm can be modified to incorporate the periodic resetting of the U-D factors according to Eq. (9.99). The resetting scheme may be used to prevent the propagation of round-off errors associated with the adaptation memory. Besides the square root Kalman filter discussed above, an alternate square root Kalman filter can be used which results in more stable convergence.[14]

The computational requirements in terms of the number of additions, multiplications, and divisions per bit for the steepest descent and square root Kalman algorithms at each iteration are listed in Table 9-1. Both algorithms are stable if the tracking parameters (Δ and ξ) are chosen correctly.

TABLE 9-1. COMPUTATIONAL REQUIREMENTS OF SQUARE-ROOT KALMAN AND STEEPEST-DESCENT ALGORITHMS

	Complex multiplications	Real multiplications	Real divisions	Real Additions
Steepest Descent Algorithm	4	2	0	11
Square Root Kalman Algorithm	7	18	2	28

9.10 COMPUTER SIMULATION FOR SPEED OF CONVERGENCE WITH MINIMUM MSE ALGORITHMS

In order to show the convergence properties of the steepest descent and square root Kalman algorithms, the actual tap weight obtained theoretically (in Section 9.7) is used as a benchmark for comparison. As a consequence, the tap weight updated bit by bit with the adaptive algorithms can be compared to the actual tap weight. For example, the tap weight given in (9.69) for $E = \frac{1}{2}$ is rewritten as

$$w_1 = \frac{\sigma_2^2 \alpha_1^*}{\sigma_2^2 |\alpha_1|^2 + \sigma_1^2 |\alpha_2|^2 + 2\sigma_1^2 \sigma_2^2} \tag{9.100}$$

Now define γ_1 and γ_2 by:

$$\gamma_1 = \frac{|\alpha_1|^2}{2\sigma_1^2}, \qquad \gamma_2 = \frac{|\alpha_2|^2}{2\sigma_2^2} \tag{9.101}$$

In Figures 9-20 and 9-21, the real part of w_1, computed according to (9.100), is shown with the corresponding tap weight updated by the steepest descent or square root Kalman algorithms as a function of the iteration number where the energy $2E$ is normalized to unity. The initial value of w_1 is set to $(0, 0)$. The straight horizontal line indicates the real part of the actual tap weight given in (9.100). The time-varying curve depicts the real part of the updated tap weight as a function of the time sample index. Similar results can be obtained for the imaginary part of the tap weight. In Figure 9-20, the convergence properties for the steepest-descent algorithm are shown for $\gamma_1 = \gamma_2 = 0$ dB, $\alpha_1 = \alpha_2 = (1.0, 0.0)$, $\sigma_1^2 = \sigma_2^2 = 0.5$, and $\Delta = 0.01$. The actual symbol I_k is used to replace the decision symbol $\tilde{I}_k$ for training purposes. The results indicate that the tap weight updated by the steepest descent algorithm will converge in

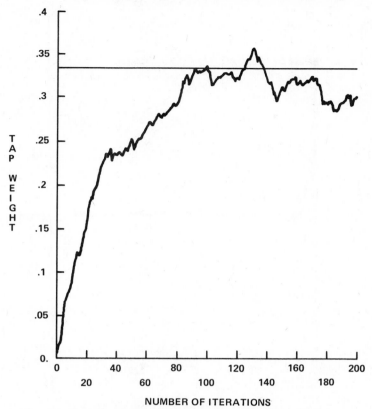

Figure 9-20.   Tap weight convergence for steepest descent algorithm.

about 200 iterations. In Figure 9-21, the convergence properties for the square root Kalman algorithm are shown for $\gamma_1 = \gamma_2 = -3$ dB, $\alpha_1 = (0.707, 0.707)$, $\alpha_2 = (1.0, 0.0)$, $\sigma_1^2 = \sigma_2^2 = 1.0$, $\xi = 0.1$, and a threshold, $T_r = 0.1$. In this figure, a training symbol (I_k) is used for the first 15 iterations and the decision-directed symbol estimate $(\tilde{I}_k)$ is then used for the rest of the iterations. In order to reduce the effects of errors in the decision-directed mode, a threshold (T_r) is set so that the signals with symbol estimate $|\hat{I}_k| < T_r$ are not utilized for updating the weights. The results depicted in Figure 9-21 indicate that the tap weight updated by the square root Kalman algorithm will converge in about 10 iterations if proper parameters ξ and T_r are chosen. The convergence rate of the square root Kalman algorithm shown in Figure 9-21 is typical and is not sensitive to signal, noise and channel parameters. Consequently, the diversity combining technique implemented with the square root Kalman algorithm will achieve an order of magnitude faster convergence than the steepest descent algorithm. Both algorithms are quite stable under poor signal-to-noise ratio (SNR ≤ 0 dB) conditions.

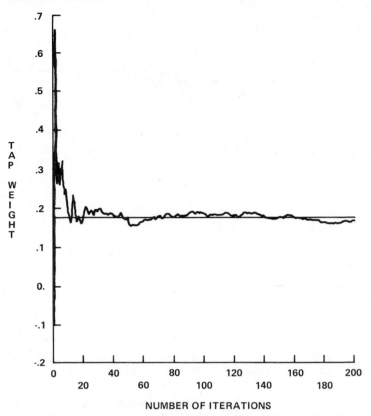

Figure 9-21. Tap weight convergence for square root Kalman algorithm.

APPENDIX: MINIMUM MEAN SQUARE ERROR (MSE) DIVERSITY ANTENNA COMBINING TECHNIQUE WITH JOINTLY STATIONARY GAUSSIAN NOISE PROCESSES

In the minimum MSE criterion an estimate of I_k is formed by combining the received signal samples $y_1(k)$ and $y_2(k)$ so that the mean square error is minimum. The estimate of the information symbol I_k is

$$D = \text{Re}\{w_1 y_1(k) + w_2 y_2(k)\} \qquad (9A.1)$$

where

$$y_1(k) = 2E\alpha_1 I_k + n_1(k) \qquad (9A.2)$$

$$y_2(k) = 2E\alpha_2 I_k + n_2(k) \qquad (9A.3)$$

are complex quantities, and w_1 and w_2 are the weights to be determined.

Assume that $I_k = 1$ and that the weights w_1 and w_2 are chosen to minimize the MSE defined as

$$E(w) = \mathscr{E}\left[|2E - (w_1 y_1(k) + w_2 y_2(k))|^2\right]$$

$$= \mathscr{E}\left[|2E - 2Ew_1\alpha_1 - 2Ew_2\alpha_2 - w_1 n_1 - w_2 n_2|^2\right] \qquad (9A.4)$$

Minimization of ε^2 with respect to w_1 and w_2 leads to

$$\left(E|\alpha_1|^2 + \sigma_1^2\right)w_1^* + \left(E\alpha_1\alpha_2^* + \bar{r}\sigma_1\sigma_2\right)w_2^* = E\alpha_1 \qquad (9A.5)$$

$$\left(E\alpha_2\alpha_1^* + \bar{r}\sigma_1\sigma_2\right)w_1^* + \left(E|\alpha_2|^2 + \sigma_2^2\right)w_2^* = E\alpha_2 \qquad (9A.6)$$

where $n_1(k)$ and $n_2(k)$ are jointly stationary processes with:

$$\mathscr{E}\left[|n_1(k)|^2\right] = 4E\sigma_1^2,$$

$$\mathscr{E}\left[|n_2(k)|^2\right] = 4E\sigma_2^2,$$

$$\mathscr{E}\left[n_1(k)n_2^*(k)\right] = 4E\sigma_1\sigma_2\bar{r} \qquad (9A.7)$$

The solutions for w_1 and w_2 from (9A.5) and (9A.6) are given by

$$w_1 = c\left(\beta_2^2\alpha_1^* - \beta_2\bar{r}\alpha_2^*\right) \qquad (9A.8)$$

$$w_2 = c\left(\alpha_2^* - \beta_2\bar{r}\alpha_1^*\right) \qquad (9A.9)$$

where

$$c = \frac{E\sigma_1^2}{Den} \qquad (9A.10)$$

$$Den = E\sigma_2^2|\alpha_1|^2 + E\sigma_1^2|\alpha_2|^2 + \sigma_1^2\sigma_2^2(1 - \bar{r}^2) - 2E\bar{r}\sigma_1\sigma_2\mathrm{Re}\left(\alpha_1\alpha_2^*\right) \qquad (9A.11)$$

$$\beta_2 = \frac{\sigma_2}{\sigma_1} \qquad (9A.12)$$

After extensive manipulation, the mean and variance of D given $I_k = 1$ can be shown to be

$$m_D = \mathscr{E}[D]$$

$$= 2Ec\left\{\beta_2^2|\alpha_1|^2 + |\alpha_2|^2 - 2\beta_2\bar{r}\,\mathrm{Re}\left(\alpha_1\alpha_2^*\right)\right\} \qquad (9A.13)$$

$$\sigma_D^2 = \mathscr{E}\left[|D - m_D|^2\right]$$

$$= 2E\sigma_1^2 c^2(1 - \bar{r}^2)\beta_2^2\left[\beta_2^2|\alpha_1|^2 + |\alpha_2|^2 - 2\beta_2\bar{r}\,\mathrm{Re}\left(\alpha_1\alpha_2^*\right)\right] \qquad (9A.14)$$

Using (9A.13) and (9A.14) leads to the error probability given in (9.94).

REFERENCES

1. L. C. Van Atta, Electromagnetic Reflection, U.S. Patent 2908002. October 6, 1959.

2. Special Issue on Active and Adaptive Antennas, *IEEE Trans. Antennas and Propagation AP-12*, March 1964.

3. F. Bryn, Optimum Signal Processing of Three-Dimensional Arrays Operating on Gaussian Signals and Noise, *J. Acoust. Soc. Am. 34*, March 1962, pp. 289–297.

4. H. Mermoz, Adaptive Filtering and Optimal Utilization of an Antenna, U.S. Navy Bureau of Ships (translation 903 of Ph.D. thesis, Institut Polytechnique, Grenoble, France), October 4, 1965.

5. S. W. W. Shor, Adaptive Technique to Discriminate Against Coherent Noise in a Narrow-Band System, *J. Acoust. Soc. Am., 39*, Jan. 1966, pp. 74–78.

6. S. P. Applebaum, Adaptive Arrays, Syracuse Univ. Res. Corp., Syracuse, N.Y., Special Projects Lab., Rep. SPL TR 66-1, August 1966.

7. B. Widrow, P. E. Mantey, L. J. Griffiths, and B. B. Goode, Adaptive Antenna Systems, *Proc. IEEE 55*, Dec. 1967, pp. 2143–2159.

8. J. Capon, R. J. Greenfield and R. J. Kolker, Multidimensional Maximum-Likelihood Processing of a Large Aperture Seismic Array, *Proc. IEEE 55*, 2, Feb. 1967, pp. 192–211.

9. L. J. Griffiths, A Simple Algorithm for Real-Time Processing in Antenna Arrays, *Proc. IEEE 57*, 10, Oct. 1969, pp. 1696–1704.

10. O. L. Frost, III, An Algorithm for Linearly Constrained Adaptive Array Processing, *Proc. IEEE 60*, No. 8, Aug. 1972, pp. 926–935.

11. R. L. Riegler and R. T. Compton, Jr., An Adaptive Array for Interference Rejection," *Proc. IEEE 61*, 6, June 1973, pp. 748–758.

12. H. C. Lin, Spatial Correlations in Adaptive Arrays, *IEEE Trans. Antennas and Propagation, AP-30*, 2, March 1982, pp. 212–223.

13. B. Widrow and M. E. Hoff, Jr., Adaptive Switching Circuits, IRE WESCON Conv. Rec., pt. 4, 1960, pp. 96–104.

14. F. M. Hsu, Square Root Kalman Filtering for High Speed Data Received over Fading Dispersive HF Channels," *IEEE Trans. Information Theory IT-28*, 5, Sept. 1982, pp. 753–763.

chapter
_____

10

DUAL-CHANNEL INTERFERENCE MITIGATION

Another application of least square algorithms arises in the case of multichannel communications. This application is different from that described in Chapter 6, where equalization based on least squares methods was presented to compensate for intersymbol interference or multipath distortion introduced by transmitting a desired signal over a dispersive channel. This application is also different from that described in the previous chapter, where adaptive array techniques based on least squares algorithms were used to suppress additive external interference. In the multichannel communications case the interference is caused by transmitting multiple signals with the same signaling format but with different information streams over channels that are cross-coupled. This cross-coupling causes serious distortion in the received signals. Once again least squares algorithms can be applied to compensate for the distortion.

Two specific examples of systems corrupted by interchannel interference include dual polarized digital radio and multiwire cable transmission. In digital radio applications two orthogonal polarizations are used to transmit two independent digital information streams on a single radio frequency.[1,2,3,4] Channel propagation anomalies introduce coupling between the two orthogonal transmitted polarizations, so that the polarized waves at the receiver are nonorthogonal. In a multiwire cable system, interchannel interference results from electromagnetic coupling between the cable pairs. The multiwire cable system can be modeled as a channel with fixed but unknown characteristics at the receiver. The digital radio example assumes a random slowly-time-varying channel whose characteristics must be adaptively tracked at the receiver.

10.1   DUAL-CHANNEL DIGITAL SIGNAL TRANSMISSION MODEL

In dual-channel digital signal transmission the signals are transmitted on two orthogonal channels. The complex envelope $s_i(t)$ for the ith channel can be represented as:

$$s_i(t) = \sum_{k=0}^{\infty} I_{ik} p(t - kT), \qquad i = 1, 2$$

The information sequence on the ith channel $\{I_{ik}\}$ modulates a basic transmitting filter pulse $p(t)$ at a rate $1/T$. The information sequence modulation can be M_a-ary PSK, PAM or a hybrid. The information sequences are assumed to be random and statistically independent, both from channel to channel and from symbol to symbol in the same channel.

The channel is assumed to be nondispersive and slowly time varying, and distorts the transmission by introducing symbols transmitted on one channel into the other channel.[5,6,7,8,9] Complex-valued coefficients T_{ij} for $i = 1, 2$ and $j = 1, 2$ represent the channel attenuation and phase shift. The factors T_{11} and T_{22} account for direct path attenuation and phase shift on channels 1 and 2, respectively. The factor T_{12} represents the complex attenuation introduced by channel 2 into channel 1, and T_{21} represents interference introduced in the opposite direction. The discrete time received signal on the ith channel is denoted x_{ik} for $i = 1, 2$. It is assumed that the discrete time signal is obtained by means of a discrete time communication model described in Section 6.1. On the ith channel the additive noise n_{ik} is assumed to be zero mean, white and Gaussian. The noise variance on both channels is the same and is equal to $2\sigma_n^2$. The received signals can then be represented by

$$x_{1k} = T_{11}I_{1k} + T_{12}I_{2k} + n_{1k}$$

$$x_{2k} = T_{21}I_{1k} + T_{22}I_{2k} + n_{2k}$$

10.2   REDUCTION OF INTERCHANNEL INTERFERENCE BY THE MSE ALGORITHM

In the MSE algorithm estimates of I_{1k} and I_{2k} are formed by linearly combining the received signal samples by using weights derived from the MSE criterion.[10,11] The estimates of the information symbols I_{1k} and I_{2k} are denoted by $\hat{I}_{1k}$ and $\hat{I}_{2k}$ respectively:

$$\hat{I}_{1k} = w_{11}x_{1k} + w_{12}x_{2k}$$

$$\hat{I}_{2k} = w_{21}x_{1k} + w_{22}x_{2k}$$

where $\{w_{ij}\}$ are the weights to be determined using the MSE algorithm.

The mean square error can now be defined as

$$E(W) = \mathscr{E}\left\{ \sum_{m=1}^{2} |I_{mk} - \hat{I}_{mk}|^2 \right\}$$

where the error signal ε_{mk} is expressed as

$$\varepsilon_{mk} = I_{mk} - \hat{I}_{mk}, \qquad m = 1, 2$$

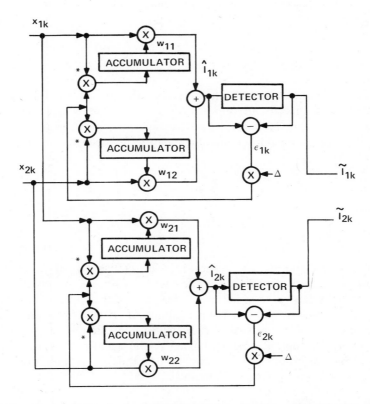

Figure 10-1. Baseband adaptive linear MSE estimation algorithm. (All arithmetic operations shown are complex. The * adjacent to the multiply indicates that the complex conjugate of the input is used as a multiplier.) (© 1977 IEEE Nichols, et al., reference 10-1).

and W is the vector of weights defined as $W^t = (w_{11}, w_{12}, w_{21}, w_{22})$. Use of the orthogonality principle or direct minimization of the mean square error results in the following equations for the weights (see appendix at the end of this chapter):

$$\begin{pmatrix} A & 0 \\ 0 & A \end{pmatrix} W = T^*$$

where $T^t = (T_{11}, T_{21}, T_{12}, T_{22})$, $\mathscr{E}\{|I_{ik}|^2\}$ is assumed to be unity, $\mathscr{E}(I_{mk}I_{lk}) = 0$ for $m \neq l$, and the matrix A is defined by

$$A = \begin{bmatrix} |T_{11}|^2 + |T_{12}|^2 + 2\sigma_n^2 & T_{21}T_{11}^* + T_{12}^*T_{22} \\ T_{21}^*T_{11} + T_{12}T_{22}^* & |T_{21}|^2 + |T_{22}|^2 + 2\sigma_n^2 \end{bmatrix}$$

The solution for the weights can be obtained directly by matrix inversion or, in the slowly varying channel case, iteratively by means of the steepest descent algorithm. The receiver structure which incorporates an adaptive linear MSE algorithm is shown in Figure 10-1.[10,11]

10.3   PERFORMANCE RESULTS

To present illustrative results, the representation of the received signals are simplified to a special but important case. First, the phase angles of the coefficients T_{11} and T_{22} are assumed to be zero with magnitudes satisfying $|T_{11}| = |T_{22}|$. Equal magnitude interchannel interference is postulated, so that

$$\frac{T_{12}}{T_{11}} = re^{j\phi_1}$$

$$\frac{T_{21}}{T_{22}} = re^{j\phi_2}$$

The phases ϕ_1 and ϕ_2 are assumed to be statistically independent, with uniform probability densities over the interval $(-\pi, \pi)$. A conditional probability of error can be obtained which depends on the values of the phases ϕ_1 and ϕ_2. Averaging the conditional probability of error over the phase distribution results in a probability of error which is a function of the signal-to-noise ratio, γ, and the single channel parameter r. The signal-to-noise ratio is defined as the average received signal power per bit on the two channels divided by the average noise power on the two channels:

$$\gamma = \frac{\frac{1}{2}\left(|T_{11}|^2 + |T_{12}|^2 + |T_{21}|^2 + |T_{22}|^2\right)}{\mathscr{E}\left[|n_{1k}|^2 + |n_{2k}|^2\right]}$$

or

$$\gamma = \frac{|T_{11}|^2 (1 + r^2)}{4\sigma_n^2}$$

The above model has been found to be a reasonable representation for depolarization introduced by heavy rainfall when two circularly polarized transmitted waves are received as two nonorthogonal elliptically polarized waves.[1,2]

Error rate performance for the MSE algorithm is shown in Figure 10-2 for various values of the crosstalk parameter $XPD = 20 \log r$, assuming perfect knowledge of the channel characteristics. Both simulated and theoretical

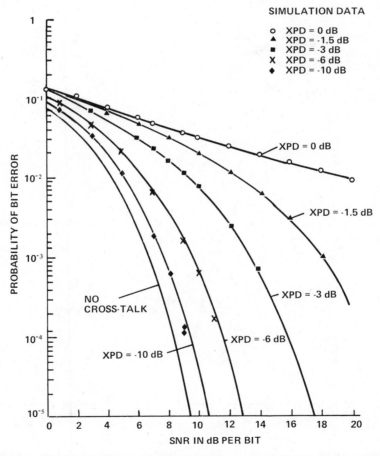

Figure 10-2. Error performance of algorithm using MSE criterion. (Simulation data is indicated by symbols.) (© 1977 IEEE, Nichols, et al., reference 10-1).

performance results have been obtained and are shown in the figure.[10,11] The solid lines correspond to theoretical results and the symbols represent simulated results. For comparison, a maximum likelihood detection (MLD) algorithm used to mitigate interchannel interference is shown in Figure 10-3 for $XPD = -3$ dB. This latter algorithm is nonlinear and results from maximizing the probability of the received signal samples on the two channels over all possible transmitted symbol pairs.[10,11] Note that the MLD algorithm offers improved performance over the MSE algorithm and that both algorithms offer substantial improvement over the case where no crosstalk correction is applied.

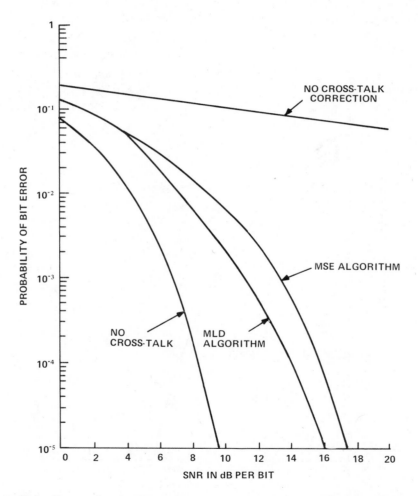

Figure 10-3. Error performance comparison for XPD = -3.0 dB (© 1977 IEEE, Nichols, et al., reference 10-1).

APPENDIX: DERIVATION OF WEIGHTS FOR INTERCHANNEL INTERFERENCE MITIGATION

In this application the orthogonality principle can be expressed as:

$$\mathscr{E}\left[\varepsilon_{mk}x_{ik}^*\right] = 0, \qquad m = 1, 2, \quad i = 1, 2$$

where

$$\varepsilon_{mk} = I_{mk} - (w_{m1}x_{1k} + w_{m2}x_{2k})$$

and

$$x_{ik} = T_{i1}I_{1k} + T_{i2}I_{2k} + n_{ik}$$

Combining the above expressions and expanding results in

$$\mathscr{E}\left\{\left[I_{mk} - (w_{m1}x_{1k} + w_{m2}x_{2k})\right]x_{ik}^*\right\} = 0, \qquad m = 1, 2; \, i = 1, 2$$

or

$$\mathscr{E}\left(I_{mk}x_{ik}^*\right) = w_{m1}\mathscr{E}\left(x_{1k}x_{ik}^*\right) + w_{m2}\mathscr{E}\left(x_{2k}x_{ik}^*\right), \qquad m = 1, 2; \, i = 1, 2$$

The individual moments can be written as

$$\mathscr{E}\left(I_{mk}x_{ik}^*\right) = \mathscr{E}\left[I_{mk}(T_{i1}I_{1k} + T_{i2}I_{2k} + n_{ik})^*\right]$$

$$= \begin{cases} T_{i1}^*, & m = 1, \quad i = 1, 2 \\ T_{i2}^*, & m = 2, \quad i = 1, 2 \end{cases}$$

$$\mathscr{E}\left(x_{1k}x_{ik}^*\right) = \mathscr{E}\left[(T_{11}I_{1k} + T_{12}I_{2k} + n_{1k})(T_{i1}I_{1k} + T_{i2}I_{2k} + n_{ik})^*\right]$$

$$= \begin{cases} |T_{11}|^2 + |T_{12}|^2 + 2\sigma_n^2, & i = 1 \\ T_{11}T_{21}^* + T_{12}T_{22}^*, & i = 2 \end{cases}$$

$$\mathscr{E}\left(x_{2k}x_{ik}^*\right) = \mathscr{E}\left[(T_{21}I_{1k} + T_{22}I_{2k} + n_{2k})(T_{i1}I_{1k} + T_{i2}I_{2k} + n_{ik})^*\right]$$

$$= \begin{cases} T_{21}T_{11}^* + T_{22}T_{12}^*, & i = 1 \\ |T_{21}|^2 + |T_{22}|^2 + 2\sigma_n^2, & i = 2 \end{cases}$$

where $\mathscr{E}\{I_{mk}I_{ik}^*\}$ is one for $m = i$ and zero otherwise.

Substituting these moments into the expanded version of the orthogonality principle given above results in:

$$T_{11}^* = w_{11}\left(|T_{11}|^2 + |T_{12}|^2 + 2\sigma_n^2\right) + w_{12}\left(T_{21}T_{11}^* + T_{12}^*T_{22}\right), \qquad m = 1, \quad i = 1$$

$$T_{21}^* = w_{11}\left(T_{11}T_{21}^* + T_{12}T_{22}^*\right) + w_{12}\left(|T_{21}|^2 + |T_{22}|^2 + 2\sigma_n^2\right), \qquad m = 1, \quad i = 2$$

$$T_{12}^* = w_{21}\left(|T_{11}|^2 + |T_{12}|^2 + 2\sigma_n^2\right) + w_{22}\left(T_{21}T_{11}^* + T_{12}^*T_{22}\right), \qquad m = 2, \quad i = 1$$

$$T_{22}^* = w_{21}\left(T_{21}^*T_{11} + T_{12}T_{22}^*\right) + w_{22}\left(|T_{21}|^2 + |T_{22}|^2 + 2\sigma_n^2\right), \qquad m = 2, \quad i = 2$$

The above equations can be expressed in the matrix notation given in Section 10.2.

REFERENCES

1. T. S. Chu, Restoring the Orthogonality of Two Polarizations in Radio Communication Systems, I," *Bell Systems Tech. J. 50*, Nov. 1971, pp. 3063–3069.
2. T. S. Chu, Restoring the Orthogonality of Two Polarizations in Radio Communication Systems, II, *Bell Systems Tech. J. 52*, March 1973, pp. 319–329.
3. R. W. Kreutel, The Orthogonalization of Polarized Fields in Dual Polarized Radio Transmission Systems, *COMSAT Tech. Rev. 3*, fall 1973, pp. 375–386.
4. S. K. Barton, Polarization Distortion in Systems Employing Orthogonal Polarization Frequency Reuse, *Marconi Rev. 38*, fourth quarter 1975, pp. 153–168.
5. S. K. Barton, Methods of Adaptive Cancellation for Dual Polarization Satellite Systems, *Marconi Rev. 39*, first quarter 1976, pp. 1–24.
6. M. J. Saunders, Cross Polarization at 18 and 30 GHz Due to Rain, *IEEE Trans. Antennas and Propagation AP-19*, March 1971, pp. 273–277.
7. D. T. Thomas, Cross Polarization Distortion in Microwave Radio Transmission Due to Rain, *Radio Science*, vol. 6, Oct. 1971, pp. 833–839.
8. T. Oguchi and Y. Hosoya, Scattering Properties of Oblate Raindrops on Cross-Polarization of Radio Waves Due to Rain (Part II): Calculation at Microwave and Millimeter Wave Region, *J. Radio Research Lab.* 21, 105, 1974, pp. 191–259.
9. T. S. Chu, Rain-Induced Cross-Polarization at Centimeter and Millimeter Wavelengths, *Bell Systems Tech. J. 53*, Oct. 1974, p. 1557.
10. H. E. Nichols, A. A. Giordano and J. G. Proakis, MLD and MSE Algorithms for Adaptive Detection of Digital Signals in the Presence of Interchannel Interference, *IEEE Trans. Information Theory IT-23*, 5, Sept. 1977, pp. 563–575.
11. H. E. Nichols, Adaptive Algorithms for Reduction of Interchannel and Intersymbol Interference in Digital Communications, Ph.D. Dissertation, Northeastern University, Boston, Mass., Nov. 1978.

chapter

11

LEAST SQUARES TECHNIQUES IN DIGITAL SPEECH PROCESSING

11.1 INTRODUCTION

In this chapter, least square estimation methods are applied to digital speech processing, with limited reference to the related field of image processing. The principal emphasis presented here addresses predictive coding of speech signals in the time domain. Predictive coding is effective in the analysis and synthesis of speech signals by exploiting the redundancy inherent in speech signals. Other aspects of digital speech processing are considered only on a limited basis for completeness in addressing the complex topic of digital speech processing. Topics which are not treated in detail but are important in the field of digital speech processing include vocoders, formant analysis, frequency domain waveform coding methods, transform coding methods, maximum likelihood methods, Kalman methods, unconstrained least square structures, pitch extraction, channel transmission errors, and implementation structures.

11.2 SPEECH PRODUCTION MODEL
AND SPEECH CHARACTERISTICS

Speech is produced when air is expelled from the lungs, passes through a nonuniform acoustic tube known as the vocal tract, and is released from the lips. When speech occurs, an acoustic pressure wave is induced in the vocal tract. The acoustic source that excites the vocal tract is usually considered to produce two types of sounds: voiced and unvoiced. Voiced sounds occur when the vocal tract is excited by pulses of air pressure in a nearly periodic fashion. The spacing between these pulses is termed the fundamental period or the pitch period. Unvoiced sounds are produced when a constriction is created in the vocal tract, causing turbulent air flow and thus a random source of acoustic noise. A simple model can then be produced by decoupling the sound source from the vocal tract, as shown in Figure 11-1.[1,2,3] This figure illustrates that the sound source consists of two component sources, that is, a random noise source for unvoiced sounds and a pulse source for voiced sounds. A switch is used to select the excitation provided at the input to the vocal tract. The vocal tract is modeled as a time-varying filter which spectrally shapes the sound source. Using this model, the speech analysis problem reduces to estimating the parameters of the vocal tract filter, the pitch period, whether the source is voiced or unvoiced, and the amplitude of the excitation. Characteristics of the speech waveform are now described.

A fundamental property of the speech waveform is that it is bandlimited. Typical voice communications circuits utilize a transmission system with a passband which extends from 200 to 3200 Hz. For digital speech transmission, sampling the speech waveform at greater than the Nyquist rate results in a typical sampling rate of 8000 Hz. In addition, predictive coding techniques described subsequently assume that the speech waveform is quantized in a discrete amplitude representation as well as sampled in time.

Depending on the observed time interval, the speech waveform can be perceived as locally stationary or nonstationary.[1,2] In intervals of one-half second or more, considerable variation in the speech waveform arises from the presence of voiced and unvoiced sounds, silent periods and variations in

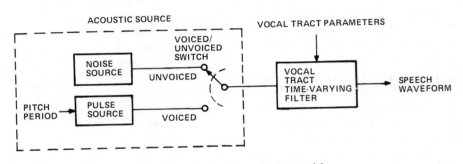

Figure 11-1. Speech production model.

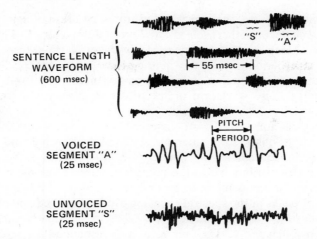

Figure 11-2. Examples of "long time" and "short time" speech segments (© 1983 IEEE Crochiere and Flanagan, reference 11-1).

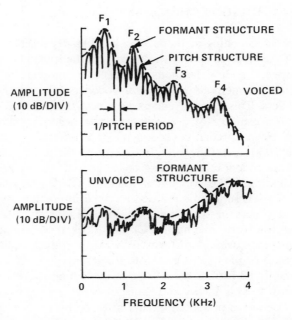

Figure 11-3. Spectral models for voiced and unvoiced speech (© 1983 IEEE Crochiere and Flanagan, reference 11-1).

345

amplitude. Over 20–40 ms, the speech waveform is locally stationary and may consist of segments of either voiced or unvoiced waveforms. Figure 11-2 illustrates typical long- and short-time speech waveforms taken from reference 1. Note that the periodicity in the quasiperiodic waveform of the voiced segment for the spoken letter A is labeled as the pitch period. No period is discernible in the noise waveform of the unvoiced segment for the spoken letter S.

Spectral models for voiced and unvoiced speech segments can be determined for short observation intervals. Examples extracted and modified from reference 1 are depicted in Figure 11-3. The shape of the spectrum envelope is directly attributable to the vocal tract filter. Over the observation interval the filter can be assumed to be time invariant. In the voiced segment case, four major peaks in the spectral envelope can be observed and are labeled F_1, F_2, F_3, and F_4. The interval between peaks in the fine structure is the reciprocal of the pitch period. The four vocal tract resonances are referred to as formants and the frequencies F_1, F_2, F_3, and F_4 are termed formant frequencies. The unvoiced speech segment does not have specific resonances, as indicated in Figure 11-3.

In the next section, speech coding techniques for digital transmission are described. Subsequently, emphasis is placed on waveform coding in the time domain.

11.3   SPEECH CODING TECHNIQUES

A variety of speech encoding methods are available, depending on the fidelity requirements of the received speech waveform and the transmission rate. Three categories of speech coding methods are vocoders, waveform coding, and hybrid or parametric methods.[1] The term vocoders is a contraction of the words voice coders. This class of coders uses available knowledge about the speech source. An example of a specific vocoder described subsequently in Section 11.4 is linear predictive coding (LPC). Waveform coders can be implemented in both the time and frequency domains. Examples of time domain waveform coders include pulse code modulation (PCM), differential pulse code modulation (DPCM), delta modulation (DM) and adaptive predictive coding (APC). These techniques are described in greater detail in Section 11.5. Waveform coders in the frequency domain include sub-band coding (SBC), adaptive transform coding (ATC) and block transformation methods such as the discrete cosine transform. Waveform coding in the frequency domain and hybrid or parametric methods are described in references 1 and 2 and are not considered further here. Prior to describing specific techniques, a brief discussion of speech fidelity and transmission rate is provided.

Quantitative measures of speech quality attained by various speech coding techniques are not readily available. Waveform coding techniques utilize signal-to-noise ratio as a simple measure. More typical measures, however,

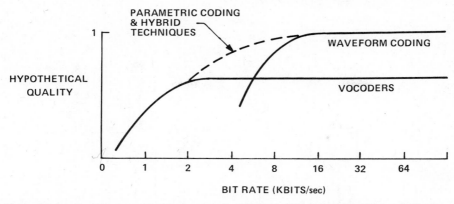

Figure 11-4. Quality versus bit rate for speech coding (© 1983 IEEE Crochiere and Flanagan, reference 11-1).

include word intelligibility or speaker recognition tests performed by a listener. Nevertheless, it is useful to compare the quality attained by various speech encoding methods as a function of transmission rate. Figure 11-4, taken from reference 1, illustrates a hypothetical quality measure. The vertical scale is a subjective quality measure in which the value 1 indicates that the method produces a signal that is close to the original speech waveform and the value 0 produces a signal that is a poor representation of the original. Reducing the transmission rate leads to more sophisticated signal processing techniques which require increasingly more accurate speech models. Note that waveform coding techniques outperform vocoders at transmission rates exceeding about 6 Kbps. Telephone-quality speech can be achieved at rates exceeding 16 Kbps. Highly intelligible quality speech can be produced at rates between 8 and 16 Kbps. Synthetic quality speech with reduced intelligibity is achievable with rates which are approximately 2Kbps.

11.4 LINEAR PREDICTION OF SPEECH

Linear prediction of speech has been extensively investigated.[3,4,5,6] A summary of linear prediction methods for stationary and nonstationary signals with application to speech communications, geophysics, etc., is provided in reference 7. Linear prediction methods permit the representation of speech signals in terms of a small number of slowly time-varying parameters. Use of linear prediction methods allows both the analysis and synthesis of human speech to be performed by exploiting the inherent redundancy in the waveform. Several linear prediction methods are described in this section.

It is now assumed that samples of the speech waveform taken at the Nyquist rate are available. Analysis of the speech samples begins by predicting the present speech sample as a linear combination of the previous samples. A set

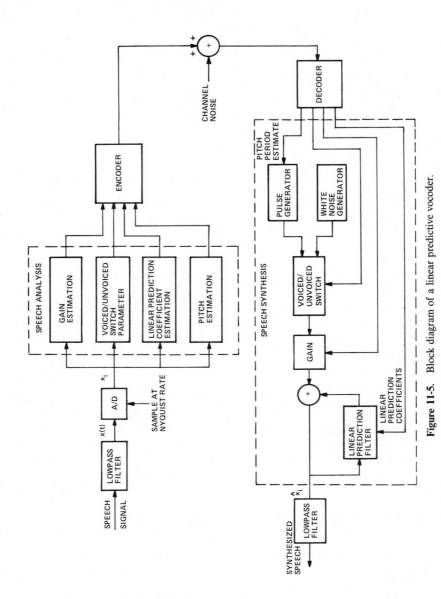

Figure 11-5. Block diagram of a linear predictive vocoder.

of predictor coefficients are then obtained by minimizing the mean square error between the actual and predicted values of the speech samples. An analysis of the speech samples then results in a set of parameters which include the predictor coefficients, the pitch period, a binary parameter indicating whether the speech is voiced or unvoiced, and the gain or rms value of the speech samples. In a speech communication system, each of these parameters is encoded and transmitted along with the quantized error signal to the receiver, where a synthesizer is employed to reconstitute the speech waveform from the estimates of the received parameters and the quantized error signal. During speech production, of course, the vocal tract shape changes continuously. Thus the set of speech parameters must be frequently updated. The update period is referred to as a frame period and is usually about 5 to 10 ms. A block diagram of a linear prediction vocoder based on this procedure is shown in Figure 11-5.

A mathematical description of the linear prediction of speech is now described. The speech sample, x_i, of the linear filter at the ith sampling instant is given by

$$x_i = \sum_{m=1}^{M} b_m x_{i-m} + e_i \qquad (11.1)$$

where the predictor coefficients $\{b_m\}$ account for the filtering action of the vocal tract, M, is the number of predictor coefficients, and e_i represents the ith sample of the error signal or excitation. Note that Eq. (11.1) is equivalent to the linear prediction formulation given by (3.2) and (3.4). This representation of the speech signal is illustrated in the block diagram depicted in Figure 11-6. The number of predictor coefficients depends on the length of the vocal tract and is typically about $M = 10$. Ordinarily, two coefficients are needed to characterize a single formant, so that ten coefficients more than adequately characterize four formants.

The linear prediction model provided above can also be described in the frequency domain by use of Z transforms. The two-sided Z transform of a discrete sequence x_n can be defined as (see Appendix Section A.3)

$$X(Z) = \sum_{n=-\infty}^{\infty} x_n Z^{-n} \qquad (11.2)$$

Applying Eq. (11.2) to (11.1) results in

$$X(Z) = X(Z) \sum_{m=1}^{M} b_m Z^{-m} + E(Z)$$

where $E(Z)$ is the Z transform of the error signal e_i. Rearranging the above equation results in $H(Z)$, the transfer function of the linear prediction filter:

$$H(Z) = \frac{X(Z)}{E(Z)} = \frac{1}{1 + \sum_{m=1}^{M} a_m Z^{-m}} \qquad (11.3)$$

where $a_m = -b_m$ for $m = 1, \ldots, M$. Note that Eq. (11.3) represents a transfer function with all poles. These poles determine the location of the formant frequencies. The power spectrum, $P(f)$, associated with the transfer function $H(Z)$ can be computed by a discrete Fourier transform representation obtained by letting $Z = e^{j2\pi fT_s}$ where T_s is the sampling interval. Thus the power spectrum is computed from

$$P_H(f) = |H(f)|^2$$

$$= \frac{1}{|1 + \sum_{m=1}^{M} a_m e^{-j2\pi fT_s}|^2} \qquad (11.4)$$

The poles of $H(Z)$ can be determined by setting the denominator of (11.3) to zero and computing the roots of the polynomial equation in Z. A formant then consists of a pair of complex conjugate poles. Several investigators have utilized formant analysis to characterize speech.[3,6,8]

11.4.1 Speech Analysis

In this section an outline of the mathematical formulation of the speech analysis problem is presented. The key analysis algorithm is linear prediction, as illustrated in Figure 11-6, where a speech sample from a voiced segment is

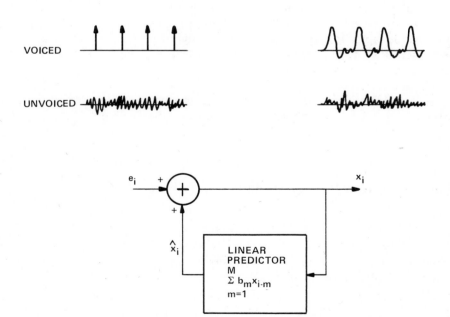

Figure 11-6. Block diagram of a functional model of speech production based on the linear prediction representation of the speech wave.

linearly predicted from the past M speech samples. Computationally, a variety of methods can be used to determine the prediction coefficients. Specific methods presented here include the covariance method[3,7] and the auto-correlation method.[3,7] Lattice structure methods using forward and backward prediction are not as common, but are available in the literature.[3,9] Stability considerations are an important element in the prediction coefficient computation and are briefly discussed.

To complete the speech analysis description, pitch extraction, gain computation, and methods of identifying voiced and unvoiced speech signals are presented.

11.4.1.1   *Estimation of Predictor Coefficients by Covariance Method*

Assume that a real sequence of N speech samples $\{x_i, \ i = 1, 2, \ldots, N\}$ is available. The covariance method is defined by setting a window, $M + 1 \leq i \leq N$, such that the error is minimized only over the interval $[M + 1, N]$. The covariance matrix element ρ_{ij} is then computed by (see Eq. (3.122)).

$$\rho_{ij} = \sum_{m=M+1}^{N} x_{m-i} x_{m-j} \tag{11.5}$$

Since the covariance matrix is not a Toeplitz matrix, the predictor coefficients cannot be solved by using the Durbin algorithm. Nevertheless, the covariance matrix is symmetric (i.e., $\rho_{ij} = \rho_{ji}$), so that the predictor coefficients are given by the normal equations (see Eq. (3.90)):

$$\sum_{m=1}^{M} b_m \rho_{ml} = \rho_{0l}, \qquad l = 1, 2, \ldots, M \tag{11.6}$$

where the time average autocorrelation estimate ρ_{mj} replaces the ensemble average autocorrelation coefficients r_{mj}. These equations can be solved by the Cholesky decomposition algorithm provided in Chapter 3.

Several techniques can be used to reduce the number of computations in evaluating the autocorrelation coefficients, ρ_{ij}. FFT methods are computationally efficient when the number of prediction coefficients, M, is on the order of the number of samples, N, used in the transform. In typical speech processing applications 4–12 coefficients are determined for a frame time of 10–30 ms, using an 8 KHz sampling rate. Thus, in a 20 ms frame there are $N = 160$ samples per frame. In this case N is much larger than M. Therefore, FFT methods are not computationally efficient, and other techniques are used.

One method for reducing the required number of multiplications in evaluating the autocorrelation coefficients utilizes a factoring procedure. The factors in Eq. (11.5) can be identified by expressing (11.5) as follows:

$$\rho_{ij} = x_{M+1-i} x_{M+1-j} + x_{M+2-i} x_{M+2-j} + \cdots$$

$$+ x_{M+l+1-i} x_{M+l+1-j} + \cdots + x_{N-i} x_{N-j} \tag{11.7}$$

Common terms in (11.7) can be identified by letting $j = l + i$. In this case, (11.7) can be rewritten as

$$\rho_{i,l+i} = x_{M+1-i}(x_{M+1-l-i} + x_{M+l+1-i})$$

$$+ x_{M+2-i}(x_{M+2-l-i} + x_{M+l+2-i}) + \cdots$$

As a result of factoring, the number of multiplications is reduced significantly.

11.4.1.2 Estimation of Prediction Coefficients by Correlation Method

The correlation method is defined by setting a window $-\infty \leq i \leq \infty$, and defining $x_i = 0$ for $i < 0$ and $i > N$. These limits allow ρ_{ij} to be simplified as (see Eq. (3.121))

$$\rho_{ij} = \sum_{m=1}^{N+M} x_{m-i} x_{m-j} \tag{11.9}$$

Note that for the correlation method

$$\rho_{ij} = \rho_{i-j} = \rho_{j-i}$$

Thus, the Durbin recursive algorithm can be used to compute the predictor coefficients from the Yule-Walker equations with time averages replacing ensemble averages (see Eq. (3.10))

$$\sum_{m=1}^{M} b_m \rho_{l-m} = \rho_l, \qquad l = 1, 2, \ldots, M \tag{11.10}$$

The Durbin algorithm is summarized as follows (see Eqs. (3.24), (3.29), and (3.31)):

$$b_{m,k} = b_{m-1,k} - b_{m,m} b_{m-1,m-k}, \qquad k = 1, \ldots, M - 1 \tag{11.11}$$

$$b_{m,m} = \frac{\rho_m - (\rho_1 \rho_2, \ldots, \rho_{m-1}) \begin{pmatrix} b_{m-1,m-1} \\ \vdots \\ b_{m-1,1} \end{pmatrix}}{P_{m-1_{\min}}} \tag{11.12}$$

$$P_{m_{\min}} = P_{m-1_{\min}}(1 - b_{m,m}^2) \tag{11.13}$$

$$P_{0_{\min}} = \rho_0 \tag{11.14}$$

for $m = 1, 2, \ldots, M$.

The coefficients $b_{m,m}$ in (11.12) are referred to as the reflection coefficients or partial correlation coefficients, with the acronym PARCOR.[3,7,9,10] The partial correlation refers to the use of a subset of the available correlation coefficients. The term reflection coefficients is borrowed from transmission line theory, in which $b_{m,m}$ represents the reflection coefficients between two sections with different impedances. A lattice structure formulation for forward and backward errors based on autocorrelation coefficients rather than the data samples (Burg method) is described in the literature for the PARCOR formulation.[3] Reference 11 describes a lattice structure for adaptive linear prediction based on the data samples.

11.4.1.3 Stability Considerations in the Correlation Method

Speech processing is typically performed with finite-word-length microprocessors or special-purpose digital computers. As a result of the short frame times, computational speed for accomplishing calculations in real time or near real time is critical. In addition to the requirement for high-speed processing, numerical accuracy must be carefully addressed. Thus, in the linear prediction of speech the stability of the prediction filter obtained with finite precision arithmetic must be considered. To present the stability problem, a brief mathematical analysis, treating finite precision effects in the computation of the PARCOR coefficients, is now provided.

The Yule-Walker equations given by (11.10) can be expressed in matrix form by:

$$\overline{R}_M \overline{B}_M = \overline{F}_M \tag{11.15}$$

where

$$\overline{B}_M^t = (b_1, b_2, \ldots, b_M) \tag{11.16}$$

$$\overline{F}_M^t = (\rho_1, \rho_2, \ldots, \rho_M) \tag{11.17}$$

and $\overline{R}_M$ is the correlation matrix given by (3.20) with all real elements and ensemble averages replaced by time averages.

The coefficients $\{b_m\}$ can be obtained from (11.15) as

$$\overline{B}_M = \overline{R}_M^{-1} \overline{F}_M \tag{11.18}$$

with the assumption that $\overline{R}_M^{-1}$, the inverse of $\overline{R}_M$, exists. The existence of the inverse requires a nonzero determinant, $|\overline{R}_M|$. To compute the determinant, the correlation matrix, $\overline{R}_M$, is diagonalized as follows:

$$D \equiv A^t \overline{R}_M A \tag{11.19}$$

where the matrix A is defined by

$$A = \begin{pmatrix} 1 & a_{11} & a_{21} & \cdots & a_{M-1,1} \\ 0 & 1 & a_{22} & \cdots & a_{M-1,2} \\ 0 & 0 & 1 & \cdots & a_{M-1,3} \\ \vdots & & & & \\ 0 & 0 & 0 & & 1 \end{pmatrix} \tag{11.20}$$

Iterative versions of the Yule-Walker equations have been determined and are provided in Eqs. (3.25) and (3.30), where $a_{M,0} = 1$ and $a_{M,m} = -b_{M,m}$ for $m = 1, \ldots, M$. Replacing the index M by m permits Eqs. (3.25) to be expressed as

$$-a_{m,m}\rho_0 - (\rho_1, \rho_2 \cdots \rho_{m-1}) \begin{pmatrix} a_{m,m-1} \\ \vdots \\ a_{m,1} \end{pmatrix} = \rho_m$$

or

$$\sum_{j=0}^{m} a_{m,j}\rho_{m-j} = 0, \qquad m = 1, \ldots, M \tag{11.21}$$

Similarly, replacing the index M by m in Eq. (3.30) for real signal samples leads to

$$\sum_{j=0}^{m} a_{m,j}\rho_j = P_{m_{\min}}, \qquad m = 1, \ldots, M \tag{11.22}$$

Note that the minimum error power can be recursively computed from (3.31) with the index M replaced by m:

$$P_{m_{\min}} = P_{m-1_{\min}}(1 - a_{m,m}^2), \qquad m = 1, \ldots, M \tag{11.23}$$

Completing the matrix multiplication in (11.19) using (11.20), (11.21), and (11.22) results in

$$D = \begin{pmatrix} P_{0_{\min}} & & & \phi \\ & P_{1_{\min}} & & \\ & & \ddots & \\ \phi & & & P_{M-1_{\min}} \end{pmatrix} \tag{11.24}$$

The determinant of $\overline{R}_M$ can be computed from the determinant of D:

$$|D| = |A^t \overline{R}_M A| = |A^t||\overline{R}_M||A| = |\overline{R}_M|$$

$$= \prod_{m=0}^{M-1} P_{m_{\min}} \tag{11.25}$$

where $\prod$ denotes the product. From (11.25) it can be seen that the determinant of $\overline{R}_M$ is related to the product of minimum mean square error terms $P_{m_{\min}}$ for $m = 0, 1, \ldots, M - 1$. Thus $\overline{R}_M$ is singular and does not have an inverse if any

$P_{m_{\min}}$ (for $m = 0$ to $M - 1$) equals zero. This result is also evident from the computation of the PARCOR coefficients, $b_{m,m}$, given in (11.12), where $P_{m-1_{\min}}$ is used in the denominator. Consequently, it is expected that a small value of $P_{m_{\min}}$ can amplify numerical errors and result in large errors in the PARCOR coefficients. In the solution procedure, $P_{m_{\min}}$ should be inspected at each recursion step. If $P_{m_{\min}}$ is negative or zero an error has been made and the process should be terminated. As a practical matter, if the computation is performed with n significant digits, the computation should cease if $P_{m_{\min}}/P_0$ is less than about 10^{1-n} (See reference 3). Note that another check on the stability of the PARCOR coefficients requires that (See (11.23))

$$-1 < b_{m,m} < 1, \qquad m = 1, 2, \ldots, M$$

The above condition is a necessary and sufficient condition in the generation of a stable filter.

11.4.1.4 Pitch Estimation

A variety of methods exist for estimating pitch period.[12,3,13,14,8] No method is completely satisfactory for all applications. Part of the problem can be attributed to the simplistic speech production model, which requires a separation into voiced and unvoiced segments. Furthermore, the voiced segments only approximately have a periodic pulse sequence. Nevertheless, a method of pitch estimation is required to produce good quality speech transmission. Sample methods described here are based on correlation techniques.

Assume that speech samples $\{x_i\}$ are available over the interval $1 \leq i \leq N$. The autocorrelation function can then be defined by

$$\rho_j = \rho_{-j} = \sum_{i=1}^{N-j} x_i x_{i+j}, \; j = 0, 1, \ldots, N - 1$$

and

$$\rho_j = 0, |j| \geq N.$$

The autocorrelation function exhibits a cyclic behavior as a result of the periodic nature of the voiced segment. Thus, the separation between adjacent peaks in the autocorrelation function is an estimate of the pitch period.

An alternate method of estimating the pitch period utilizes an FFT approach. In this case zeros are appended to the N speech data samples for $i = N + 1, \ldots, L$, where L is selected to be a power of two. Then an L point FFT of the appended data sequence is computed to produce the spectral sequence $\{X_k\}$. The magnitude squared of each sample is computed, resulting in a power spectral estimate (see Chapter 7). Computing the inverse Fourier transform of the squared spectral sequence provides an estimate of the autocorrelation function, which leads to an estimate of the pitch period as described above.

11.4.1.5   *Voiced/Unvoiced Switch Estimation*

The selection of voiced and unvoiced segments of speech is closely related to the pitch extraction problem. Typical methods of selection are based on examining the density of zero crossings in the speech signal and the peak value of the autocorrelation function.

Suppose that ρ_p represents the peak autocorrelation function value for a sample $p > 0$. Note that p is, in fact, an estimate of the pitch period. A specific voicing threshold, T, can be used to decide whether a voiced or unvoiced segment exists. The decision rule is to choose a voiced segment if $\rho_p/\rho_0 \geq T$ and to choose an unvoiced segment otherwise. A typical value for the threshold is $T = 0.18$ (See reference 3).

11.4.1.6   *Gain Computation*

The final parameter required for speech analysis is the gain or rms level of the error signal. Recall that the error signal is given by

$$e_i = x_i - \sum_{m=1}^{M} b_m x_{i-m} \tag{11.26}$$

The gain, g, can then be defined as

$$g = \sqrt{\mathscr{E}\left(e_i^2\right)} \tag{11.27}$$

Once the predictor coefficients are obtained, Eq. (3.14) can be used to estimate the gain, g:

$$g = \sqrt{P_{M_{\min}}} = \sqrt{\sum_{m=0}^{M} a_m \rho_m} \tag{11.28}$$

11.4.2   Speech Synthesis

The speech signal is synthesized by means of the same parameters used in speech analysis. A block diagram of the speech synthesizer is depicted in Figure 11-7. In addition to the error signal the control parameters supplied to the synthesizer are the pitch period, the binary voiced-unvoiced parameter, the gain or rms value and the M PARCOR coefficients. In this formulation the effect of channel transmission errors has been neglected. If channel transmission errors are not negligible, then e_i in Figure 11-7 is replaced by $e_i + n_i$, where n_i is the noise introduced by the channel. The linear prediction synthesizer shown in Figure 11-7b is obtained by using Eqs. (3.145) and (3.146)

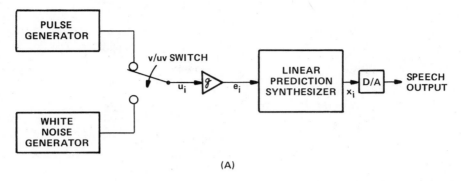

(A)

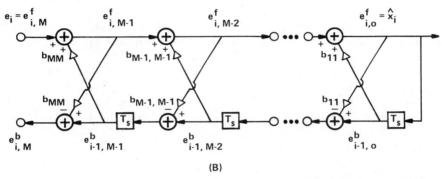

(B)

Figure 11-7. Block diagram of the speech synthesizer. (A) general block diagram; (B) linear prediction synthesizer using two multipliers per lattice section.

with M replaced by m:

$$e^f_{i,m-1} = e^f_{i,m} + b_{m,m} e^b_{i-1,m-1} \qquad (11.29)$$

$$e^b_{i,m} = e^b_{i-1,m-1} - b_{m,m} e^f_{i,m-1} \qquad (11.30)$$

where T_s is the sampling interval. The input and output forward error signals are determined by $e^f_{i,M} = e_i$ and $e^f_{i,0} = e^b_{i,0} = \hat{x}_i$, respectively. Use of the PARCOR coefficients for synthesizing the speech signal allows the stability of the filter to be checked directly. The pulse generator produces a pulse of unit amplitude at the beginning of each pitch period. The white noise generator produces uncorrelted random samples with unit variance. The selection between the pulse generator and the white noise generator is made by the voiced/unvoiced (v/uv) switch. The amplitude of the excitation signal is adjusted by the amplifier with gain, g. The synthesized speech samples $\hat{x}_i$ are finally digital to analog (D/A) converted to form the continuous speech wave. The synthesizer control parameters are reset to their new values at the beginning of each pitch period for voiced speech and once every few ms for unvoiced speech.

11.5   TIME DOMAIN WAVEFORM CODING OF SPEECH SIGNALS

An analog source can be digitally encoded by use of various classical techniques such as pulse code modulation (PCM), differential pulse code modulation (DPCM) and delta modulation (DM). To perform the source coding, samples of the analog waveform are taken at the Nyquist sampling rate or higher. Each sample is usually quantized to a discrete amplitude value by selecting one of a set of several discrete values. This quantization process introduces an error, which is modeled as additive noise. Efficient digital encoding schemes decrease the quantization noise by exploiting the redundancy present in the digital bit stream. Linear prediction techniques described in the previous section provide a useful method to minimize the redundancy. For signals with stationary characteristics, linear predictors with fixed coefficients are suitable. For signals with nonstationary characteristics, such as speech signals, alternative methods, which can follow changes in both the spectral envelope and signal periodicities, are required. These alternative methods are referred to as adaptive predictive coding (APC) techniques. In APC methods, both the quantizer and the predictor are made adaptive in order to provide efficient source coding.

In the following subsections short descriptions of PCM, DPCM, DM and APC are given. An overview of these methods is provided in reference 2, and a good comparison of the first three methods applied to speech processing is available in references 15 and 16. APC methods are treated in numerous references, including 1, 2, 4, 15, 17, 18, and 19.

11.5.1   Pulse Code Modulation (PCM)

Consider a bandlimited analog source, $x(t)$, with bandwidth, W, to be sampled at a rate exceeding the Nyquist rate, that is, $1/T_s > 2W$, where T_s is the sampling interval (See Appendix A.1.18). The samples, x_n, are each quantized to one of 2^B discrete levels, where B is the number of binary digits used to represent each sample. The digital bit rate from the source is therefore B/T_s bits per second. The quantized value of the sample x_n is denoted by $\tilde{x}_n$ and is related to x_n by

$$\tilde{x}_n = x_n + q_n \tag{11.31}$$

where q_n is an additive noise term or error introduced by the quantization process. The amount of quantization noise power is related to the amount of redundancy present in the digital bit stream. Efficient source coding techniques reduce the redundancy and decrease quantization noise. A measure of the efficiency of the source coding technique is thus provided by the ratio of signal to quantization noise power denoted by SNR. Mathematically, this SNR is defined by

$$\text{SNR} = 10 \log \frac{\mathscr{E}(x_n^2)}{\mathscr{E}(q_n^2)} \tag{11.32}$$

where the quantities $\mathscr{E}[x_n^2]$ and $\mathscr{E}[q_n^2]$ correspond to the signal and quantization noise powers, respectively, assuming that both the signal and quantization mean levels are each zero.

A common quantization model is to assume that the noise is uniformly distributed over the step size, Δ. In this case, $p(q_n)$, the probability density function (pdf) of the quantization error, q_n, is given by

$$p(q_n) = \begin{cases} \dfrac{1}{\Delta}, & -\dfrac{\Delta}{2} \le q_n \le \dfrac{\Delta}{2} \\ 0 & \text{otherwise} \end{cases} \tag{11.33}$$

In this case the quantization noise power is given by

$$\mathscr{E}(q_n^2) = \int_{-\infty}^{\infty} q_n^2 p(q_n)\, dq_n$$

$$= \int_{-\Delta/2}^{\Delta/2} \frac{q_n^2}{\Delta}\, dq_n$$

$$= \frac{\Delta^2}{12} \tag{11.34}$$

Now assume that the amplitude range of the signal x_n is represented by R_x. For 2^B discrete amplitude levels, the step size Δ can be expressed as

$$\Delta = \frac{R_x}{2^B} \tag{11.34a}$$

The SNR can now be computed from (11.32), (11.34), and (11.34a):

$$\text{SNR} \simeq 6B + 10.8 + 10 \log \frac{\mathscr{E}(x_n^2)}{R_x^2} \tag{11.35}$$

Thus the SNR improves 6 dB for each additional bit used in the quantizer.

Further evaluation of (11.35) can be accomplished by assuming that the amplitude range is ± 4 times the RMS value of the signal.

$$R_x = 8\sqrt{\mathscr{E}\left(x_n^2\right)} \tag{11.36}$$

Substituting (11.36) in (11.35) yields

$$\text{SNR} \simeq 6B - 7.3 \tag{11.37}$$

Equation (11.37) is referred to as standard PCM with 4-sigma loading. An input-output characteristic for a 3-bit uniform quantizer is shown in Figure 11-8.

Alternate quantizer characteristics have been investigated that utilize non-uniform quantization.[2,15,20] This form of quantization is particularly useful in speech processing, where small signal amplitudes, which occur more frequently, can be assigned a greater number of quantization levels than the larger signal amplitudes. Assume that the input and output voltages of the

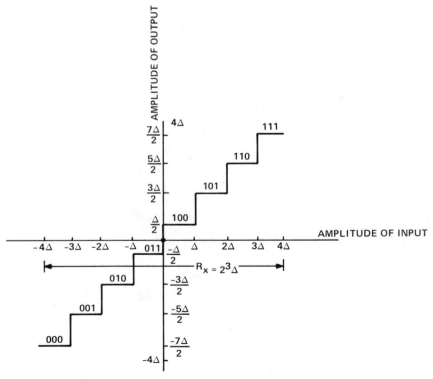

Figure 11-8.   Input-output characteristic for a 3-bit uniform quantizer.

quantizer are denoted by v and u, respectively, where the maximum value of the input voltage is normalized to unity. Two frequently used PCM quantizer characteristics in commercial telephone applications are the μ-law and A-law compression techniques. The A-law is defined as

$$u = \begin{cases} \dfrac{Av}{1 + \ln A}, & 0 \le v \le A^{-1} \\[2mm] \dfrac{1 + \ln Av}{1 + \ln A}, & A^{-1} \le v \le 1 \end{cases}$$

where   is a compression factor. A second law, which is also based on a nonur.form quantization of the input signal in accordance with a logarithmic scale, is referred to as the μ-law, defined by

$$u = \begin{cases} \dfrac{\ln(1 + \mu v)}{\ln(1 + \mu)}, & 0 \le v \le 1 \\[2mm] \dfrac{-\ln(1 - \mu v)}{\ln(1 + \mu)}, & -1 \le v \le 0 \end{cases} \tag{11.38}$$

In (11.38), μ is the compression factor, which is often set to 255. To provide a better understanding of logarithmic compression characteristics, Eq. (11.38) is plotted in Figure 11-9 for various values of μ. Note that $\mu = 0$ results in no

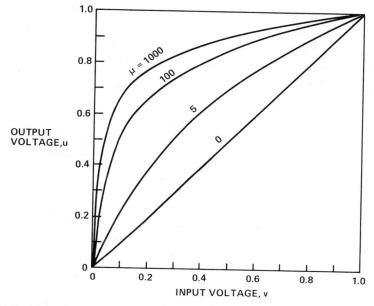

Figure 11-9. Nonuniform quantizer compression characteristic (© 1974 IEEE Jayant, reference 11-15).

compression. Using the compression characteristic like that in (11.38) in cascade with a uniform quantizer results in a nonuniform quantizer, referred to as a log-quantizer. It has been shown that a 7-bit log-quantizer provides similar performance to an 11-bit uniform quantizer which, according to (11.37), corresponds to a 24 dB improvement in SNR by use of the log-quantizer.[15]

Implementation of a nonuniform quantizer is usually accomplished by passing the signal through a nonlinear device that compresses the signal amplitude in accordance with (11.38) and is then followed by a uniform quantizer. To reconstruct the signal from the quantized samples, the inverse logarithmic relation is used to expand the signal amplitude. The combined characteristics of compression followed by expansion are linear and are referred to as a compander.[20]

Nonuniform quantization is one solution used to accommodate the wide dynamic range in voice communications systems. Alternate quantization characteristics used in speech processing are adaptive to variations in the input signal. As a result the nonstationary or time-varying nature of the speech signal can be followed. One method used in adaptive quantization utilizes a quantizer characteristic which contracts or expands in accordance with the short term signal power estimate. In this method the step size varies with the variance of prior signal samples. One such algorithm utilizes a step size that depends only on the previous sample and is adjusted recursively according to

$$\Delta_{n+1} = \Delta_n M(|\tilde{x}_n|) \tag{11.39}$$

where Δ_{n+1} is the step size used for the input x_{n+1}, Δ_n is the previous step size, and $M(|\tilde{x}_n|)$ is a multiplier which depends only on the magnitude of $\tilde{x}_n$, the quantizer output for sample x_n.[15] Note that there is no need to transmit the step size information to the receiver since it can be computed directly from the quantizer output.

11.5.2 Differential Pulse Code Modulation (DPCM)

A block diagram of a differential pulse code modulation (DPCM) system is shown in Figure 11-10. In this case the error signal, e_n, is given by

$$e_n = x_n - \hat{\tilde{x}}_n \tag{11.40}$$

where

$$\hat{\tilde{x}}_n = \sum_{m=1}^{M} b_m \tilde{x}_{n-m} \tag{11.41}$$

The quantization error, q_n, is defined by

$$q_n = \tilde{e}_n - e_n \tag{11.42}$$

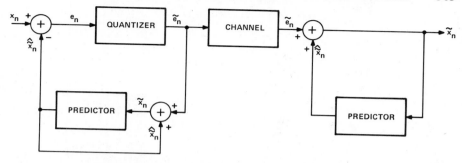

Figure 11-10. Block diagram of a DPCM communication system.

Since

$$\tilde{x}_n = \hat{\tilde{x}}_n + \tilde{e}_n \qquad (11.43)$$

Eq. (11.42) can be equivalently expressed as

$$q_n = \tilde{x}_n - x_n \qquad (11.44)$$

Note that in the absence of channel transmission errors, knowledge of the prediction coefficients and the error signal permits the quantized signal samples to be computed at the receiver.

The block diagram shown in Figure 11-10 is in fact a linear predictive communication system with a quantizer at the transmitter placed within the prediction loop. Elimination of the quantizer results in a linear predictive coding scheme described in Section 11.4. A heuristic description of DPCM is now provided, following a discussion given by Jayant.[15]

DPCM exploits the redundancy inherent in adjacent samples of the input signal. Thus, more efficient source coding is possible using the DPCM technique. Define the difference between successive signal samples by D_n:

$$D_n = x_n - x_{n-1} \qquad (11.45)$$

As a result of the signal correlation the variance of the difference signal is smaller than the variance of the input signal. Assuming x_n has zero mean, the variance of D_n is given by

$$\mathscr{E}(D_n^2) = \mathscr{E}\left[(x_n - x_{n-1})^2\right]$$
$$= \mathscr{E}(x_n^2) + \mathscr{E}(x_{n-1}^2) - 2\mathscr{E}(x_n x_{n-1}) \qquad (11.46)$$

For stationary signal statistics with an autocorrelation function $r_n = \mathscr{E}(x_i x_{i+n})$, (11.46) can be expressed as

$$\mathscr{E}(D_n^2) = 2r_0 - 2r_1 \qquad (11.47)$$

If the normalized autocorrelation is defined as

$$r_n' = \frac{r_n}{r_0} \tag{11.48}$$

Eq. (11.47) can be expressed as

$$\mathscr{E}(D_n^2) = 2r_0(1 - r_1') \tag{11.49}$$

Note that $-1 \leq r_n' \leq 1$. If r_1', the normalized correlation between adjacent samples, is greater than 0.5, then the variance of the difference signal is smaller than the variance of the input signal, x_n. Due to the significant correlation in adjacent samples it is preferable to quantize the difference signal, D_n, rather than the input signal. In other words, use of the difference signal instead of the input signal results in a lower quantization noise power and thus a higher SNR for the same number of quantization levels.

A generalization of the above result can be obtained by returning to the linear prediction case. Consider the error signal in the absence of quantization:

$$e_n = x_n - \hat{x}_n$$

$$= x_n - \sum_{m=1}^{M} b_m x_{n-m}$$

In Section 3.1.1, the mean square error was minimized, resulting in Eqs. (3.10) and (3.14), which are repeated here for convenience:

$$\sum_{m=1}^{M} b_m r_{l-m} = r_l, \qquad l = 1, \ldots, M \tag{3.10}$$

$$P_{M_{\min}} = \sum_{m=0}^{M} a_m r_{-m} \tag{3.14}$$

Note that for real signals $r_m = r_{-m}$. Since $a_0 = 1$ and $a_m = -b_m$ for $m \neq 0$, Eq. (3.14) can be expressed as

$$P_{M_{\min}} = r_0 - \sum_{m=1}^{M} b_m r_m$$

Expressing the above equation in terms of the normalized autocorrelation given in (11.48) results in

$$P_{M_{\min}} = r_0\left(1 - \sum_{m=1}^{M} b_m r_m'\right) \tag{11.50}$$

Similarly, (3.10) can be rewritten in terms of the normalized autocorrelation:

$$\sum_{m=1}^{M} b_m r'_{l-m} = r'_l, \qquad l = 1, \ldots, M \tag{11.51}$$

If it can be shown that $\sum_{m=1}^{M} b_m r'_m > 0$, then the error signal has a smaller variance than the input signal.

Define the covariance matrix R'_M and the vectors B and V by

$$B = \begin{pmatrix} b_M \\ \vdots \\ b_1 \end{pmatrix}$$

$$V = \begin{pmatrix} r'_M \\ \vdots \\ r'_1 \end{pmatrix}$$

$$R'_M = \begin{pmatrix} r'_0 & r'_1 & \cdots & r'_{M-1} \\ \vdots & & \ddots & \vdots \\ r'_{M-1} & \cdots & & r'_0 \end{pmatrix}$$

Equation (11.51) can now be expressed in matrix form by

$$R'_M B = V \tag{11.52}$$

The solution for the optimal coefficients is then given by

$$B = R'^{-1}_M V \tag{11.53}$$

The expression $\sum_{m=1}^{M} b_m r'_m$ can also be expressed in matrix form:

$$B'V = \sum_{m=1}^{M} b_m r'_m$$

Using (11.53) in the above expression leads to

$$B'V = \left(R'^{-1}_M V \right)' V$$

$$= V'\left(R''_M \right)^{-1} V \tag{11.54}$$

Since R'_M is a real covariance matrix, $R'_M = R''_M$, so that (11.54) becomes

$$B'V = V'R'^{-1}_M V$$

It is well known that a covariance matrix is positive semidefinite, that is (see Appendix Section A.2.5):

$$Y'R'_M Y \geq 0$$

where Y is an arbitrary M dimensional vector. Reference 21 proves the less well-known fact that the inverse of a positive semidefinite matrix is also positive semidefinite:

$$V'R'^{-1}_M V \geq 0 \qquad (11.55)$$

Thus it has been shown that

$$\sum_{m=1}^{M} b_m r'_m \geq 0 \qquad (11.56)$$

As an example, assume $M = 1$. Then from (11.51) and (11.48)

$$b_1 = \frac{r'_1}{r'_0} = r'_1$$

so that (11.50) becomes

$$P_{M_{min}} = r_0 (1 - r'^2_1)$$

For any $r'_1 > 0$, where r'_1 is of course restricted to the range $-1 < r'_1 \leq 1$, then $P_{M_{min}} < r_0$.

Using the above results it is now shown that DPCM offers an improvement in SNR over the SNR obtained from PCM. At this point it is convenient to attach subscripts to SNR, that is, SNR_{PCM} and SNR_{DPCM} are used to denote the SNR in the PCM and DPCM cases, respectively. The SNR in the PCM case can be written from (11.32) as

$$\text{SNR}_{PCM} = 10 \log \frac{\mathscr{E}(x_n^2)}{\mathscr{E}(q_n^2)} \qquad (11.57)$$

Figure 11-11 illustrates a PCM quantizer and only the quantizer portion of the DPCM system. From Figure 11-11 it can be seen that, with regard to the quantizer, the difference in the two cases is the input signal, that is, the input signal is the error signal e_n in the DPCM case and x_n in the PCM case. Thus the SNR in the DPCM case can be written from (11.32) as

$$\text{SNR}_{DPCM} = 10 \log \frac{\mathscr{E}(x_n^2)}{\mathscr{E}(q_n^2)} \qquad (11.58)$$

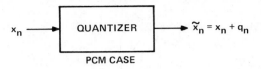

PCM CASE

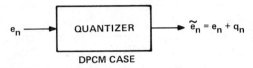

DPCM CASE

Figure 11-11.   Quantization in PCM and DPCM cases.

It is now possible to express the SNR_{DPCM} in terms of SNR_{PCM} and a gain factor. Rewriting (11.58) yields

$$\text{SNR}_{\text{DPCM}} = 10 \log \left[\frac{\mathscr{E}(x_n^2)}{\mathscr{E}(e_n^2)} \frac{\mathscr{E}(e_n^2)}{\mathscr{E}(q_n^2)} \right]$$

$$= 10 \log \frac{\mathscr{E}(x_n^2)}{\mathscr{E}(e_n^2)} + 10 \log \frac{\mathscr{E}(e_n^2)}{\mathscr{E}(q_n^2)}$$

Comparing the PCM and DPCM cases in Figure 11-11, it can be seen that the second term in the above equation is in fact the signal-to-noise ratio in the PCM case where the input signal is e_n rather than x_n. Therefore the above equation can be written as

$$\text{SNR}_{\text{DPCM}} = \text{SNR}_{\text{PCM}} + 10 \log \frac{\mathscr{E}(x_n^2)}{\mathscr{E}(e_n^2)} \tag{11.59}$$

As a result of the above analysis the $\mathscr{E}(e_n^2)$ for minimum MSE is given by Eq. (11.50):

$$P_{M_{\min}} = r_0 G^{-1} \tag{11.60}$$

where the factor G is defined by

$$G = \left(1 - \sum_{m=1}^{M} b_m r'_m \right)^{-1} \tag{11.61}$$

Since $\mathscr{E}(x_n^2) = r_0$, Eq. (11.59) can be rewritten using (11.60):

$$\text{SNR}_{\text{DPCM}} = \text{SNR}_{\text{PCM}} + 10 \log G \tag{11.62}$$

The quantity, G, defined by (11.61), represents the SNR improvement of DPCM over PCM. This result has been previously given in references 2 and 15.

Plots of the DPCM gain G have been obtained as a function of M, the order of the prediction filter.[2] Figures 11-12 and 11-13, extracted from reference 2, illustrate the gain for low-pass filtered and bandpass filtered speech, respectively, assuming the autocorrelation functions given in Figure 11-14. The autocorrelation function in Figure 11-14 results from speech sampled at 8 KHz and long term averaged over 55 seconds. The low-pass filtered curves shown in the upper portion of Figure 11-14 correspond to the maximum, minimum, and average over four speakers in the band from 0 Hz to 3400 Hz. The bandpass filtered curves shown in the lower portion of Figure 11-14 correspond to the maximum, minimum, and average over four speakers in the band from 200 Hz to 3400 Hz. Note that in Figures 11-12 and 11-13, three curves, corresponding

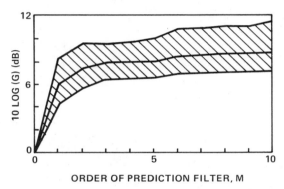

Figure 11-12.   DPCM gain, G (in dB) versus order of prediction filter for low-pass filtered speech (© 1979 IEEE Flanagan, et al., reference 11-2).

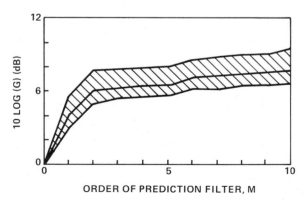

Figure 11-13.   DPCM gain, G (in dB) versus order of prediction filter for bandpass filtered speech (© 1979 IEEE Flanagan, et al., reference 11-2).

to the maximum, minimum, and average over the four speakers, are produced. For small prediction filters, that is, M equals 2 or 3, the principal benefit from DPCM is attained and indicates significant improvement in SNR over PCM. In the time-varying case, adaptive prediction will result in further enhancements.

Effects which result from the presence of the quantizer in Figure 11-10 are now considered. Suppose the quantizer were placed outside the predictor loop. The transmitter portion of Figure 11-10 would then be redrawn, as shown in Figure 11-15. Figure 11-15 is in fact a linear predictor in cascade with the quantizer. Assume no channel transmission errors. In Figure 11-15 the input to the prediction filter is x_n at the transmitter and $\tilde{x}_n$ at the receiver. In Figure 11-10 the input to the prediction filter is $\tilde{x}_n$ at both the transmitter and receiver. As a result, small quantization errors are accumulated in the receiver prediction filter, using the diagram shown in Figure 11-15. In Figure 11-10, the only input to the prediction filter is the quantized signal, so that the same predicted sequence is produced at both the transmitter and the receiver. As a result of (11.44), the quantized sample differs from the original sample by only the quantization error in the case of Figure 11-10, independent of the predictor used. In Figure 11-15 the difference between the quantized sample $\tilde{x}_n$ and the

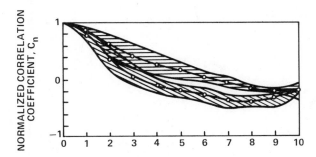

Figure 11-14. Long-time autocorrelation function for low-pass-filtered speech (upper curve) and bandpass-filtered speech (lower curve) (© 1979 IEEE Flanagan, et al., reference 11-2).

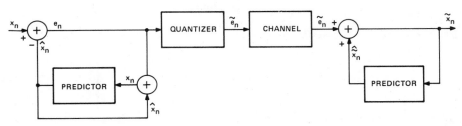

Figure 11-15. Quantizer placed external to the prediction loop.

input sample x_n depends on the prediction filter used, so that quantization errors tend to accumulate.

It should be noted that the development of the SNR improvement formula given by (11.62) proceeded by neglecting the quantization noise. Thus, the gain in SNR is independent of the number of bits, B. For sufficiently fine quantization this result is fairly accurate.

Thus far it has been shown that DPCM is a more efficient source coding technique than PCM. An interesting question in the study of source coding techniques is how efficient DPCM is in comparison with the optimum source coding technique. This question has been answered by O'Neal[22,23] for the special case of a Gaussian noise source, using a rate bound due to Shannon: [24]

$$R \geq W \log_2 \frac{Q_s}{\mathscr{E}(q_n^2)} \tag{11.63}$$

where R is the information rate, W is the signal bandwidth, Q_s is the entropy power of the signal, and $\mathscr{E}(q_n^2)$ is the quantization noise power. O'Neal[22,23] points out that in the case of linear prediction, the prediction process has whitened the error sequence. Furthermore, the Gaussian source produces a Gaussian error sequence so that the samples of the error sequence are independent. In this case the entropy power equals the mean square value of the error sequence, that is, $Q_s = \mathscr{E}(e_n^2)$. Thus the bound in (11.63) can be written as

$$R \geq W \log_2 \frac{\mathscr{E}(e_n^2)}{\mathscr{E}(q_n^2)} \tag{11.64}$$

Recall that the SNR is given by (11.32):

$$\text{SNR} = 10 \log \frac{\mathscr{E}(x_n^2)}{\mathscr{E}(q_n^2)} \tag{11.32}$$

Equation (11.32) can be written in the alternate form

$$\text{SNR} = 10 \log \frac{\mathscr{E}(x_n^2)}{\mathscr{E}(q_n^2)} \frac{\mathscr{E}(e_n^2)}{\mathscr{E}(e_n^2)} \tag{11.65}$$

From (11.59)–(11.62), it can be seen that (11.65) can be expressed as

$$\text{SNR} = 10 \log G + 10 \log \frac{\mathscr{E}(e_n^2)}{\mathscr{E}(q_n^2)} \tag{11.66}$$

where G is the gain factor given by (11.61). If the source is sampled at the Nyquist rate with a sampling interval, T_s, then $1/T_s = 2W$. The source information rate in bits per second is $R = B/T_s$ where B is the number of bits/sample. Therefore $R/2W = B$ and the bound in (11.64) can be written as

$$B \geq \tfrac{1}{2} \log_2 \frac{\mathscr{E}(e_n^2)}{\mathscr{E}(q_n^2)}$$

Changing the above inequality from base 2 to base 10 with $2/\log_2 10 \simeq 0.6$ yields

$$6B \geq 10 \log \frac{\mathscr{E}(e_n^2)}{\mathscr{E}(q_n^2)} \tag{11.67}$$

Substituting (11.66) into (11.67) results in

$$\text{SNR} \leq 6B + 10 \log G \tag{11.68}$$

The inequality in (11.68) represents the best SNR which can be achieved by reducing the redundancy in a Gaussian source with a linear predictor for a given bit rate and signaling bandwidth.

To compare the bound to the SNR achieved by DPCM, recall that (11.62) provides SNR_{DPCM} in terms of SNR_{PCM}:

$$\text{SNR}_{\text{DPCM}} = \text{SNR}_{\text{PCM}} + 10 \log G \tag{11.62}$$

A formula for SNR_{PCM} using 4-sigma loading was provided by (11.37) and could be used in (11.62). An alternate formula for SNR_{PCM} can be obtained following the derivation in O'Neal[22] for a Gaussian source. In this case an optimum quantizer derived by Panter and Dite[25] is available which permits the quantization noise to be expressed as

$$\mathscr{E}(q_n^2) = \frac{\sqrt{3}}{2} \frac{\pi}{2^{2B}} \mathscr{E}(x_n^2) \tag{11.69}$$

From (11.69) the quantity SNR_{PCM} can be written as

$$\text{SNR}_{\text{PCM}} = 10 \log \frac{\mathscr{E}(x_n^2)}{\mathscr{E}(q_n^2)}$$

$$= 10 \log \frac{2}{\sqrt{3}} \frac{2^{2B}}{\pi}$$

$$= 6B - 4.35 \tag{11.70}$$

Thus, (11.70) represents the SNR for a PCM system with a quantizer designed to optimally operate with a Gaussian input.

TABLE 11-1. MULTIPLIERS FOR PCM AND DPCM IN THE
ADAPTIVE STEP SIZE ALGORITHM

	PCM			DPCM		
$M(\|\tilde{x}_n\|)$	$B = 2$	$B = 3$	$B = 4$	$B = 2$	$B = 3$	$B = 4$
$M(\|\tilde{x}_1\|)$	0.60	0.85	0.80	0.80	0.90	0.90
$M(\|\tilde{x}_2\|)$	2.20	1.00	0.80	1.60	0.90	0.90
$M(\|\tilde{x}_3\|)$	—	1.00	0.80	—	1.25	0.90
$M(\|\tilde{x}_4\|)$	—	1.50	0.80	—	1.75	0.90
$M(\|\tilde{x}_5\|)$	—	—	1.20	—	—	1.20
$M(\|\tilde{x}_6\|)$	—	—	1.60	—	—	1.60
$M(\|\tilde{x}_7\|)$	—	—	2.00	—	—	2.00
$M(\|\tilde{x}_8\|)$	—	—	2.40	—	—	2.40

Substituting (11.70) into (11.62) yields

$$\text{SNR}_{\text{DPCM}} = 6B - 4.35 + 10 \log G \qquad (11.71)$$

By comparing (11.71) with (11.68), it can be seen that DPCM is only 4.35 dB below the bound for a Gaussian source.

Other bounds have been obtained by O'Neal[23] for nonstationary source statistics. The results have been applied to both speech and television signals.

For time-varying source statistics DPCM can be made adaptive and is referred to as ADPCM. The recursive step size algorithm in (11.39) presented for PCM can also be used in the DPCM case. Typical multiplier values for 2-, 3- and 4-bit adaptive quantizers for both PCM and DPCM using a first order predictor are presented in Table 11-1 and are taken from Jayant.[15] The multipliers in the DPCM case are slightly different from those in the PCM case as a result of the reduced correlation present in DPCM.

11.5.3 Delta Modulation

Delta modulation is a variation of DPCM in which a two-level or 1-bit quantizer is used in conjunction with a fixed first order predictor and a scale factor which multiplies the quantized error signal. This scheme is typically adopted when the source is oversampled to increase adjacent sample correlation, so that simple quantizer and prediction algorithms can be utilized. Two common forms of delta modulation will be described: linear DM (LDM) and adaptive DM (ADM). In LDM the input is approximated by a series of linear segments with a constant slope and a fixed step size. In ADM the step size is varied in accordance with the variation in the slope of the input signal.

An LDM block diagram is depicted in Figure 11-16. In the figure the error signal, e_n, is given by

$$e_n = x_n - \hat{\tilde{x}}_n \qquad (11.72)$$

where $\hat{\tilde{x}}_n$ is the current estimate of the input signal x_n. Note that the current estimate is related to the previous receiver output $\tilde{x}_n$ according to

$$\hat{\tilde{x}}_n = \tilde{x}_{n-1} \qquad (11.73)$$

The quantized error signal, $\tilde{e}_n$, is the sum of the error signal and the quantization noise, q_n:

$$\tilde{e}_n = e_n + q_n \qquad (11.74)$$

From the block diagram it can be seen that

$$\tilde{x}_n = \tilde{x}_{n-1} + \Delta\tilde{e}_n \qquad (11.75)$$

where the parameter, Δ, is a scale factor or step size. Using (11.73) and (11.75) results in the equation

$$\hat{\tilde{x}}_n = \hat{\tilde{x}}_{n-1} + \Delta\tilde{e}_{n-1}$$

Substituting (11.72) and (11.74) into the above equation yields

$$\hat{\tilde{x}}_n = (1 - \Delta)\hat{\tilde{x}}_{n-1} + \Delta x_{n-1} + \Delta q_{n-1} \qquad (11.76)$$

For $\Delta = 1$, Eq. (11.76) implies that the estimated value of the current sample x_n is the previous sample x_{n-1} added to the quantization noise, q_{n-1}. Thus, in this case, the receiver output is the current input corrupted by the quantization noise.

Figure 11-16 is usually drawn in the form of Figure 11-17, where the dotted portion is replaced by an accumulator. To identify the equivalence between Figures 11-16 and 11-17, the recursive form in (11.75) can be expressed as an accumulator:

$$\tilde{x}_n = \sum_{k=1}^{n} \Delta\tilde{e}_k \qquad (11.77)$$

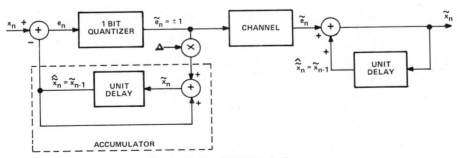

Figure 11-16.   LDM block diagram.

The selection of the step size value Δ must be made by considering the two types of delta modulation quantizing errors, referred to as slope-overload distortion and granular noise. Figure 11-18 illustrates both conditions. In the left-hand portion of the figure, the input signal increases steeply and the step size, Δ, is too small to follow the change in signal amplitude. In the right-hand portion of the figure, the input is relatively flat and the step size, Δ, is too large, so that the estimated signal hunts in the vicinity of the input signal and produces significant quantization noise. Thus, the step size must be optimally selected to achieve a compromise between the slope overload distortion and

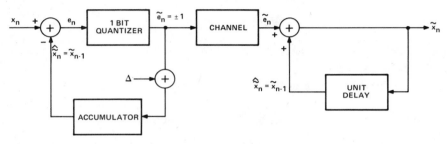

Figure 11-17.   Typical form of LDM block diagram.

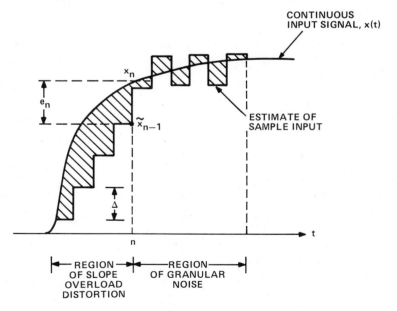

Figure 11-18.   Illustration of slope overload distortion and granular noise in LDM.

granular noise. It is possible to select a size to minimize the total error power. Typical results for this case are shown in Jayant.[15] However, the nonstationary characteristics of speech suggest that an adaptive scheme is preferable.

Before considering ADM, the SNR for the LDM, denoted by $\mathrm{SNR}_{\mathrm{LDM}}$ is provided in accordance with Flanagan, et al.:[2]

$$\mathrm{SNR}_{\mathrm{LDM}} = 10 \log k_1 + 30 \log f_s$$

where f_s is the sampling frequency and k_1 is a constant which depends on the input power spectrum and the performance of the single-bit quantizer. The above equation illustrates that the SNR is proportional to the cube of the sampling frequency. This relationship demonstrates the importance of over-sampling the input signal in attaining acceptable LDM performance.

Adaptive delta modulation (ADM) is accomplished by adjusting the step size in accordance with the variation in the input signal. This procedure improves the dynamic range of the source coder and reduces quantization noise. A common rule obtained by Jayant[15] is given by

$$\Delta_n = \Delta_{n-1}(P)^{\tilde{e}_n \tilde{e}_{n-1}} \tag{11.78}$$

where P is a parameter greater than 1. The optimum value of P lies in the range from 1 to 2. If the quantization error power for speech encoding is minimized, then the optimum value of P is 1.5.

The rule given by (11.78) can be explained as follows: If the input signal varies rapidly and slope overload distortion occurs, the error signal product in (11.78) will be positive, resulting in an increase in the step size used for the next iteration. If the input signal is relatively flat and a granular noise

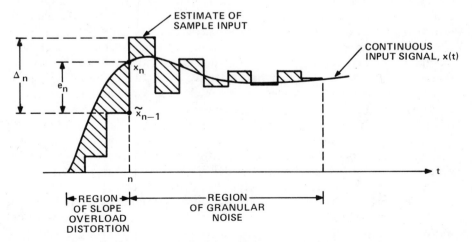

Figure 11-19.   Illustration of slope overload distortion and granular noise in ADM.

condition exists, the error signal product is negative, resulting in a decrease in the step size used for the next iteration. These effects are illustrated in Figure 11-19.

Another common ADM technique is called continuously variable slope delta modulation (CVSD). In CVSD the step size parameter is recursively computed according to

$$\Delta_n = \alpha \Delta_{n-1} + \Delta_0 k \tag{11.79}$$

where α, Δ_0 and k are constants. Typically, $\alpha = 0.996$, Δ_0 is the input step size, and k is equal to one or zero according to the following rule

$$k = \begin{cases} 1, & \text{if } |\tilde{e}_n + \tilde{e}_{n-1} + \tilde{e}_{n-2} + \tilde{e}_{n-3}| = 4 \\ 0, & \text{otherwise} \end{cases}$$

Variations of the above rule using fewer quantized error samples or other delayed error samples have been used.[2]

11.5.4 Adaptive Predictive Coding (APC)

In Section 11.4, linear prediction of speech was presented in which the error signal and a set of parameters were transmitted to the receiver. The parameters, which include the predictor coefficients, the pitch period, the voiced/unvoiced switch parameter, and the gain, were computed and transmitted to the receiver once per frame period. In Section 11.4, quantizer effects were neglected. Thus far in Section 11.5, only linear prediction with fixed coefficients was considered. In this case the coefficients are assumed to be available at both the transmitter and the receiver. Adaptive quantizer operation was considered for PCM, DPCM, and DM, and was accomplished by transmitting only the quantized error signal. To achieve satisfactory performance, the methods provided in Section 11.5 are usually restricted to data rates in excess of 16 Kbps. For lower rates, adaptive techniques must be considered in which both the quantizer and the predictor are changing to accommodate the nonstationary nature of the speech signal. Adaptive predictive coding (APC) techniques then combine the linear prediction method, in which a set of parameters are transmitted every frame period, with adaptive quantization. Various methods will subsequently be described which permit the predictor to follow both the changing short-time spectral envelope of the speech signal and the varying periodicities in the voiced speech. Algorithms to be presented are therefore generalizations of the techniques provided in Sections 11.4 and 11.5.1–11.5.3.

Descriptions of APC systems focus on prediction techniques required for both the short-time spectral envelope and the spectral fine structure in voiced speech.[2,17,26] Prediction techniques based on the spectral envelope were described in Section 11.4.1, and require the estimation of predictor coefficients, b_k, for $k = 1, \ldots, M$. Both covariance and correlation methods were described

in Section 11.4.1 for estimating the prediction coefficients. A common representation of the prediction filter is given by its Z transform presented in
Section 11.4 (see Appendix Section A.3):

$$P_B(Z) = \sum_{k=1}^{M} b_k Z^{-k} \tag{11.80}$$

where $P_B(Z)$ is the Z transform of the tapped delay line whose delay elements
are spaced at the sampling interval and whose multipliers, b_k, $k = 1, \ldots, M$,
are the prediction coefficients (see Figure 11-6). The zeros of the prediction
filter determine the location of the formant frequencies, as described in Section
11.4. Note that the transfer function of the receiver filter, $H(Z)$, given by
(11.3), can then be expressed as

$$H(Z) = \frac{1}{1 - P_B(Z)} \tag{11.81}$$

Errors due to quantization or channel noise can cause instabilities in the
receiver filter if the prediction coefficients are used. Thus the PARCOR
coefficients described in Section 11.4.1 are transmitted to the receiver to ensure
the stability of the receiver prediction filter. For acceptable quality speech
sampled at 8 KHz, typically 10 prediction coefficients are computed and
updated every frame period which corresponds to about 10–30 ms.

Prediction techniques based on spectral fine structure have been described
in Section 11.4.1.4. In this case a simple predictor for detecting periodicities in
the error signal given by (11.1) can be represented in Z transform notation by

$$P_p(Z) = \beta Z^{-L} \tag{11.82}$$

where $P_p(Z)$ is the Z transform of the spectral fine structure prediction filter
(pitch predictor), L is a delay between successive peaks in the autocorrelation
of the speech samples,* and β is a gain parameter. The delay parameter, L, is
in the range 2–20 ms, and corresponds to the pitch period or an integral
number of pitch periods. The gain parameter β, sometimes referred to as the
pitch gain, can be estimated by minimizing the mean square error given by

$$\sum_{k=1}^{K} (x_k - \beta x_{k-L})^2 \tag{11.83}$$

where K represents the number of samples in the frame period. The minimization then results in[2]

$$\hat{\beta} = \frac{\sum_{k=1}^{K} x_k x_{k-L}}{\sum_{k=1}^{K} x_{k-L}^2} \tag{11.84}$$

---

*L is assumed to be an integer.

It should be noted that for unvoiced speech the parameter β is small and the delay L is random. Improved spectral fine-structure estimation can be obtained by using additional samples preceding and following the pitch delay, resulting in a third order predictor given by

$$P_p(Z) = \beta_1 Z^{-L+1} + \beta_2 Z^{-L} + \beta_3 Z^{-L-1} \tag{11.85}$$

where β_1, β_2, and β_3 are amplitude coefficients which can be estimated by MSE minimization. It is known that periodicities in voiced speech are not as prevalent at high frequencies as they are at low frequencies.[2,26] Thus, the three amplitude coefficients provide a frequency-dependent gain factor in the pitch prediction loop.

The two types of prediction filters for spectral envelope and spectral fine structure can be combined in a variety of ways to provide a prediction gain. In Section 11.5.2 the prediction gain for DPCM was computed and was given by Eq. (11.61). In a similar manner, the prediction gain using two prediction filters can be computed and results in additional gain over a single prediction filter. For two predictors in cascade the first predictor provides a prediction gain of approximately 13–14 dB, and the second provides typically an additional 3 dB.[2]

The two prediction filters can be placed in series in either order. Here order is important since the predictors are time varying. A block diagram of an APC receiver system is shown in Figure 11-20. Note that the overall transfer function, $H_T(Z)$, for the diagram is expressed in terms of the combined predictor, $P(Z)$, by

$$H_T(Z) = \frac{1}{1 - P(Z)} \tag{11.86}$$

where

$$P(Z) = P_B(Z) + P_p(Z)\big[1 - P_B(Z)\big] \tag{11.87}$$

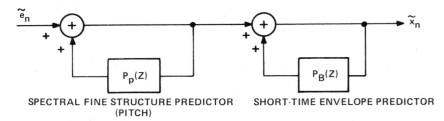

SPECTRAL FINE STRUCTURE PREDICTOR   SHORT-TIME ENVELOPE PREDICTOR
(PITCH)

Figure 11-20.   APC receiver system.

or

$$P(Z) = P_p(Z) + P_B(Z)\left[1 - P_p(Z)\right] \qquad (11.88)$$

The combined predictor, given by (11.88), has the implementation structure shown in Figure 11-21, and corresponds to the case where short-time envelope prediction is performed on the speech signal after pitch prediction.[4] The combined predictor, given by (11.87), corresponds to the case where pitch prediction is performed after the short-term spectral envelope prediction. Atal[26] suggests that it is preferable to perform short-term spectral envelope prediction first.

Quantization effects are now considered. To minimize speech distortion in the reconstructed signal, the spectrum of the quantization noise must be considered in terms of its relationship to the spectrum of the speech signal. For example, since the speech signal energy is large in the vicinity of the formant frequencies, the noise energy can also be larger in this region. One method used to control the quantization noise power spectrum is termed D*PCM.[2] In this method, the speech signal, x_n, is prefiltered by a time-varying filter with a Z transform, $1 - F(Z)$, to produce a new signal y_n. A combined predictor, $P(Z)$, such as that described by (11.87), is then applied to the signal, $\tilde{y}_n$, to provide prediction for both the short-time spectral envelope and the spectral fine structure. Figure 11-22 illustrates the D*PCM technique. Note that to

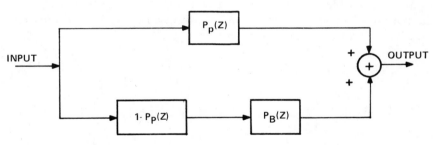

Figure 11-21. Combined predictor P(Z) for equation (11.87).

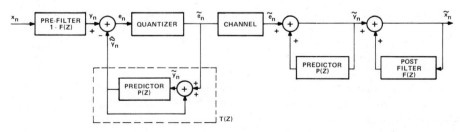

Figure 11-22. D*PCM method using pre- and post-filtering to control quantization noise spectrum.

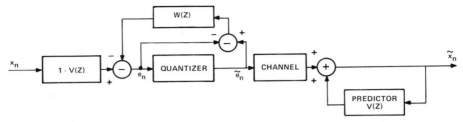

Figure 11-23. Predictive coding with quantization noise spectral shaping.

reconstruct the speech signal, the receiver structure first uses the predictor, $P(Z)$, to estimate $\tilde{y}_n$, and subsequently uses the post filter with Z transform, $F(Z)$, to estimate $\tilde{x}_n$. For sufficiently fine quantization, selection of the filter characteristic, $F(Z)$, allows a specified quantization noise spectrum to be generated.[26]

Another predictive coder capable of producing any desired spectrum for the quantization noise is shown in Figure 11-23. In this figure the filter with Z transform, $V(Z)$, is the predictor for the speech signal, and the filter with Z transform, $W(Z)$, is used to shape the quantization noise. The quantization noise shaping filter, $W(Z)$, can be selected to reduce noise in regions where the speech signal is low, which once again increases the noise in the vicinity of the formant frequencies.

It is now shown that, under certain conditions, Figures 11-22 and 11-23 are equivalent. To identify the relationships required to achieve the equivalence, the output of the speech encoder, $\tilde{e}_n$, is expressed in both cases. First refer to Figure 11-22 and define the transfer function, $T(Z)$, by

$$T(Z) = \frac{\hat{\hat{Y}}(Z)}{\tilde{E}(Z)} \tag{11.89}$$

where $\hat{\hat{Y}}(Z)$ and $\tilde{E}(Z)$ are the Z transform of $\hat{\hat{y}}_n$ and $\tilde{e}_n$, respectively. From Figure 11-22 it can be seen that

$$\hat{\hat{Y}}(Z) = \left(\hat{\hat{Y}}(Z) + \tilde{E}(Z) \right) P(Z) \tag{11.90}$$

From (11.89) and (11.90), the transfer function, $T(Z)$ can be written as

$$T(Z) = \frac{P(Z)}{1 - P(Z)} \tag{11.91}$$

Let the Z transforms of x_n and e_n be defined by $X(Z)$ and $E(Z)$, respectively. From Figure 11-22, the speech encoder error signal can be related to the speech encoder input:

$$E(Z) = X(Z)[1 - F(Z)] - \hat{\hat{Y}}(Z) \tag{11.92}$$

Using (11.89) and (11.91) in (11.92) results in

$$E(Z) = X(Z)[1 - F(Z)] - \frac{P(Z)}{1 - P(Z)}\tilde{E}(Z) \qquad (11.93)$$

Rewriting (11.93) yields a relationship between the speech encoder output and its input:

$$\tilde{E}(Z) = \frac{(1 - P(Z))(1 - F(Z))}{P(Z)}X(Z) - \left(\frac{1 - P(Z)}{P(Z)}\right)E(Z) \quad (11.94)$$

Now refer to Figure 11-23. The speech encoder output can be related to the speech signal input in Z transform notation as follows:

$$E(Z) = X(Z)(1 - V(Z)) - W(Z)[\tilde{E}(Z) - E(Z)] \qquad (11.95)$$

Rewriting (11.95) yields

$$\tilde{E}(Z) = \left[\frac{1 - V(Z)}{W(Z)}\right]X(Z) - \left[\frac{1 - W(Z)}{W(Z)}\right]E(Z) \qquad (11.96)$$

From the receiver structure in Figures 11-22 and 11-23 it can be seen that

$$\frac{1}{1 - P(Z)}\frac{1}{1 - F(Z)} = \frac{1}{1 - V(Z)}$$

or

$$(1 - P(Z))(1 - F(Z)) = 1 - V(Z) \qquad (11.97)$$

Comparing (11.94) and (11.96) with the condition specified by (11.97) implies that equivalence is achieved by requiring

$$P(Z) = W(Z) \qquad (11.98)$$

Therefore, the relationships given by (11.97) and (11.98) provide the conditions under which the block diagrams in Figures 11-22 and 11-23 are functionally equivalent.

The diagrams in Figures 11-22 and 11-23 are not the only general predictive coding schemes which permit quantization noise spectral shaping. Another form functionally equivalent to those in Figures 11-22 and 11-23 is provided in Figure 11-24. Additional generalized predictive coding schemes exist and are available in the literature.[2] These predictive coders are all functionally equivalent and differ principally in their implementation structure. From a practical viewpoint it has been shown that the performance of the various predictive coding schemes is essentially the same.[26]

11.5.5 Comparison of Speech Encoding Techniques

Table 11-2 was constructed from data available in Flanagan, et al.[2] and Proakis.[27] The transmission rate needed to achieve a given speech quality is provided. In the table, "toll communications" refers to the quality comparable to analog speech, which has a frequency range extending from 200 Hz to 3200 Hz, a signal-to-noise ratio in excess of 30 dB, and harmonic distortion less than 2–3 percent. Communications-quality speech results in a highly intelligible signal with reduced quality and poorer speaker recognition. Synthetic speech is unnatural and machinelike, without a substantial in-speaker recognition. The column in Table 11-2 on implementation complexity is subjective and breaks down into complex structures such as APC and LPC and simple structures.

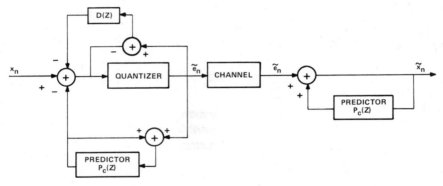

Figure 11-24. Predictive coding with quantization noise spectral shaping.

TABLE 11-2. COMPARISON OF SPEECH ENCODING TECHNIQUES

Encoding Method	Quantizer Type	Number of Quantizer Bits	Transmission Rate (bps)	Speech Quality	Implementation Complexity
PCM	Uniform	12	96,000	Toll	Simple
Log PCM	Nonuniform	7–8	56,000	Toll	Simple
DPCM	Nonuniform	4–6	32,000	Toll	Simple
ADPCM	Adaptive	3–4	{ 32,000 { 16,000	Toll Communications	Simple
LDM	Binary	1	64,000	Toll	Simple
ADM	Adaptive	1	{ 40,000 { 24,000	Toll Communications	Simple
LPC	—	—	2,400	Synthetic	Complex
APC	Adaptive	1–2	{ 16,000 { 7,200	Toll Comunications	Complex

The complex structures typically require a number of logic counts, which exceed the number required in the simple structures by an order of magnitude. Complexity differences not indicated in the table exist. For example, APC is simpler to implement than LPC, and ADM is perhaps the simplest scheme shown.

11.5.6   Additional Linear Prediction Algorithms

It is interesting to consider the application of the unconstrained least square lattice structures described in Chapter 4 and the Kalman algorithms presented in Chapter 5 to the linear prediction of speech signals. Studies involving the use of the Kalman algorithm, provided in Table 5-1, have been investigated.[28,29] Studies utilizing the unconstrained least square lattice structure provided in Section 4.1.6 have been considered by Honig, et al.[30] In both cases the results indicate that the complexity introduced by these algorithms is not offset by any substantial performance gain.

11.6   IMAGE PROCESSING

PCM, DPCM, and ADPCM techniques presented for speech processing can be applied in two dimensions to image processing.[31,32] In this case an image is subdivided into picture elements, referred to as pixels. In the simplest scheme, which uses PCM, the sampled image is quantized on a pixel-by-pixel basis. To provide images of reasonable quality and to avoid false contouring effects, PCM requires 6–8 bits/pixel. In general, PCM provides unacceptable image reconstruction with fewer than 6–8 bits/pixel as a result of the inability of PCM to exploit the spatial correlation between pixel values. DPCM is one method that can be used to improve PCM performance in a manner similar to that obtained for speech processing.

Let the two dimensional image signal be represented by $x_{i,j}$, where the index pair $\{i, j\}$ defines a specific pixel. It is common to represent the signal by a two dimensional autoregressive model in the form

$$x_{i,j} = \sum_k \sum_l a_{k,l} x_{i-k,j-l} + W_{i,j}, \qquad i, j \geq 0, \quad \left(k, l \in D'_a \right) \quad (11.99)$$

where $a_{k,l}$ are image structure coefficients in a specified region of space $\{D'_a\}$, and $W_{i,j}$ is a two-dimensional, zero mean sequence of independent, identically distributed noise variables with a fixed variance σ_w^2. A block diagram of the image encoding/decoding system is shown in Figure 11-25. This diagram is the two-dimensional version of the DPCM diagram shown in Figure 11-10 for speech processing. The estimated signal, $\hat{\tilde{x}}_{i,j}$, is given by

$$\hat{\tilde{x}}_{i,j} = \sum_{k=1}^{K} \sum_{l=1}^{L} b_{k,l} \tilde{x}_{i-k,j-l} \qquad (11.100)$$

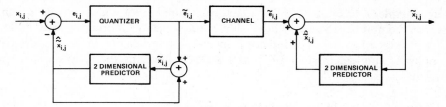

Figure 11-25.   Two-dimensional DPCM encoding/decoding system.

where $b_{k,l}$ are the two-dimensional prediction coefficients for $k = 1, \ldots, K$ and $l = 1, \ldots, L$. Equation 11.100 represents the two-dimensional predictor in the transmitter and receiver portions of Figure 11-25. Following the procedure outlined for speech processing, an error signal at the input to the quantizer can be defined. Then the mean square can be minimized by ignoring the quantizer, resulting in an expression for the prediction coefficients. The two-dimensional DPCM system provides good quality image reconstruction using 2–3 bits/pixel, and represents a substantial improvement over PCM. In most cases image statistics are nonstationary, which again suggests the use of an adaptive quantizer. In the latter case good performance can be achieved even at a rate as low as 1 bit/pixel. The interested reader can examine references 31 and 32 for further information on this complex subject.

REFERENCES

1. R. E. Crochiere and J. L. Flanagan, Current Perspective in Digital Speech, *IEEE Communications Magazine*, January 1983, pp. 32–40.

2. J. L. Flanagan, M. R. Schroeder, B. S. Atal, R. E. Crochiere, N. S. Jayant and J. M. Tribolet, Speech Coding, *IEEE Trans. Communications COM-27* 4, April 1979, pp. 710–737.

3. J. D. Markel and A. H. Gray, Jr., *Liner Prediction of Speech*, Springer-Verlag, New York, 1976.

4. B. S. Atal and M. R. Schroeder, Adaptive Predictive Coding of Speech Signals, *Bell System Tech. J. 49*, 6, Oct. 1970, pp. 1973–1986.

5. B. S. Atal and S. L. Hanauer, Speech Analysis and Synthesis by Linear Prediction of the Speech Wave, *J. Acoustic Society America 50*, 2, 1971, pp. 637–655.

6. J. Makhoul, Spectral Analysis of Speech by Linear Prediction, *IEEE Trans. Audio Electroacoust. AU-21*, June 1973, pp. 140–148.

7. J. Makhoul, Linear Prediction: A Tutorial Review, *Proc. IEEE 63*, April 1975, pp. 561–580.

8. R. W. Schafer and J. D. Markel, Eds., *Speech Analysis*, IEEE Press, 1979.

9. F. Itakura and S. Saito, On the Optimum Quantization of Feature Parameters in the PARCOR Speech Synthesizer, In *Proc. IEEE 1972 Conference on Speech Communications and Process*, April 1972, pp. 434–437.

10. H. Wakita, Estimation of the Vocal Tract Shape by Optimal Inverse Filtering and Acoustic/Articulatory Conversion Methods, SCRL Monograph No. 9, Speech Communications Research Laboratory, Santa Barbara, Calif., 1972.

11. J. I. Makhoul and L. K. Cosell, Adaptive Lattice Analysis of Speech, *IEEE Trans. Acoustics, Speech, and Signal Processing*, ASSP-29, 3, June 1981, pp. 654–659.

12. L. R. Rabiner, M. J. Cheng, A. E. Rosenberg, and C. A. McGonegal, A Comparative Performance Study of Several Pitch Detection Algorithms, *IEEE Trans. Acoustics, Speech, and Signal Processing*, October 1976.

13. M. J. Ross, H. L. Schaffer, and A. Cohen, R. Freudberg and H. J. Manley, Average Magnitude Difference Function Pitch Extractor, *IEEE Trans. Acoustics, Speech, and Signal Processing*, October 1974.

14. A. M. Noll, Cepstrum Pitch Determination, *J. Acoustical Society of America*, February 1976.

15. N. Jayant, Digital Coding of Speech Waveforms: PCM, DPCM, and DM Quantizers, *Proc. IEEE 62* May 1974.

16. R. W. Schafer and L. R. Rabiner, Digital Representations of Speech Signals, *Proc. IEEE*, 63, April 1975, pp. 662–677.

17. A. J. Goldberg and H. L. Schaffer, A Real Time Adaptive Predictive Coder Using Small Computers, *IEEE Trans. Communications COM-23*, 12, Dec. 1975, pp. 1443–1451.

18. M. Honig and D. C. Messerschmitt, Comparison of Adaptive Linear Prediction Algorithms in ADPCM, *IEEE Trans. Communications COM-30*, 7, July 1982, pp. 1775–1785.

19. B. Atal, Predictive Coding of Speech at Low Bit Rates, *IEEE Trans. Communications COM-30*, 4, April 1982, pp. 600–614.

20. P. F. Panter, *Modulation, Noise and Spectral Analysis*, McGraw-Hill, New York, 1965.

21. P. B. Liebelt, *An Introduction to Optimal Estimation*, Addison-Wesley, Reading, Mass, 1967, p. 27.

22. J. B. O'Neal, Jr., A Bound on Signal-to-Quantizing Noise Ratios for Digital Encoding Systems, *Proc. IEEE*, 55, March 1967, pp. 287–292.

23. J. B. O'Neal, Jr., Bounds on Subjective Performance Measures for Source Encoding Systems, *IEEE Trans. Information Theory*, IT-17, 3, May 1971, pp. 224–231.

24. C. E. Shannon and W. Weaver, *The Mathematical Theory of Communication*, Univ. of Illinois Press, Urbana 1962.

25. P. F. Panter and W. Dite, Quantization Distortion in PCM with Non-Uniform Spacing of Levels, *Proc IRE*, Jan. 1951.

26. B. S. Atal, Predictive Coding of Speech at Low Bit Rates, *IEEE Trans. Communications COM-30*, 4, April 1982, pp. 600–614.

27. J. G. Proakis, *Digital Communications*, McGraw-Hill, New York, 1983, p. 89.

28. S. F. Boll, A Priori Digital Speech Analysis, Ph.D. Dissertation, Dept. Electrical Engineering, Utah University, March 1973.

29. J. D. Gibson, V. P. Berglund and L. C. Sauter, Kalman Backward Adaptive Predictor Coefficient Identification in ADPCM with PCQ, *IEEE Trans. Communications COM-28*, 3, March 1980, pp. 361–371.

30. M. L. Honig and D. C. Messerschmitt, Comparison of Adaptive Linear Prediction Algorithm in ADPCM, *IEEE Trans. Communications COM-30*, 7, July 1982, pp. 1775–1785.

31. D. G. Daut and J. W. Modestino, Two-Dimensional DPCM Image Transmission Over Fading Channels, *IEEE Trans. Communications COM-31*, 3, March 1983.

32. J. W. Modestino, V. Bhaskaran and J. B. Anderson, Tree Encoding of Images in the Presence of Channel Errors, *IEEE Trans. Information Theory IT-27*, 6, Nov. 1981.

appendix

PRINCIPAL MATHEMATICAL RESULTS

A.1 PROBABILITY RESULTS[1,2]

A.1.1 Definition of Probability

Let S be defined as the samples of all outcomes, ζ, of an experiment. Define an event, A, as a subset of the sample space. Then, S denotes the certain event and ϕ denotes the impossible event. Let $\cup$ and $\cap$ denote union and intersection, respectively. If A, B, and A_i for $i = 1, 2, \ldots$ denote events, the probability of an event A is a number $P(A)$ assigned to A such that

1. $0 \le P(A) \le 1$, for any A contained in S.
2. $P(S) = 1$.
3. $P(\phi) = 0$.
4. If $A \cap B = \phi$, then $P(A \cup B) = P(A) + P(B)$ and A and B are termed mutually exclusive.
5. $P(A \cup B) = P(A) + P(B) - P(A \cap B) \le P(A) + P(B)$.
6. If $A_i \cap A_j = \phi$ for $i \ne j$, then $P(A_1 \cup A_2 \cup \ldots \cup A_n \cup \ldots) = \sum_{i=1}^{\infty} P(A_i)$ and A_i are termed pairwise disjoint.

A.1.2 Conditional Probability

Given events A, M with $P(M) > 0$, the conditional probability of A given M is denoted by $P(A|M)$ and is defined by

$$P(A|M) = \frac{P(A \cap M)}{P(M)}$$

A.1.3 Bayes Rule

$$P(A|B) = \frac{P(B|A)P(A)}{P(B)}$$

If $A_i \cap A_j = \phi$ for $i \neq j$ and $S = A_1 \cup A_2 \cup \cdots \cup A_N$, then

$$P(A_i|B) = \frac{P(B|A_i)P(A_i)}{\sum_{i=1}^{N} P(B|A_i)P(A_i)}$$

where $P(A_i)$ are termed a priori probabilities and $P(A_i|B)$ is termed the a posteriori probability.

A.1.4 Independent Events

Two events A and B are independent if and only if

$$P(A \cap B) = P(A)P(B)$$

Note that N events A_i, $i = 1, \ldots, N$, are independent if and only if

$$P(A_1 \cap A_2 \cap \cdots \cap A_N) = \prod_{i=1}^{N} P(A_i)$$

A.1.5 Random Variables

A random variable (rv), $X(\zeta)$, is a number assigned to the specific outcome, ζ. Alternatively stated, $X(\zeta)$ is a rule of correspondence between elements ζ of the sample space, S, and the set of real numbers.
  Notes:

1. An rv can be discrete or continuous.
2. The event $\{X \leq x\}$ is the set consisting of all outcomes, ζ, such that $X(\zeta) \leq x$ where x is a specific value.

3. A complex rv Z is defined in terms of two real rv's X and Y by $Z = X + jY$.

A.1.6 Cumulative Distribution Function (cdf)

Given rv's X, Y and X_i, $i = 1, \ldots N$, the cdf of X, $F(x)$, is $F(x) = P\{X \leq x\}$. The joint cdf of X and Y, $F(x, y)$, is $F(x, y) = P\{X \leq x, Y \leq y\}$. The joint cdf of $X_1, \ldots, X_N$, $F(x_1, \ldots, x_N)$ is

$$F(x_1, \ldots x_N) = P\{X_1 \leq x_1, \ldots X_N \leq x_N\}$$

The conditional cdf of X, given an event M with probability $P\{M\}$ is

$$F(X|M) = P\{X \leq x|M\} = \frac{P\{X \leq x, M\}}{P\{M\}}$$

Note:

1. For a discrete rv X with values x_i, the cdf of X is

$$F(x) = \sum_i P\{X = x_i\} \text{ where } i \text{ corresponds to all } x_i \leq x$$

A.1.7 Probability Density Function (pdf)

The pdf of X, $p(x)$, is $p(x) = (dF(x)/dx)$. The joint pdf of X and Y, $p(x, y)$, is

$$p(x, y) = \frac{\partial^2 F(x, y)}{\partial x \, \partial y}.$$

The joint pdf of $X_1, \ldots, X_N$, $p(x_1, \ldots, x_N)$ is

$$p(x_1, \ldots, x_N) = \frac{\partial^N F(x_1, \ldots, x_N)}{\partial x_1, \ldots, \partial x_N}$$

The conditional pdf of X given an event M, $p(x|M)$, is

$$p(x|M) = \frac{dF(x|M)}{dx}$$

Notes:

1. $p(x) \geq 0$, $p(x, y) \geq 0$, $p(x_1, \ldots, x_N) \geq 0$.
2. $\int_{-\infty}^{\infty} p(x) \, dx = 1$, $\int_{-\infty}^{\infty} \int_{-\infty}^{\infty} p(x, y) \, dx \, dy = 1$,

$$\int_{-\infty}^{\infty} \cdots \int_{-\infty}^{\infty} p(x_1, \ldots, x_N) \, dx_1 \ldots, dx_N = 1$$

3. $P\{x_1 \leq X \leq x_2\} = \int_{x_1}^{x_2} p(x)\,dx.$
4.   Marginal pdf's are given by

$$\int_{-\infty}^{\infty} p(x, y)\,dy = p(x)$$

$$\int_{-\infty}^{\infty} p(x_1, \ldots, x_N)\,dx_N = p(x_1, \ldots, x_{N-1})$$

$$\vdots$$

5.   The pdf of a discrete rv can be written in continuous form:

$$p(x) = \sum_i P\{X = x_i\}\delta(x - x_i) \qquad \text{for } i \text{ such that } x_i \leq x$$

A.1.8   Expected Value

The expected value of an rv, X, is denoted $\mathscr{E}(X)$ and is

$$\mathscr{E}(X) = \int_{-\infty}^{\infty} xp(x)\,dx$$

The conditional expected value of a rv, X, given a rv, Y, is

$$\mathscr{E}(X|Y) = \int_{-\infty}^{\infty} xp(x|y)\,dx$$

Notes:

1.   The expected value is also known as the mean, average, and first moment and is often denoted by $\eta = \mathscr{E}(X)$.
2.   Let $Z = g(X, Y)$ where g is an arbitrary function and Z, X, and Y are rv's. Then

$$\mathscr{E}(Z) = \int_{-\infty}^{\infty}\int_{-\infty}^{\infty} g(x, y)p(x, y)\,dx\,dy$$

3.   Given rv's X and Y and constants a and b, if $Y = aX + b$, then $\mathscr{E}(Y) = a\mathscr{E}(X) + b$. For arbitrary functions $g_i(x)$ for $i = 1, \ldots, N$,

$$\mathscr{E}\left\{\sum_{i=1}^{N} g_i(X)\right\} = \sum_{i=1}^{N} \mathscr{E}\{g_i(X)\}$$

4. For a discrete rv, X, with values x_i,

$$\mathscr{E}\{x\} = \sum_i x_i p\{X = x_i\}$$

where the sum includes all values, x_i.

A.1.9 Variance

The variance of a real rv X is denoted by $V(X)$ and is

$$V(X) = \int_{-\infty}^{\infty} (x - \eta)^2 p(x)\, dx$$

Notes:

1. The variance is also known as the dispersion or second central moment and is often denoted by $\sigma^2 = V(X)$ where σ is referred to as the standard deviation of X.
2. $\sigma^2 = \mathscr{E}[(X - \eta)^2] = \mathscr{E}(X^2) - [\mathscr{E}(X)]^2$.
3. Given rv's X and Y and real constants a and b, if $Y = aX + b$, then $V(Y) = a^2 V(X)$.
4. For a discrete rv, X, with values x_i

$$V(X) = \sum_i (x_i - \eta)^2 P\{X = x_i\}$$

where the sum includes all values, x_i.
5. For a complex rv Z

$$V(Z) = \mathscr{E}\{|Z - \mathscr{E}[Z]|^2\}$$

Thus, if $Z = aX + b$, where all quantities are complex, then

$$V(Z) = |a|^2 V(X).$$

A.1.10 Covariance

The covariance of real rv's X and Y with means η_X and η_Y, respectively is denoted by $\text{COV}(X, Y)$ and is

$$\text{COV}(X, Y) = \mathscr{E}[(X - \eta_X)(Y - \eta_Y)]$$

Notes:

1. $\text{COV}(X, Y) = \mathscr{E}(XY) - \eta_X \eta_Y$.
2. The correlation coefficient between X and Y is denoted ρ. If X and Y have standard deviations σ_X and σ_Y, respectively, then

$$\rho = \frac{\text{COV}(X, Y)}{\sigma_X \sigma_Y} \qquad \text{with} \qquad -1 \le \rho \le 1$$

3. The rv's X and Y are orthogonal if $COV(X, Y) = 0$. The rv's X and Y are uncorrelated if $\mathscr{E}(XY) = \eta_X \eta_Y$.

4. For complex rv's Z and W with means η_Z and η_W, respectively,

$$COV(Z, W) = \mathscr{E}\{[Z - \eta_Z][W^* - \eta_W^*]\}$$

A.1.11   Moments

The moments of a rv X, m_k, are

$$m_k = \mathscr{E}[X^k] = \int_{-\infty}^{\infty} x^k p(x)\, dx$$

The central moments of a rv X, μ_k, are

$$\mu_k = \mathscr{E}[(X - \eta)^k] = \int_{-\infty}^{\infty} (x - \eta)^k p(x)\, dx$$

Notes:

1. $m_0 = 1$, $m_1 = \eta$.
2. $\mu_0 = 1$, $\mu_1 = 0$, $\mu_2 = \sigma^2$.

A.1.12   Characteristic Function (cf)

The cf of an rv X, $\Phi(w)$, is

$$\Phi(w) = \mathscr{E}[e^{jwx}] = \int_{-\infty}^{\infty} e^{jwx} p(x)\, dx$$

The joint cf of rv's $X_1, \ldots, X_N$, $\Phi(w_1, \ldots, w_N)$ is

$$\Phi(w_1, \ldots, w_N) = \mathscr{E}[e^{j(w_1 X_1 + \cdots + w_N X_N)}]$$

Notes:

1. $p(x) = (1/2\pi)\int_{-\infty}^{\infty}\Phi(w)e^{-jwx}\, dw$.
2. $\mathscr{E}(X^n) = (-j)^n(d^n\Phi(w)/dw^n)$ evaluated at $w = 0$.

A.1.13   Independent Random Variables

Assume $X_1, \ldots, X_N$ are statistically independent. Then the joint pdf $p(x_1, \ldots, x_N) = \prod_{i=1}^{N} p(x_i)$

Notes:

1. For arbitrary functions g_i with $i = 1, \ldots N$

$$\mathscr{E}\left\{\prod_{i=1}^{N} g_i(X_i)\right\} = \prod_{i=1}^{N} \mathscr{E}\{g_i(X_i)\}$$

2. The variance of the sum of N independent rv's X_i is

$$V\left\{\sum_{i=1}^{N} X_i\right\} = \sum_{i=1}^{N} V(X_i)$$

3. The cf at the sum, S, of N independent rv's, X_i, is the product of the cf's of the individual rv's. The pdf of S is the N fold convolution of the X_i: Let $S = \sum_{i=1}^{N} X_i$ where $\Phi_S(w_1, \ldots, w_N)$ is the cf of S, and $\Phi_{X_i}(w_i)$ is the cf of X_i, then

$$\Phi_S(w_1, \ldots, w_N) = \prod_{i=1}^{N} \Phi_{X_i}(w_i)$$

If $p_S(s)$ is the pdf of S and $p_{X_i}(x_i)$ is the pdf of X_i, then

$$p_S(s) = p_{X_1}(x_1) * p_{X_2}(x_2) * \cdots * p_{X_N}(x_N)$$

where the notation $*$ denotes convolution. For independent rv's X and Y with pdf's $p_X(x)$ and $p_Y(y)$, respectively, the sum S has a pdf, $p_S(s)$ given by

$$p_S(s) = \int_{-\infty}^{\infty} p_X(s - y) p_Y(y) \, dy$$

Also

$$\Phi_S(w) = \Phi_X(w) \Phi_Y(w)$$

where $\Phi_S(w)$, $\Phi_X(w)$, and $\Phi_Y(w)$ are the cf's of S, X, and Y, respectively.

A.1.14   Normal Distribution and Density Function

Assume X is normally (or Gaussian) distributed with mean η and variance σ^2. Then

$$p(x) = \frac{1}{\sqrt{2\pi\sigma^2}} e^{-(x-\eta)^2/2\sigma^2}$$

$$F(x) = \frac{1}{2} + \frac{1}{2} \mathrm{erf}\left(\frac{x - \eta}{\sqrt{2}\,\sigma}\right)$$

where the error function erf t is defined in reference 2 by:

$$\mathrm{erf}\, t = \frac{2}{\sqrt{\pi}} \int_0^t e^{-x^2} \, dx$$

The complementary error function, erfc t is defined by

$$\text{erfc } t = 1 - \text{erf } t = \frac{2}{\sqrt{\pi}} \int_t^\infty e^{-x^2} \, dx$$

Notes:

1. The cf of a normal rv X is

$$\Phi(w) = \exp\left\{ jw\eta - \tfrac{1}{2}w^2\sigma^2 \right\}$$

2. Assume N jointly distributed normal random variables X_i, $i = 1, \ldots, N$ with means η_i, $i = 1, \ldots, N$, variances, σ_i^2, $i = 1, \ldots, N$ and covariances, $\mu_{ij} = \mathscr{E}[(X_i - \eta_i)(X_j - \eta_j)]$, $i, j = 1, \ldots, N$. Let $\bar{\eta}$ be an $N \times 1$ column vector of the means; M denotes an $N \times N$ covariance matrix with elements μ_{ij}; $\det(M)$ denotes the determinant of M, and $\bar{X}$ is an $N \times 1$ column vector of the random variables $X_1, \ldots, X_N$. Then

$$p(x_1, \ldots, x_N) = \frac{1}{(2\pi)^{N/2}[\det M]^{1/2}} \exp\left\{ -\tfrac{1}{2}(\bar{X} - \bar{\eta})^t M^{-1}(\bar{X} - \bar{\eta}) \right\}$$

Denoting $\bar{w}$ as an $N \times 1$ column vector for $w_1, \ldots, w_N$, the corresponding cf is

$$\Phi(w_1, \ldots, w_N) = \exp\left\{ j\bar{\eta}^t \bar{w} - \tfrac{1}{2}\bar{w}^t M \bar{w} \right\}$$

3. If $\bar{X}$ is an $N \times 1$ column vector of the jointly normal rv's $X_1, \ldots, X_N$, and A is an $N \times N$ matrix, then the linear transformation $\bar{Y} = A\bar{X}$ is an $N \times 1$ column vector of jointly normal rv $Y_1, \ldots, Y_N$, with mean $\bar{\eta}_Y = A\bar{\eta}_X$ and covariance matrix $Q = AMA^t$.

4. If X_1, X_2, X_3, and X_4 are real zero mean jointly normal rv's,[2] then

$$\mathscr{E}\{ X_1 X_2 X_3 X_4 \} = \mathscr{E}\{ X_1 X_2 \}\mathscr{E}\{ X_3 X_4 \} + \mathscr{E}\{ X_1 X_3 \}\mathscr{E}\{ X_2 X_4 \}$$

$$+ \mathscr{E}\{ X_1 X_4 \}\mathscr{E}\{ X_2 X_3 \}$$

If Z_1, Z_2, Z_3, and Z_4 are complex zero mean jointly normal rv, then

$$\mathscr{E}\{ Z_1 Z_2^* Z_3 Z_4^* \} = \mathscr{E}\{ Z_1 Z_2^* \}\mathscr{E}\{ Z_3 Z_4^* \} + \mathscr{E}\{ Z_1 Z_4^* \}\mathscr{E}\{ Z_2^* Z_3 \}$$

A.1.15   Central Limit Theorem

Assume $X_1, X_2, \ldots, X_N$ are independent, identically distributed rv's. Then, if $S = \sum_{i=1}^N X_i$ with mean $\eta_S = \sum_{i=1}^N \mathscr{E}(X_i)$ and variance $\sigma_S^2 = \sum_{i=1}^N V(X_i)$, as N

increases without bound the pdf of S approaches a normal pdf with mean η_S and variance σ_S^2.

Notes:

1. The means and variances must exist and no one term in the sum can dominate to ensure convergence.

2. Under certain conditions the sum will be normally distributed even if the rv's are not identically distributed and, with further restrictions, are not statistically independent.

A.1.16 Statistical Properties of Estimators[3]

A.1.16.1 Unbiased Estimate

Assume rv's $X_1, \ldots, X_N$, with pdf $p(x_1, \ldots, x_N; \theta)$ where θ is a parameter. An estimate $d(x_1, \ldots, x_N)$ is an unbiased estimate of the parameter θ if

$$\mathscr{E}\{d(X_1, \ldots, X_N)\} = \theta$$

Notes:

1. Assume rv's $X_1, \ldots, X_N$ are independent identically distributed with mean η and variance σ^2. The sample mean, $\overline{X}$, is defined by

$$\overline{X} = \frac{1}{N} \sum_{i=1}^{N} X_i$$

Since $\mathscr{E}(\overline{X}) = \eta$, $\overline{X}$ is an unbiased estimate of η.

2. The sample variance, S_X^2, is defined by

$$S_X^2 = \frac{1}{N} \sum_{i=1}^{N} (X_i - \overline{X})^2$$

Since $\mathscr{E}(S_X^2) = \sigma^2$, S_X^2 is an unbiased estimate of σ^2.

A.1.16.2 Consistent Estimate

An estimate $d(x_1, \ldots, x_N)$ is a consistent estimate of a parameter G if the probability distribution of $d(x_1, \ldots, x_N)$ concentrates on the parameter θ as $N \to \infty$, that is, $d(x_1, \ldots, x_N)$ is a consistent estimate of θ if

$$P[\theta - \varepsilon \leq d(X_1, \ldots, X_N) \leq \theta + \varepsilon] \to 1 \quad \text{as} \quad N \to \infty \text{ for small } \varepsilon$$

Notes:

1. Assume rv's $X_1, \ldots, X_N$ are independent identically distributed with mean η and variance, σ^2. Then

$$V(\overline{X}) = \frac{\sigma^2}{N}$$

As N increases, the distribution of $\overline{X}$ becomes more concentrated about its mean, η.

A.1.16.3 Efficient Estimate

An efficient estimate is a best unbiased estimate that achieves the information limit to the variance (the equality condition in the Cramer-Rao bound corresponding to an estimate with minimum variance), and is therefore an efficient estimate.

   Notes:

1. If $X_1, \ldots, X_N$ are independent rv's with the same pdf $p(x_i; \theta)$ and the estimate $d(x_1, \ldots, x_N)$ is an unibased estimate of θ:

$$\mathscr{E}[d(X_1, \ldots, X_N)] = \theta$$

then the Cramer-Rao inequality becomes[3]

$$V(d(X_1, \ldots, X_N)) \geq \frac{-1}{\mathscr{E}\left\{\left[\frac{\partial^2}{\partial\theta^2}\log p(x_1, \ldots, x_N; \theta)\right]\right\}}$$

Assume X_i $i = 1, \ldots, N$ are each independent normally distributed with mean η and variance σ^2. If $d(x_1, \ldots, x_N) = \overline{X}$ with $\mathscr{E}[\overline{X}] = \eta$ and $\theta = \eta$, then $V(d(X_1, \ldots, X_N)) \geq \sigma^2/N$ and $d(x_1, \ldots, x_N)$ is an efficient estimate of η.

A.1.16.4 Sufficient Statistic

A statistic $d(x_1, \ldots, x_N)$ is a sufficient statistic if the conditional distribution of the rv's $X_1, \ldots, X_N$, given the value of $d(x_1, \ldots, x_N)$, is independent of the parameters.

   Notes:

1. Given parameters $\theta_1, \theta_2, \ldots$, then $d(x_1, \ldots, x_N)$ is a sufficient statistic if

$$p(x_1, \ldots, x_N; \theta_1, \theta_2, \ldots) = g(d; \theta_1, \theta_2, \ldots)h(x_1, \ldots, x_N)$$

where g depends only on the statistic d and the parameters $\theta_1, \theta_2, \ldots,$ and h is independent of the parameters $\theta_1, \theta_2, \ldots$.

2. From any unbiased estimate not based on a sufficient statistic, an improved estimate can be obtained which is based on a sufficient statistic. The improved estimate has smaller variance and is obtained by averaging with respect to the conditional distribution, given the sufficient statistic.

3. The minimum variance unbiased estimate of a parameter is the unbiased estimate that is based on a sufficient statistic.

A.1.17   Stochastic Process[1,2]

A stochastic (or random) process, $X(t)$, is defined here as an ensemble or set of time functions such that for every outcome, ζ, a time function $X(t, \zeta)$ can be assigned. A stochastic process $X(t)$ is statistically determined if the nth order distribution functions are known for any n and $t_1, \ldots, t_n$:

$$F(x_1, \ldots, x_n; t_1, \ldots, t_n) = P\{ X(t_1) \leq x_1, \ldots, X(t_n) \leq x_n \}$$

Notes:

1. For t fixed, $X(t)$ is a rv.
2. For ζ fixed (no probability measure), $X(t)$ is a single time function.

A.1.17.1   Moments of a Stochastic Process

The mean of a stochastic process $X(t)$ with first-order pdf $p(x; t)$ is

$$\eta(t) = \mathscr{E}\{ X(t) \} = \int_{-\infty}^{\infty} xp(x; t)\, dx$$

The autocorrelation $R(t_1, t_2)$ of $X(t)$ is

$$R(t_1, t_2) = \mathscr{E}\{ X(t_1) X(t_2) \}$$

The autocovariance $C(t_1, t_2)$ of $X(t)$ is

$$C(t_1, t_2) = \mathscr{E}\{ (X(t_1) - \eta(t_1))(X(t_2) - \eta(t_2)) \}$$

Notes:

1. $C(t_1, t_2) = R(t_1, t_2) - \eta(t_1)\eta(t_2)$.
2. The variance of $X(t)$ is $C(t, t) = R(t, t) - \eta^2(t)$.

A.1.17.2   Stationary Stochastic Process

A stochastic process is stationary if its statistics are not affected by a shift in the time origin:

$$p(x_1, \ldots, x_n; t_1, \ldots, t_n) = p(x_1, \ldots, x_n; t_1 + \varepsilon, \ldots, t_n + \varepsilon) \qquad \text{for any } \varepsilon$$

Notes:

1. $p(x; t) = p(x)$ independent of time.
2. $\mathscr{E}[X(t)] = \eta$ (a constant).
3. The autocorrelation $R(\tau) = \mathscr{E}\{X(t)X(t + \tau)\}$.
4. For complex stochastic processes $R(\tau) = \mathscr{E}\{X^*(t)X(t + \tau)\}$.
5. The covariance of $X(t)$ with mean η is

$$C(\tau) = \mathscr{E}\{[X(t + \tau) - \eta][X^*(t) - \eta^*]\} = R(\tau) - |\eta|^2$$

6. The crosscovariance of two stochastic processes $X(t)$ and $Y(t)$ with means η_X and η_Y, respectively, is

$$C(\tau) = \mathscr{E}\{[X(t + \tau) - \eta_X][Y^*(t) - \eta_Y^*]\}$$

$$= R_{XY}(\tau) - \eta_X \eta_Y^*$$

where

$$R_{XY}(\tau) = \mathscr{E}\{X(t + \tau)Y^*(t)\}$$

A.1.17.3 Power Spectrum of A Stochastic Process

The power spectrum or power spectral density (PSD) of a stationary process $X(t)$ with autocorrelation $R(\tau)$ is denoted by $P(w)$ and is

$$P(w) = \int_{-\infty}^{\infty} R(\tau)e^{-jw\tau}\,d\tau$$

Notes:

1. $R(\tau) = \dfrac{1}{2\pi}\displaystyle\int_{-\infty}^{\infty} P(w)e^{jw\tau}\,dw.$
2. The average power of a complex stochastic process, $X(t)$, is

$$\mathscr{E}(|X(t)|^2) = R(0) = \frac{1}{2\pi}\int_{-\infty}^{\infty} P(w)\,dw \geq 0$$

3. The cross power spectrum of two complex stochastic processes $X(t)$ and $Y(t)$ is

$$P_{XY}(w) = \int_{-\infty}^{\infty} R_{XY}(\tau)e^{-jw\tau}\,d\tau$$

where

$$R_{XY}(\tau) = \frac{1}{2\pi}\int_{-\infty}^{\infty} P_{XY}(w)e^{jw\tau}\,dw$$

4. A stationary process $X(t)$ is ergodic if time averages equal ensemble averages. Then

$$P(w) = \lim_{T \to \infty} \mathscr{E} \left\{ \frac{\int_{-T}^{T} X(t) e^{-jwt} \, dt}{2T} \right\}^2 = \int_{-\infty}^{\infty} R(\tau) e^{-jw\tau} \, d\tau$$

5. If $X(t)$ is a white noise process, then $\mathscr{E}(X(t)) = 0$, $R(\tau) = N_0 \delta(\tau)$, and $P(w) = N_0$ where N_0 is a constant PSD.

A.1.17.4 *Linear Time Invariant Systems* [2]

Assume a linear time invariant system with input $X(t)$, output $Y(t)$, and impulse response $h(t)$. If $X(t)$ is a complex stochastic process, then $Y(t)$ is a complex stochastic process with

$$Y(t) = \int_{-\infty}^{\infty} h(\sigma) X(t - \sigma) \, d\sigma$$

If $X(t)$ is stationary with mean η_X and autocorrelation $R_X(\tau)$, then

$$\mathscr{E}\{Y(t)\} = \eta_X \int_{-\infty}^{\infty} h(\sigma) \, d\sigma$$

$$R_Y(\tau) = \int_{-\infty}^{\infty} \int_{-\infty}^{\infty} h(\alpha) h(\beta) R_X(\tau + \alpha - \beta) \, d\alpha \, d\beta$$

$$R_{YX}(\tau) = \int_{-\infty}^{\infty} h(\alpha) R_X(\tau - \alpha) \, d\alpha$$

$$P_Y(w) = |H(w)|^2 P_X(w)$$

$$P_{XY}(w) = H^*(w) P_X(w)$$

where

$$H(w) = \int_{-\infty}^{\infty} h(t) e^{-jwt} \, dt$$

A.1.18 Sampling Theorem [1]

Assume the stochastic process $X(t)$ has a bandlimited PSD $P(w)$, that is, $P(w) = 0$ for $|w| > w_c$ where $w_c = 2\pi f_c$ and f_c represents the highest frequency. Then

$$X(t) = \sum_{n=-\infty}^{\infty} X(nT_s) \frac{\sin(w_c t - n\pi)}{w_c t - n\pi}$$

where the sampling interval $T_s = 1/2f_c$. Note that $X(nT_s)$ represents samples of the process $X(t)$ taken at the Nyquist rate, $f_s = 2f_c = 1/T_s$. Using this theorem, all results quoted for continuous rv's apply in the discrete case.

A.2 MATRIX RESULTS[4,5,6]

A.2.1 Transpose Properties of a Matrix

Given two matrices A and B where t denotes transpose,

$$(A + B)^t = A^t + B^t$$

$$(AB)^t = B^t A^t$$

If A is invertible, then

$$(A^t)^{-1} = (A^{-1})^t$$

A.2.2 Inverse of a Product of Matrices

Given two invertible matrices A and B,

$$(AB)^{-1} = B^{-1}A^{-1}$$

If the columns of A are linearly independent, then $A^{t*}A$ is invertible.

A.2.3 Trace of a Matrix

Given an $N \times N$ matrix A with elements a_{ij},

$$A = \begin{bmatrix} a_{11} & a_{12} & \cdots & a_{1N} \\ a_{21} & & & \\ \vdots & & \ddots & \vdots \\ a_{N1} & \cdots & & a_{NN} \end{bmatrix}$$

The trace of A denoted by $\text{Tr}(A)$ is then defined as the sum of the diagonal elements:

$$\text{Tr}(A) = \sum_{i=1}^{N} a_{ii}$$

A.2.4 Symmetric Matrix

A matrix A is symmetric if $A^t = A$.
   Note:

1. $\frac{1}{2}(A + A^t)$ is a symmetric matrix.

A.2.5 Quadratic Form

Let x be a real vector with N components: $x^t = (x_1, \ldots, x_N)$. Let A be a real $N \times N$ matrix with elements a_{ij}. Then a quadratic form in x is

$$x^t A x = \sum_{i=1}^{N} \sum_{k=1}^{N} a_{ik} x_i x_k$$

Notes:

1. $x^t A x = x^t A_1 x$ where $A_1 = \frac{1}{2}(A + A^t)$ and is referred to as the symmetric part of A.
2. A real symmetric matrix, A (i.e., $A = A^t$), is positive definite if $x^t A x > 0$ for any nonzero vector x. The matrix is positive semidefinite if $x^t A x \geq 0$ for any vector x and is zero for some nonzero vector.
3. Given a real symmetric quadratic form, a linear transformation exists which transforms the matrix into a diagonal form.
4. A real covariance matrix is positive semidefinite.
5. Let x be a complex vector with N components. For an $N \times N$ matrix A with elements a_{ij}, a quadratic form, q, in x is

$$q = x^{t*} A x = \sum_{i=1}^{N} \sum_{k=1}^{N} a_{ik} x_i^* x_k$$

   A Hermitian matrix A is defined by $A^{t*} = A$. If A is Hermitian, the quadratic form q is real, that is, $q = q^*$.[7]
6. A covariance matrix, R, consisting of complex variances, is Hermitian, positive semidefinite.[7]

A.2.6 Similarity Transform and Matrix Diagonalization

Two square matrices, A and $\tilde{A}$, are similar if a nonsingular transformation matrix T exists which results in a similarity transformation:[6]

$$\tilde{A} = T^{-1} A T$$

   Note:

1. Assume I is an $N \times N$ identity matrix, A is an $N \times N$ square matrix with elements a_{ij} for $i = 1, \ldots N$ and $j = 1, \ldots, N$, and λ is a scalar

parameter. The eigenvalues of A are those values of λ which cause the matrix $(A - \lambda I)$ to be singular. The eigenvalues are determined by the characteristic equation of A computed from the determinant (det) of $A - \lambda I$:

$$\det[A - \lambda I] = \begin{vmatrix} a_{11} - \lambda & a_{12} & \cdots & a_{1N} \\ a_{21} & a_{22} - \lambda & & a_{2N} \\ \vdots & & \ddots & \vdots \\ a_{N1} & a_{N2} & \cdots & a_{NN} - \lambda \end{vmatrix} = 0$$

2. The diagonal elements of $\tilde{A}$ are eigenvalues of A, and each eigenvalue λ_i of A occurs $m_i \geq 1$ times as a diagonal element. If λ_i occurs m_i times for $i = 1, \ldots, L$, then $\sum_{i=1}^{L} m_i = N$, and $\tilde{A}$ has a canonical upper triangular form known as the Jordan form. For distinct eigenvalues, $\tilde{A}$ is diagonal with the N eigenvalues along the diagonal.

3. A unitary matrix, U, is a matrix with orthonormal columns: $U'^{*}U = I$. An $N \times N$ Hermitian matrix, A, has real eigenvalues and can be diagonalized by a unitary matrix, U:

$$U'^{*}AU = \Lambda$$

where Λ is a diagonal matrix with eigenvalues λ_i, $i = 1, \ldots, N$. If the matrix A is positive definite, its eigenvalues are positive real.[8]

4. A complex covariance matrix R can be diagonalized by a unitary matrix U: $U'^{*}RU = \Lambda$.[7] The determinant of U is unity so that $\det R = \prod_{i=1}^{N} \lambda_i$. Also

$$R^{-1} = U\Lambda^{-1}U'^{*}$$

Both R and R^{-1} are positive definite if $\det R > 0$. Since the eigenvalues of λ are positive real, Λ can be factored; for example, a square root factorization is

$$\Lambda = \gamma'^{*}\gamma$$

where γ is a diagonal matrix with elements $\lambda_i^{1/2}$. Therefore the covariance matrix and its inverse can be factored:

$$R = (U'^{*})^{-1}\Lambda U$$

$$= (U'^{*})^{-1}\gamma'^{*}\gamma U$$

$$= (\gamma U^{-1})'^{*}\gamma U^{-1}$$

$$= SS'^{*}$$

where

$$S = (\gamma U^{-1})'^{*}$$

A.2.7 Matrix Identities[5]

The following identities are frequently used:

1. $(C - B^{t*}A^{-1}B)^{-1} = C^{-1} - C^{-1}B^{t*}(BC^{-1}B^{t*} - A)^{-1}BC^{-1}$

2. $\begin{bmatrix} A & B \\ B^{t*} & C \end{bmatrix}^{-1}$

$$= \begin{bmatrix} (A - BC^{-1}B^{t*})^{-1} & -(A - BC^{-1}B^{t*})^{-1}BC^{-1} \\ -C^{-1}B^{t*}(A - BC^{-1}B^{t*})^{-1} & C^{-1}B^{t*}(A - BC^{-1}B^{t*})^{-1}BC^{-1} + C^{-1} \end{bmatrix}$$

3. $\begin{bmatrix} A & B \\ B^{t*} & C \end{bmatrix}^{-1}$

$$= \begin{bmatrix} A^{-1} + A^{-1}B(C - B^{t*}A^{-1}B)^{-1}B^{t*}A^{-1} & -A^{-1}B(C - B^{t*}A^{-1}B)^{-1} \\ -(C - B^{t*}A^{-1}B)^{-1}B^{t*}A^{-1} & (C - B^{t*}A^{-1}B)^{-1} \end{bmatrix}$$

A.2.8 Toeplitz Matrix[8]

Assume an $N \times N$ matrix which has identical elements along symmetrical diagonals and is conjugate symmetric:

$$\begin{pmatrix} a_0 & a_1 & \cdots & a_{N-1} \\ a_1^* & a_0 & \ddots & \vdots \\ \vdots & \ddots & \ddots & a_1 \\ a_{N-1}^* & \cdots & a_1^* & a_0 \end{pmatrix}$$

This matrix is called a Toeplitz matrix. The form of this matrix permits a simple recursive solution to linear equations.

A.3 TRANSFORM RESULTS[9, 10]

A.3.1 Fourier Transform

A time function, $s(t)$, has a Fourier transform pair defined by

$$S(f) = \int_{-\infty}^{\infty} s(t)e^{-j2\pi ft} \, dt$$

$$s(t) = \int_{-\infty}^{\infty} S(f)e^{j2\pi ft} \, df$$

where $S(f)$ is known as the Fourier transform.

A.3.2　Discrete Fourier Transform

The discrete Fourier transform pair for a set of samples s_n for $n = 1, \ldots, N$ is given by

$$S_k = \sum_{n=1}^{N} s_n e^{-j2\pi nk/N}, \qquad k = 1, \ldots, N$$

$$s_n = \frac{1}{N} \sum_{k=1}^{N} S_k e^{j2\pi nk/N}$$

A.3.3　*Z* Transform (*Z*-Sited)

The Z transform pair for a set of samples s_n is given by

$$S(Z) = \sum_{n=-\infty}^{\infty} s_n Z^{-n}$$

$$s_n = \frac{1}{2\pi j} \oint_C S(Z) Z^{n-1} dZ$$

where the integral is a contour integral in the Z plane over any closed path, C, in the region of convergence encompassing the origin in the Z plane.

REFERENCES

1.　A. Papoulis, *Probability, Random Variables and Stochastic Processes*, McGraw-Hill, New York, 1965.

2.　A. D. Whalen, *Detection of Signals in Noise*, Academic, New York, 1971.

3.　D. A. S. Fraser, *Statistics, An Introduction*, Wiley, New York, 1958.

4.　G. Korn and T. Korn, *Mathematical Handbook for Scientists and Engineers*, McGraw-Hill, New York, 1961.

5.　P. Liebelt, *An Introduction to Optimal Estimation*, Addison-Wesley, Reading, Mass., 1967.

6.　G. Strang, *Linear Algebra and Its Applications*, Academic, New York, second edition, 1980.

7.　M. Schwartz, W. R. Bennett and S. Stein, *Communications Systems and Techniques*, McGraw-Hill, New York, 1966, Appendix B.

8.　G. Arfken, *Mathematical Methods for Physicists*, Academic, New York, 1966, pp. 161–172.

9.　G. M. Jenkins and D. M. Watts, *Spectral Analysis and Its Applications*, Holden-Day, San Francisco, 1969.

10.　A. V. Oppenheim and R. W. Schafer, *Digital Signal Processing*, Prentice-Hall, Englewood Cliffs, N.J., 1975.

Index